FUTURE FRONTIERS IN AGRICULTURAL MARKETING RESEARCH

Agricultural

FUTURE FRONTIERS IN

Marketing Research

EDITED BY

PAUL L. FARRIS

PROFESSOR, DEPARTMENT OF AGRICULTURAL ECONOMICS
PURDUE UNIVERSITY

IOWA STATE UNIVERSITY PRESS / AMES, IOWA

First edition, 1983

Library of Congress Cataloging in Publication Data
Main entry under title:

Future frontiers in agricultural marketing research.

1. Farm produce—Marketing—Research.   I.   Farris, Paul L.
HD9000.5.F873   1983   630′68′8   82–12684
ISBN 0-8138-0560-0

C O N T E N T S

PREFACE

THIS BOOK was prepared to help meet the agricultural marketing research challenge of the 1980s and beyond. New marketing problems are emerging, and several persistent issues of the past are continuing to appear in new ways. How well forthcoming marketing problems are solved will have an important bearing on the productivity and performance of agriculture and the general economy. Success in solving future marketing problems will also affect public attitudes toward various forms of market organization and exchange processes that will be relied upon to allocate resources and distribute income.

The volume is the result of an initiative by R. J. Aldrich, Adminstrator of the Cooperative State Research Service (CSRS), U.S. Department of Agriculture, who appointed a steering committee to plan the work. Members were Richard G. Garner, CSRS; Howard W. Ottoson, University of Nebraska, Chairman of the Marketing Subcommittee of ESCOP (Experiment Station Committee on Organization and Policy); and Paul L. Farris, Purdue University, who served as chairman. R. J. Hildreth, Farm Foundation, served with the committee. CSRS provided funds to help support the work.

The steering committee launched the venture by meeting with a few persons with considerable knowledge of agricultural marketing research. These were Ben C. French; John E. Lee, Jr.; William T. Manley; Willard F. Mueller; Don Paarlberg; and James D. Shaffer. The group prepared an outline for the volume, with a suggested title, focus, and author for each chapter. The authors met several times to review and critique outlines, manuscript drafts, and revised chapters.

The book is the individual and collaborative work of a group of professional agricultural economists who see major challenges in agricultural marketing research. Their ideas, perspectives, and suggestions are expected to be of interest to persons who seek ways to strengthen and improve research related to important issues involving marketing, agriculture, and the general economy in the years ahead.

Paul L. Farris

FUTURE FRONTIERS IN
AGRICULTURAL MARKETING RESEARCH

PAUL L. FARRIS

1

AGRICULTURAL MARKETING
RESEARCH IN PERSPECTIVE

THIS INTRODUCTORY CHAPTER highlights agricultural marketing research developments in the United States and indicates some emerging influences that may affect research priorities in the future. Succeeding chapters deal with individual subject matter areas and topics for research. Several research suggestions are especially timely at this stage in the growth of the U.S. economy. Others are less time dependent. Ongoing issues include marketing in relation to economic productivity, technological progress, and social well-being. Public policies relating to these issues are of continuing significance, along with policy aspects of economic organization, decision making, and market processes.

GROWTH OF MARKETING ISSUES
As the U.S. economy developed from a stage primarily of agricultural self-sufficiency to high degrees of specialization and industrialization, functions of marketing, exchange, and coordination expanded and grew more complex. Advances in the nation's physical and institutional infrastructure contributed to and affected the performance of the marketing system and productivity of the economy. Marketing activities are continuing to evolve as U.S. agriculture and the economy as a whole progress toward further stages of economic development.

A variety of questions related to marketing accompanied growth and development of the economy. Early concerns dealt primarily with market exchange and fair dealing. Whether marketing activities were economically productive was also given considerable attention by some early economic writers (see Cassels 1936). Vestiges of that concern are still expressed from time to time, particularly with reference to futures markets (Gray and Rutledge 1971).

PAUL L. FARRIS is Professor of Agricultural Economics, Purdue University, West Lafayette, Indiana.

3

Later issues involved competition, adequacy of information, efficiency, productivity, progress, income distribution, and other dimensions of market performance. Agricultural price and income issues of the late nineteenth and early
twentieth century included several marketing-type questions.

Precise boundaries of the domain of marketing have not been agreed
upon, but agricultural marketing has come to encompass off-farm activities on
both the product and input sides of farm production, including buying and
selling by farmers, and institutions and arrangements that affect exchange and
exchange relationships in the food and fiber sector.

Agricultural marketing research in the United States evolved slowly prior
to the 1930s. During and immediately following the Great Depression, several
factors coincided and interacted to emphasize marketing as a subject needing
research attention. Most important, particularly with regard to urgency and
timing, were economic circumstances in agriculture. Coming at about the same
time were advances in economic theory, methods of analysis, and data.
Marketing problems were being seen in a new light and with enough
understanding to encourage commitment of resources and people to solve
identified marketing problems. Public recognition of the need for marketing
research, education, and service was expressed in a substantial way in the
Research and Marketing Act of 1946.

Circumstances in Agriculture. Conditions within agriculture have had a
major influence on the amounts and kinds of emphasis given to marketing
problems. These conditions were in turn affected significantly by underlying
developments in the general economy. Growth of industrialization and urbanization in the latter part of the nineteenth century gave rise to a rapidly
changing market for agricultural commodities and to a larger, more complex
and varied processing, handling, and distribution system. Many local markets
became regional, national, and international in scope.

Technological developments also led to increasing requirements for purchased inputs and to a rapidly expanding farm input sector. Industrial concentration increased greatly in several sectors of the U.S. economy, including
some closely interrelated with agriculture such as transportation, processing of
farm-produced commodities, merchandising of grains, food distribution, and
manufacture of machinery and chemicals used in production.

The changing structure of the nonfarm economy alongside the relatively
unchanging structure of farm production intensified disquiet in the farm community and gave rise to increased political activity that farmers believed would
improve their lot. They supported inflationary policies in the general
economy, lobbied for antitrust legislation and other measures designed to limit
the perceived market power in the industrial sector, and established
cooperatives and group ventures in marketing and purchasing (Galbraith
1956). Farmers' activities in support of the greenback movement after the Civil
War, regulation of the railroads under the Interstate Commerce Act of 1887,
and control of monopoly under the Sherman Antitrust Act of 1890 were
related to unfavorable economic conditions in agriculture.

As farm conditions began to improve in the early years of the twentieth century, farm unrest subsided. Then, following relatively favorable conditions in agriculture during and immediately after World War I, farm prices broke sharply in 1920 and 1921. This helped provide the impetus for passage of the Capper-Volstead Act in 1922 (Benedict 1955). Farm prices generally declined during the 1920s and sank very low in the early 1930s. Political activities of farmers intensified during the 1920s, and increased efforts were made to pass national legislation designed to help farmers. The prevailing idea in that legislation was to raise prices on farm products for the domestic market and channel the excess abroad at lower world prices. But no such legislation was enacted. In 1929 the Agricultural Marketing Act became law, and a revolving fund of $500 million was provided to make loans to cooperatives, primarily for the purchase of commodities to maintain prices. In the face of rapidly deteriorating prices, the venture failed (Benedict 1955).

The new administration that came to power in 1933 began to experiment with a variety of ideas and approaches to deal with national economic hardship. The mood of the public was receptive, and Congress passed a wide range of legislation designed to address the economic ills of the nation. Several programs were tried, modified, adjusted, and/or discarded. Some were declared unconstitutional. Agricultural legislation during this period was oriented principally toward production control, and it thrust the government into more direct participation in agricultural affairs than ever before.

Marketing challenges seen in the 1930s were primarily those that would raise prices and incomes of farmers. Following the largely unsuccessful ventures of cooperatives in raising farm prices in the 1920s and the Federal Farm Board experience in the early 1930s, the Agricultural Marketing Agreement Act (which provided for establishment of marketing orders and agreements for milk and selected commodities) was enacted in 1937. Marketing and purchasing cooperatives were supported. The emphasis on outlook, which had its beginning in the 1920s, was continued. The marketing margin concept was refined and empirical measures devised. Other programs and services were undertaken to regulate competition, increase market information, enhance standardization and grading of agricultural commodities, and increase efficiency. A major objective of many of these programs was to reduce marketing costs as a way of benefiting producers and consumers.

The recovery of farm prices at the beginning of World War II shifted the emphasis in farm policy. Instead of the familiar burdensome agricultural supplies, weak demand, and low prices and incomes, the farm economy became buoyant. Prices rose faster than farm costs, and farmers experienced greater prosperity than they had known in decades. Still, there was the nagging fear that after the war the low farm price and income problem would return. Efforts undertaken to design government farm programs for the postwar period were accompanied by renewed attention to the problems of marketing. The Land-Grant Colleges and Universities Committee on Postwar Agricultural Policy (1944) commented that, in addition to better outlook information, two other types of improvements were needed to make prices of the free market serve both farmers and consumers well. These were greater efficiency in physical distribution of products and better operation of the pricing process. It was generally believed that if agricultural marketing problems could be successfully addressed and solved, a substantial contribution would be made to solution of the overall farm price and income problem. The environment thus turned out to be favorable for passage of the Research and Marketing Act of 1946 (Black 1947). With substantial infusion of new federal funds into marketing research, made possible by the act and subsequent legislation, there was a burst of new marketing research activity in the U.S. Department of Agriculture (USDA) and the land-grant colleges and a coincident entry of new researchers into the field.

Advances in Theory. The defined problems and research approaches gradually began to incorporate theoretical insights and methods of the 1930s and 1940s. From the turn of the century to about 1930 the ruling microeconomic ideas were based on Marshall's *Principles of Economics.* Essentially, two forms of market organization were recognized, pure competition and monopoly. Yet it was commonly observed that most economic activity was characterized by structures somewhere in between. Then almost simultaneously two major theoretical works appeared in 1933, Chamberlin's *The Theory of Monopolistic Competition* (United States), and Robinson's *The Economics of Imperfect Competition* (England). Further theoretical discussions and developments in the 1930s led to identification of two new conceptual market forms, monopolistic competition and oligopoly, to accompany the existing Marshallian forms of pure competition and monopoly (Galbraith 1948).

Economists who were interested in marketing problems welcomed the new theoretical ideas. However, it was not easy to apply them in research. Stewart and Dewhurst (1939) held that much of the distribution sector fitted the monopolistic competition model and concluded that distribution did cost too much, not because firms were making excess profits, but because the nature of competition allowed underemployed resources to be tied up in the industry on a continuous basis. Waugh provided important leadership for much marketing work in the USDA. For example, in 1938 he illustrated how establishing and maintaining different net prices for different parts of a crop might benefit producers and society. The idea later evolved into the Food Stamp Act. The Temporary National Economic Committee of the late 1930s examined a wide range of problems related to organization and competition and provided much descriptive information and perceptive observation about various sectors of the economy (see especially Hoffman 1940).

In 1940, Clark developed the idea that market forms with elements of monopoly might be considered workably competitive if they gave rise to aspects of performance considered to be satisfactory. In 1941, Nicholls extended the theoretical concepts of Chamberlin and Robinson to agricultural industries and elaborated the models to the buying and selling sides of several types of agricultural markets.

At the macro level, the major theoretical contribution of the 1930s, Keynes's *General Theory of Employment, Interest and Money,* dealt with the behavior of aggregate sectors of the economy and emphasized the possibility of equilibrium at less than full employment. Macroeconomic developments were not seen to be particularly related to microeconomic problems, and the view that these two branches of economics were separate prevailed for many years. Only relatively recently, as price-cost interrelations in the inflationary process became more evident, have we become aware of the dynamic interaction between microeconomic and macroeconomic activity. This interaction is coming to be recognized as particularly important in the agricultural economy, and it promises to be an important conceptual and empirical challenge in the future. This topic was addressed by the Cabinet Committee on Price Stability in 1968, for which W. F. Mueller served as Executive Director; it was later examined by Gordon (1981) and Okun (1981).

Thus in the late 1940s, the kit of theoretical tools available to deal with many marketing problems was still in the process of development and evaluation. The state of conceptual knowledge had a major bearing on the kinds of problems identified and the research projects given emphasis in new marketing research initiatives.

Methods of Analysis. Much early work in marketing was descriptive; Thomsen (1945) classified 82 percent of 589 marketing reports in this way. Markets were described, marketing channels delineated, marketing margins measured, and cost studies of plant operations and marketing functions undertaken. Price series were analyzed for seasonal, cyclical, and trend patterns. Regularities were sought out and elaborated for the practical user. Tables were prepared to show relationships among two or more variables. Multiple correlation techniques were developed and debated in the 1930s and 1940s (e.g., Malenbaum and Black 1937; Bean 1940; Ezekiel 1940). Techniques for handling systems of equations were just beginning to be addressed in the 1940s (Haavelmo 1944). Linear programming ideas were in their infancy. The computer had not yet arrived. Budgeting was used in studying many firm-type problems.

Data. Relevant information bearing on many important marketing problems has always been inadequate and difficult to obtain. Early data series consisted mainly of prices received by farmers and prices of selected farm cost items. Retail prices gathered by the Bureau of Labor Statistics provided a source of data on selected retail food prices. Much USDA work went into development of data series on marketing costs and margins. The Censuses of Manufactures and Distribution were useful in describing general characteristics of industries engaged in processing and distributing farm foods and supplying farm inputs.

Much early data needed to analyze firm and industry efficiency problems had to be gathered firsthand. Data gathering included such methods as using stopwatches in processing and packing plants to measure labor efficiency, selecting samples of grain from country elevators to study pricing and grading practices, establishing and supervising record-keeping systems and panels, and considerable use of special-purpose surveys. As problem conceptualizations evolved, data priority needs changed. Variability in success of obtaining different types of relevant data influenced the course and accomplishments of marketing research.

EARLY POST-WORLD WAR II EMPHASIS ON
AGRICULTURAL MARKETING RESEARCH

In addressing itself to marketing problems following World War II, the agricultural economics profession was influenced not only by the existing state of theoretical knowledge, methods, and data but also by ideas and approaches that had come out of its considerable experience in farm management and agricultural policy. Trelogan (1951) observed, "Too many research workers persist in trying to do their marketing research on farms or in confining their attention to problems at first assembly points, when most of the marketing costs are absorbed in marketing functions farther removed from the farms." Although issues of pricing efficiency, grading, marketing information, consumer preferences, and cooperative operations received increasing attention, the much emphasized problem area was firm operating efficiency. Research studies focused on topics such as labor efficiency in packing plants and measurement of various cost items. Some attention was given to efficiency aspects of performing selected marketing functions such as transportation costs in milk distribution and potential savings in alternative methods of farm product assembly. Presumably, if efficiency were improved, the marketing margin would be smaller and either farmers or consumers, or possibly both, would benefit. Input markets were virtually neglected. Marketing research was not oriented toward the predominant agricultural policy issue of the period, which was the balancing of overall supply and demand for agricultural commodities at levels that would increase returns to farmers.

The burgeoning activities in agricultural marketing were accompanied by numerous discussions at professional association gatherings, regional research committee meetings, and marketing workshops and in written commentaries (e.g., see discussions by Spencer 1947; Brownlee 1948; Nicholls 1948; Banfield 1949; Bressler 1949, 1951; Joy 1949; Shepherd 1949; Trelogan 1951). The series of national marketing workshops held during the 1950s brought together marketing researchers from the USDA and land-grant universities and served as forums and stimuli for exploration of new activities, ideas, and approaches in marketing. Expansion of regional research, in accordance with provisions of the Research and Marketing Act, brought researchers with similar interests together, increased interaction and communication, and broadened the range of marketing studies undertaken at various institutions. But as time passed, it was gradually being recognized that the product of increased research activity in marketing consisted of many fragmented studies on a variety of topics. Although they added to knowledge in particular areas, some broader questions relating to organization and performance of production and marketing systems as wholes were being neglected (Black 1954; Kohls 1957).

While a variety of marketing activities were being pursued in the 1950s, a significant organizational development, primarily in the poultry industry, caught many marketing scholars by surprise. The trade press was carrying discussions of vertical integration and raising questions for which the profession had no ready answers. There was a flurry of activity in academic circles and the USDA, seeking understanding and evaluation of this new phenomenon. Although general economists had written about vertical integration much earlier (e.g., Clark 1923) and actual integration was commonplace in many nonfarm sectors, agricultural economists were scrambling for a framework to interpret integration in agriculture (e.g., see Bottum et al. 1957; Mighell and Jones 1963). In the Foreword to the latter report, Koffsky wrote, "The present study attempts to build a more balanced image of production and market structure that will lead to the solution of economic problems associated with novel forms of coordination."

The appearance of vertical integration in the poultry industry turned out to be only a dramatic example of emerging organizational changes in agriculture. Marketing economists became increasingly sensitive to organizational developments in other agricultural industries and to questions of market power and control by processors, distributors, and input supply firms. In 1959, Bain's book *Industrial Organization* was published and accepted as a useful and relevant framework for understanding and interpreting an important set of marketing problems related to organization and competition in agricultural industries.

TRANSITION OF MARKETING RESEARCH IN THE 1960s

In recognition of the profound transformation under way in several agribusiness industries, characterized primarily by increasing plant and firm size, declining firm numbers, and expansion in various forms of integration, marketing economists turned increasingly to describing and analyzing the organizational changes that were occurring. A North Central Regional Committee, NCR-20, prepared a report on market structure research. The included article by Clodius and Mueller (1961) built upon previous writings in the area, especially by Bain, and endeavored to relate the framework to agriculturally oriented problems. Considerable emphasis was given to setting forth some potential areas for empirical research, followed by additional efforts to define high-priority problems (see Farris 1964).

The emphasis on industrial organization was seen to be relevant to emerging problems, but research on these problems was seriously hampered by lack of pertinent data. Business firms were generally not willing to provide the needed data on sales, costs, buying, selling, and pricing practices and on their investment and resource allocation decisions. The inability of marketing researchers to provide useful answers to some important questions being raised by farmers and the consuming public was an important factor contributing to establishment of the National Commission on Food Marketing in the mid-1960s.

The commission's work (1966) added to general understanding of food marketing by highlighting several emerging tendencies and implications for public policy for the food industry. By having the power of subpoena, the commission acquired some particular types of data, along with insights from business, that were previously unavailable to researchers. The research was on a scale large enough to benefit from staff interaction on concurrent and related studies. The one and one-half years of operation was too short a period, however, to analyze in depth many of the potentially promising problem areas. A working paper prepared by Shaffer (1968) suggested research organization alternatives and helped provide the stimulus for undertaking several potentially promising subsector studies.

To carry forward and build upon previous industrial organization research, a North Central Regional Committee research project was proposed in the late 1960s and activated as NC-117 in 1973. The project was organized with a core group of researchers at the University of Wisconsin and a number of collaborating researchers from the North Central region and several other states and the USDA. The workers were able to collaborate with committees of Congress, the Federal Trade Commission, and the USDA in undertaking research. NC-117 produced several studies of the changing food system and provided a basis for further understanding of important issues and research challenges.

Accompanying the work on market organization and competition in the 1960s were three major advances in methods of analysis. These were econometrics, mathematical programming, and simulation. By employing these methods, agricultural economists were able to gain greater understanding of market forces in domestic and foreign markets, determine optimum locations of plants, select optimum product lines and inputs for efficiency of plant and firm operation, and project the probable effects of alternative policies and constraints on firms and industries. Business games were developed to provide insights into business behavior and consequences of alternative managerial decisions. The games were found to be useful in educational work with students, farmers, and business people.

Research work employing methodological advances included several studies that were not designed for immediate practical application, partly because of the need to develop illustrations and ascertain potential uses and limitations of new techniques. In some instances special efforts were required to adapt new techniques to fit relevant problems, to conceptualize problems to employ the techniques effectively, and, especially, to obtain relevant and appropriate data. Some of the successes were achieved on intrafirm problems where firms provided confidential data to researchers with the understanding that the results would not be made public. While research of this type did not yield publishable results, it did help perfect techniques and train researchers in method applications that were useful for other types of problems. As experience in employing new methods was gained, the methods were applied to a widening variety of problems. Advances in computer technology and communications facilitated applications and presented opportunities for theoretical and empirical research relative to processes and types of decision making (Simon 1978). New research methods and computational capability in turn helped stimulate generation of relevant data. These developments contributed in major ways to solution of a broad range of marketing problems and began to show high promise for further applications in the future.

Various aspects of past marketing studies have been admirably documented by Breimyer (1973), French (1977), Helmberger et al. (1981), and Tomek and Robinson (1977). Freebairn's (1967) discussion of ideas about grading and the survey by Gray and Rutledge (1971) on the evolution of concepts relative to futures markets are valuable sources of knowledge in these areas. Waugh's (1954) annotated book of readings on agricultural marketing is a useful compilation of several excerpts relating to ideas and research results prior to the early 1950s. Marketing research prior to 1932 is described by the Taylors (1952). Breimyer's (1963) review of federal marketing programs from the establishment of the USDA Office of Markets in 1913 to 1963 traces the principal areas of emphasis through that half-century. The historical pattern of legislation coordination and administration of marketing research in the USDA was traced by Wiser and Bowers (1981).

QUESTIONING AND CHALLENGE OF MARKETING RESEARCH IN THE 1970s

Metzger (1973) studied marketing research for the Cooperative State Research Service (CSRS), USDA, and observed "a noticeable waning in the enthusiasm for marketing research and perhaps a decline in its effectiveness." Babb's (1977b) CSRS study of federal funding requirements for marketing shows a leveling off in the relative amount of federal funds spent on marketing research after the mid-1960s. There was a growth in marketing technology work, which experiment station administrators thought was desirable. However, heads of agricultural economics departments felt that there was need to strengthen and expand marketing research by their staffs.

The momentous shocks to the agricultural marketing system in the early 1970s placed great strain on the system and raised a number of critical questions, some longstanding and some new. For example, attention to the market and farm share of the consumer's food dollar was highlighted periodically in the 1970s by consumer groups who were concerned about rising food prices and by farmers who continued to see a lack of close correspondence between changes in retail and farm prices.

Because of a large number of substantial economic changes in the 1970s, this was a period when events and circumstances, many of which originated outside agriculture, again dominated perceived problems in marketing. Several new problem areas were related to regulation, environmental quality, changes in population age distributions, social values and life-styles, energy, changing resource price relationships, new technology, nutrition, and foreign trade.

MARKETING RESEARCH IN THE FUTURE

Growth of the U.S. economy throughout its history brought widespread changes in the role and functions of markets and marketing firms. Further marketing changes can be expected as the economy and agriculture continue to develop. As in the past, the course of change will be influenced by many factors. Among these will be the overall rate of growth, relative importance of particular sectors, and degree of stability in the general economy.

The state and development of the infrastructure of the U.S. economy will significantly affect the performance of the marketing system, kinds of marketing problems likely to be faced, and general advances in economic productivity. All dimensions of the infrastructure will be important—physical, technological, institutional, and informational. For example, various components of the transportation system, including port and storage facilities, can have a major bearing on efficiency in production, manufacturing, and distribution. The technological base of the economy, including processes and incentives for generation of new knowledge and adoption of innovations, directly influences increases in economic productivity. The legal and institutional framework, including banking and finance, can have an appreciable effect on efficiency, competition, and ability of the system to adjust and overcome impediments to progress. Communications and information are becoming increasingly important in efficient functioning of an ever more highly specialized and interrelated economic system. The general state of knowledge, including education and skills of the population, may be one of the more significant components of the U.S. infrastructure.

Although future marketing problems cannot be predicted precisely, many current tendencies can be expected to continue to exert their influences. Further theoretical developments can be anticipated, shaped in part by needs to solve new problems, advances in problem-solving methodology, and availability of data bearing on relevant issues.

As marketing problems of the future are approached, goals and values of society will continue to influence judgments of researchers as to the relative importance and priority of particular problems. Undoubtedly, longtime objectives such as the level, stability, and distribution of farm income will continue to be important. Farm input procurement will grow in importance and complexity. Greater attention to desires of consumers may be required. People are likely to be concerned with such attributes as adequacy, availability, stability, variety, and price of food. Other goals also seem to be emerging. These include food safety, quality, nutrition, and various types of subsidized food consumption. These developments imply continuing emphasis on marketing efficiency and on finding new ways to reduce marketing costs. However, emphasis on efficiency will be subject to an increasing number of constraints imposed by the desires of consumers for more particularized attributes in their food purchases.

The overall research challenge faced by the United States in overcoming declining productivity growth in the general economy may involve several marketing issues. Renewed emphasis on problems involving the more basic features, processes, and relationships of the production and distribution system will be required to help raise economic output, enhance the U.S. competitive position in world markets, and improve the satisfaction of the nation's citizens.

Other kinds of goals seem likely to assume greater significance than in the past. Important among these will be organization and operation of the system. In this connection, the issue increasingly may be whether the legal and economic framework within which the system operates provides the kinds of incentives that enhance competitive organization and behavior, which in turn will stimulate adoption of new technology and increase efficiency. Related to this issue may be the extent to which the developing organization of the system will help satisfy criteria with respect to income distribution, decentralization of decision making, distribution of economic power, freedom, and other basic goals and values of society.

A significant research policy issue involves deciding which problems should be researched by the public sector and which by the private sector. The decision criteria are not always clear and may involve more considerations than might be apparent at first glance. In general, one would not expect the private sector to finance or undertake research to produce benefits that could not be captured to some extent either directly or indirectly in a proprietary way. Similarly, the case for public support of research that would not generate benefits beyond a particular firm or group would be difficult to make. In actual practice, new knowledge has a way of spreading, albeit sometimes with a lag. Thus, in general, while research should be encouraged, support for any particular project should be subject to carefully examined priorities. Public support of research would seem appropriate where benefits would appear to become widespread, the private sector is neither motivated nor financially capable of generating research funds to solve a major intractable problem, or risks may be particularly high.

The objective of this book is to help develop perspectives about research priorities in marketing, to conceptualize marketing problems in their new economic settings, to indicate the kinds of research approaches appropriate in various situations, and to suggest the kinds of commitments of resources that will be needed to solve the emerging problems.

REFERENCES

Association of Land-Grant College and Universities. 1944. *Postwar Agricultural Policy.*
Babb, E. M. 1977a. Marketing Research at State Agricultural Experiment Stations—Problems and Possible Solutions. Purdue University Agric. Exp. Sta. Bull. 150, West Lafayette, Ind.
_____. 1977b. Impacts of Federal Funding Requirements on Marketing Research at State Agricultural Experiment Stations. Coop. State Res. Serv. USDA.
Bain, Joe S. 1959. *Industrial Organization.* New York: Wiley.
Banfield, Edward C. 1949. Planning under the Research and Marketing Act of 1946: A study in the sociology of knowledge. *J. Farm Econ.* 31:48–75.
Bean, L. H. 1940. The short-cut graphic method of multiple correlation. *Q. J. Econ.* 54:318–31.
Black, John D. 1947. Guideposts in the development of a marketing program. *J. Farm Econ.* 29:616–31.
_____. 1954. Lines of advance in research in marketing. *J. Farm Econ.* 36:1061–68.
Black, John D., and Wilfred Malenbaum. 1940. Rejoinder. *Q. J. Econ.* 54:346–58.
Bottum, J. C., N. S. Hadley, L. S. Hardin, R. L. Kohls, J. K. McDermott, and V. W. Ruttan. 1957. Vertical Integration in Agriculture. Purdue University Agric. Ext. Serv. EC-154.
Breimyer, Harold F. 1963. Fifty years of federal marketing programs. *J. Farm Econ.* 65:749–58.
_____. 1973. The economics of agricultural marketing: A survey. *Rev. Mark. Agric. Econ.* 61:115–65.
Bressler, R. G. 1949. Agricultural marketing research. *J. Farm Econ.* 31:553–62.
_____. 1951. Research in marketing efficiency. *J. Farm Econ.* 33:944–54.
Brownlee, O. H. 1948. Marketing research and welfare economics. *J. Farm Econ.* 30:55–68.
Cassels, J. M. 1936. The significance of early economic thought on marketing. *J. Mark.* 1:129–33.

Chamberlin, E. H. 1933. *The Theory of Monopolistic Competition*. Cambridge, Mass.: Harvard University Press.

Clark, J. M. 1923. *Studies in the Economics of Overhead Costs*. Chicago: University of Chicago Press.

_____. 1940. Toward a concept of workable competition. *Am. Econ. Rev.* 30:241–56.

Clodius, Robert L., and Willard F. Mueller. 1961. Market structure analysis as an orientation for research in agricultural economics. *J. Farm Econ.* 43:513–53.

Ezekial, Mordecai. 1940. Further comment. *Q. J. Econ.* 54:331–46.

Farris, Paul. L., ed. 1964. *Market Structure Research*. Ames: Iowa State University Press.

Freebairn, J. W. 1967. Grading as a market innovation. *Rev. Mark. Agric. Econ.* 35:147–62.

French, Ben C. 1977. The analysis of productive efficiency in agricultural marketing: Models, methods and progress. In L. R. Martin, ed., *A Survey of Agricultural Economics Literature*, vol. 1, pp. 91–206. Minneapolis: University of Minnesota Press.

Galbraith, J. K. 1948. Monopoly and the concentration of economic power. In H. S. Ellis, ed., *A Survey of Contemporary Economics*. Philadelphia: Blakiston.

_____. 1956. *American Capitalism*, rev. ed. Cambridge, Mass.: Riverside Press.

Gordon, Robert J. 1981. Output fluctuations and gradual price adjustment. *J. Econ. Lit.* 19:493–530.

Gray, Roger W., and David J. S. Rutledge. 1971. The economics of commodity futures markets: A survey. *Rev. Mark. Agric. Econ.* 39:51–108.

Haavelmo, T. 1944. The probability approach in econometrics, suppl. *Econometrica* 12:1–118.

Helmberger, Peter, Gerald R. Campbell, and William D. Dobson. 1981. Organization and performance of agricultural markets. In L. R. Martin, ed., *A Survey of Agricultural Economics Literature*, vol. 3. Minneapolis: University of Minnesota Press.

Hoffman, A. C. 1940. Large Scale Organization in the Food Industries. Temporary National Economic Committee, Monogr. 35, Washington, D.C.

Joy, Bernard. 1949. Marketing research under the Research and Marketing Act. *J. Farm Econ.* 31:1148–55.

Keynes, John Maynard. 1935. *The General Theory of Employment, Interest and Money*. New York: Harcourt, Brace.

Kohls, R. L. 1957. A critical evaluation of agricultural marketing research. *J. Farm Econ.* 37:1600–09.

Malenbaum, Wilfred. 1940. Concluding remarks. *Q. J. Econ.* 54:358–64.

Malenbaum, Wilfred, and John D. Black. 1937. The short-cut graphic method: An illustration of "flexible" multiple correlation techniques. *Q. J. Econ.* 52:66–112.

Marshall, Alfred. 1922. *Principles of Economics*. London: Macmillan.

Metzger, H. B. 1973. Views on strengthening marketing economics research. *Agric. Sci. Rev.* 11(1):25–29.

Mighell, Ronald L., and Lawrence A. Jones. 1963. Vertical Coordination in Agriculture. USDA, Agric. Econ. Rep. 19.

National Commission on Food Marketing. 1966. Food from Farmer to Consumer, and ten technical studies. Washington, D.C.: USGPO.

Nicholls, William H. 1941. *A Theoretical Analysis of Imperfect Competition with Special Application to the Agricultural Industries*. Ames: Iowa State College Press.

_____. 1948. Reorientation of agricultural marketing and price research. *J. Farm Econ.* 30:43–54.

Okun, Arthur M. 1981. *Prices and Quantities: A Macroeconomic Analysis*. Washington, D.C.: Brookings Institution.

Robinson, Joan. 1933. *The Economics of Imperfect Competition*. London: Macmillan.

Shaffer, James D. 1968. A Working Paper Concerning Publicly Supported Economic Research in Agricultural Marketing. USDA, ERS.

Shepherd, Geoffrey. 1949. The field of agricultural marketing research: Objectives, definition, content, criteria. *J. Farm Econ.* 31:444–55.

Simon, Herbert A. 1978. Rationality as process and product of thought. *Am. Econ. Rev.* 68:1–16.

Spencer, Leland. 1947. Marketing research under the research and marketing act of 1946 (summary of round table discussion). *J. Farm Econ.* 29:292–98.

Stewart, Paul W., and J. Frederic Dewhurst. 1939. *Does Distribution Cost Too Much?* New York: Twentieth Century Fund.

Taylor, Henry C., and Anne Dewees Taylor. 1952. *The Story of Agricultural Economics in the United States, 1840–1932*. Ames: Iowa State College Press.

Thomsen, F. L. 1945. A critical examination of marketing research. *J. Farm Econ.* 17:947–62.

Tomek, William G., and Kenneth L. Robinson. 1977. Agricultural Price Analysis and Outlook. In L. R. Martin, ed., *A Survey of Agricultural Economics Literature*, vol. 1, pp. 327–409. Minneapolis: University of Minnesota Press.

Trelogan, Harry C. 1951. Marketing research in the United States during the past five years. *J. Farm Econ.* 33:932-43.

Waugh, Frederick V. 1938. Market prorates and social welfare. *J. Farm Econ.* 22:403-16.

Waugh, Frederick V., ed. 1954. *Readings on Agricultural Marketing.* Ames: Iowa State College Press.

Wiser, Vivian, and Douglas E. Bowers. 1981. Marketing Research and Its Coordination in USDA: An Historical Approach. USDA, Agric. Econ. Rep. 475.

BRUCE W. MARION
WILLARD F. MUELLER

2

INDUSTRIAL ORGANIZATION, ECONOMIC POWER, AND THE FOOD SYSTEM

INDUSTRIAL ORGANIZATION (IO) has emerged as a field of study within economics, in which the primary focus is organization and performance of markets and industries. Although individual firms may be the unit of observation in some IO studies, the emphasis in industrial organization is mainly on understanding behavior of groups of firms that either act as competitors (a selling or buying industry) or interact as suppliers and customers.

Whereas the organizational theorist attempts to understand firm behavior by examining the internal structure, motivations, and decision rules of individual firms (Cyert and March 1963), the IO economist attempts to explain the behavior of groups of firms by basic economic conditions (e.g., demand characteristics, state of technology) and the competitive environment within which firms operate. Thus while the former focuses largely on factors internal to the firm, the latter concentrates largely on factors external to the firm. Little has been accomplished in integrating these alternative theoretical approaches to firm behavior.

Industrial organization is a branch of applied price theory. The basic IO paradigm holds that the structure (S) of a market strongly influences the competitive conduct (C) of firms within the market, which in turn strongly influences market performance (P). Within this basic paradigm there are a large number of hypotheses, some conflicting. The basic tenets of IO are covered well elsewhere (Bain 1968; Scherer 1980) and will not be dealt with here.

BRUCE W. MARION is Agricultural Economist, Economic Research Service, USDA, and Professor, Department of Agricultural Economics, University of Wisconsin, Madison. WILLARD F. MUELLER is William F. Vilas Research Professor of Agricultural Economics, Professor of Economics, and Professor in the Law School, University of Wisconsin, Madison. The helpful comments of Walter Armbruster, William Boehm, John Connor, and Vernon Sorenson on earlier drafts are gratefully acknowledged.

The *raison d'être* of IO is imperfectly competitive markets. In a world of perfect competition, industrial organization would be irrelevant. Since IO provides the only well-developed framework for examining behavior of imperfectly competitive markets, the field has grown in importance as imperfect markets (especially those involving oligopolies) have become more prevalent. Although IO remains a rather imprecise and controversial field of economics, especially when compared to the more simplistic and deterministic models of markets often emphasized in graduate theory courses, is attractive to economists seeking answers to critical questions about an increasingly complex industrial system.

DEVELOPMENT AND EVOLUTION OF INDUSTRIAL ORGANIZATION

In 1939, Mason laid out the plan for systematic study of what was to become industrial organization. His plan called for in-depth studies of a large enough number of diversely structured industries to permit generalizations about relationships of structure to performance. During the following two decades, a large number of industry case studies were conducted. These provided valuable insights into the individual industries examined but provided little basis for generalizations about structure-performance relationships.

The 1950s was a period of considerable theoretical debate among IO economists—a period during which claims of Schumpeter, Stigler, Adelman, and others that there was nothing to fear from monopoly power were countered by the theories and evidence of Bain, Means, Kaysen and Turner, and others that tight oligopolies were widespread and socially costly. By 1960 the field was rich in concepts but short on empirical testing.

Bain (1945), the author of perhaps the most comprehensive industry study, was the key figure in the demise of the industry study approach. He argued for cross-sectional analyses whereby one or more structural variables common to a number of industries were correlated with industry performance variables. In doing so, he rejected his earlier acceptance of Mason's belief that verified scientific knowledge would come only from close examination of individual firms or industries.

During the 1960s a large number of cross-sectional econometric studies were conducted with particular emphasis on industry structure–performance relationships. Verification of the basic S-C-P paradigm was emphasized along with testing alternative theories. Although poor data often marred results, the accumulated evidence pointed toward significant and costly market power problems in many parts of the economy. Market power was found to be positively related to profits and to have a deleterious effect on inventions and innovations when concentration reached very high levels. The normative case against economic concentration grew stronger as a result of the evidence developed during the 1960s.

Additional evidence on structure-performance relationships has been developed during the 1970s. This decade has involved still further shifting in emphasis within industrial organization. Greater attention has been placed on new theories, methodological issues, firms as the basic unit of analysis as opposed to industries, firm market share as a measure of market power vis-a-vis market concentration, dynamics of market structure over time, additional elements of market structure such as conglomeration and information, and effects of market power on a broader set of consequences such as distribution of income and opportunity, political power, and inflation. X-inefficiency has come to be recognized as possibly equal in importance, as a consequence of market power, to misallocation and lost innovation. (X-inefficiency, a term coined by Leibenstein 1966, refers to the tendency for costs as well as prices to be higher in noncompetitive markets as a result of lax cost controls, sheer waste, excessive advertising and promotional expenditures, excess capacity, and operating with suboptimal scale plants.) Still, one of the basic issues has persisted over the years: How much monopoly is justified by economies of scale, and what are the performance trade-offs?

Agricultural economists were slow in applying industrial organization theory to organization and performance of agricultural markets. Two important exceptions were A. C. Hoffman and William H. Nicholls. Hoffman's 1940 monograph for the Temporary National Economic Committee (TNEC) was the first comprehensive examination of the organization of food manufacturing industries. Nicholls's contributions were both theoretical and empirical. His *Imperfect Competition within Agricultural Markets* (1941) expanded the theories of Chamberlain and Robinson to the buyer side of markets as well as providing empirical evidence of the structure and conduct of various agricultural markets. His *Price Policies in the Cigarette Industry* (1951) followed in the Mason industry study tradition.

Despite these and other important contributions to the IO literature in agricultural markets, Nicholls (1948) had little success in persuading other agricultural economists to follow his lead. They largely ignored his urging the use of IO theory to examine marketing and price determination in the food sector. One of his students was persuaded, however, and coauthored a paper admonishing the profession to use industrial organization theory as an orientation for research into agricultural markets (Clodius and Mueller 1961).

IO research by agricultural economists accelerated in the 1960s, much of it stimulated by the National Commission on Food Marketing and a series of seminars and workshops sponsored by NCR-20, a North Central Regional Research Committee. Several members of NCR-20 were instrumental in creating NC-117, which invited participation by researchers throughout the nation. Organized in 1974, this project has undertaken several studies of the organization, control, and performance of the U.S. food system.

EMPIRICAL FINDINGS

Comprehensive surveys of IO empirical studies have been made by Weiss (1971), Goldschmid et al. (1974), and Scherer (1980). Industrial organization studies have largely concentrated on structure-performance relationships. Conduct, the third element of the IO paradigm, has received much less attention, partly because of measuring difficulties and partly because its theory has not been well developed. In some cases, measures of conduct such as advertising expenditures are used as surrogates of market structure (product differentiation). Measures of conduct (e.g., market rivalry, market entry and exit) have also been included in econometric models of market performance or structural change (Cotterill and Mueller 1979; Marion et al. 1979). The determinants and effects of various types of market conduct remain relatively poorly understood, however, and may depend upon future research that combines the orientation and skills of the organizational theorists with that of the IO economist.

Seller Concentration and Profit. The relationship between seller concentration and profits has received by far the most widespread and penetrating analyses of any relationships suggested by the IO paradigm. The theoretical basis for this relationship is straightforward: successful collusion (tacit or explicit) is expected to lead to results approaching joint profit maximization, and the ability to collude increases with concentration. Although the quality of data, measures of profit, level of aggregation, and empirical models have varied greatly, a significant positive relationship between industry concentration and profit has been found in the majority of these studies. After a careful analysis of 46 separate studies of concentration-profit relationships, Weiss (1974) concludes:

> To summarize, the theory of the dominant firm unequivocally points to high prices and suggests high profit rates for dominant firms. Our assorted oligopoly theories are more equivocal in their details, but all of them that have not been discredited point to higher margins in concentrated industries once more. Our massive effort to test these predictions has, by and large, supported them for "normal" years such as the period 1953–1967, though the concentration-profits relationship is weakened or may even disappear completely in periods of accelerating inflation or directly following such periods. By and large the relationship holds up for Britain, Canada, and Japan, as well as in the United States. In general the data have confirmed the relationship predicted by theory, even though the data are very imperfect and almost certainly biased toward a zero relationship.

Of the relatively few studies that have examined the effect of relative firm dominance (using some measure of firm market share) as well as industry concentration, several have examined food industries. Kelly (1969), Imel and Helmberger (1971), and Rogers (1978), in studying food manufacturing firms, all found significant positive relationships to profits for both relative firm dominance and market concentration. Marion et al. (1979) found both variables were positively and significantly related to prices and profits of food retailing firms in different metropolitan markets.

Thus there appears to be little question of the positive effect of market concentration and relative firm dominance on firm and industry profits. However, this relationship appears to vary for different stages of the business cycle. Seller concentration, for example, generally has been found to have a significant positive relationship to profits except during inflationary periods.

Weiss (1971) suggests that this is because prices and wages are sticky in oligopolistic industries and respond to inflationary pressures with a lag.

The exact nature of the structure-profit relationship also has not been defined; results of some studies have supported the notion that a threshold level of concentration must be reached before concentration and profits are positively related, while other studies indicate a positive relationship throughout the range of data. The nature of the concentration-profit relationship has particular relevance for antitrust enforcement, which must decide whether concentration has reached levels that confer substantial market power on sellers.

Monopoly and oligopoly theory deal primarily with prices, not profits. The effect of concentration on prices is expected to be greater than indicated by empirical analysis of concentration-profit relationships. Unfortunately, very few studies have had access to the necessary price data to test the relationship of prices to market concentration and relative firm dominance. A study of grocery chain prices and profits (Marion et al. 1979) supports the above expectation. Studies of prices in tax-exempt bond markets (Kessel 1971), gasoline retailing (Marvel 1978), and banking (Hester 1979) are consistent with the findings by Marion et al. concerning market structure–price relationships. These studies did not have access to profit data however, and hence were unable to test the X-inefficiency hypothesis. Prices rose more rapidly than profits as relative firm dominance and market concentration increased. At least in part, this difference may be a result of the tendency for cost-increasing nonprice forms of competition to be emphasized in concentrated markets. In part, it may also stem from X-inefficiency in concentrated markets. Although some case studies support the latter conclusion, no systematic analysis has been made of the relationship between market concentration/firm dominance and X-inefficiency.

Entry Barriers and Profit. Because of the difficulty of measuring entry barriers, relatively few studies have examined their effects on firm or industry profits. In general, researchers have attempted to measure minimum efficient plant size, absolute capital requirements, and/or advertising expenditures or have qualitatively defined entry barriers as low, medium, or high. Results have been mixed and often open to methodological question. Bain (1956) and Mann (1966) have found that high industry concentration had a similar effect on profits in industries with moderate or substantial barriers to entry but a considerably stronger effect when barriers were high.

The evidence is more consistent concerning the relationship of advertising intensity to firm profitability in consumer goods industries. Studies by Kelly (1969), Imel and Helmberger (1971), Comanor and Wilson (1974), and Rogers (1978) have found that consumer goods industries (or firms) with high levels of advertising enjoy substantially higher profits than those with low or moderate advertising levels. High advertising levels are expected to result in stronger product differentiation and higher entry barriers. For nondurable consumer goods industries, there is empirical support for the expectation that advertising levels are higher in concentrated industries. Studies involving a broader sample of industries have been less consistent in the relationship found between advertising and concentration.

Market Structure and Progressiveness. Progressiveness includes research and development (R & D), innovation, and diffusion of new processes or products. The theoretical linkage between market structure and progressiveness has been subject to continuous debate since Schumpeter (1950) set forth his hypothesis that large firm size and market power are essential for technological progress. This is countered by theories holding that technological progress is dulled by great levels of market power, since the incentive to be innovative is reduced. Much of the empirical work on progressiveness has sought to test these conflicting hypotheses.

Empirical analyses have generally examined the relationship of market structure to R & D expenditures or employment (measures of R & D input) and/or to patents issued (a measure of R & D output). Other studies have examined the origins of important inventions and innovations. Although the results have been mixed, they generally indicate a threshold relationship between market structure and R & D and innovation. Scherer (1980) concludes:

> A bit of monopoly power in the form of structural concentration is conducive to invention and innovation, particularly when advances in the relevant knowledge base occur slowly. But very high concentration has a favorable effect only in rare cases, and more often it is apt to retard progress by restricting the number of independent sources of initiative and by dampening firms' incentive to gain market position through accelerated research and development. Likewise, it seems important that barriers to new entry be kept at modest levels, and that established industry members be exposed continually to the threat of entry by technically audacious newcomers.

Studies of technology in the food manufacturing industries by Imel et al. (1972) and Mueller et al. (1979) are generally consistent with Scherer's conclusion. The latter study further found that inventions and innovations promoting efficiency in food manufacturing are largely spawned outside these industries, especially in food machinery and other industries that specialize in developing and manufacturing technology for food manufacturers. This explains the paradox that many food manufacturing industries enjoy above average increases in productivity while spending less on R & D than any other industrial group. The study concludes that while there may be a case for restructuring food manufacturing industries (either making them more or less competitive), no persuasive case can be made on grounds of technological performance because the bulk of inventions and innovations originate outside food manufacturing industries.

Diffusion studies are concerned with the speed with which industries adopt new techniques once they have been introduced in a commercially viable form. The relatively few studies that have been conducted indicate that smaller firms and competitive industries do at least as well as large firms and oligopolistic industries in adopting new techniques once they have been introduced (Mansfield et al. 1977).

Econometric analyses of progressiveness have been seriously hampered by data and measurement problems. Some of the more valuable insights into the innovative process have resulted from case studies of individual industries. These are difficult to generalize, however.

Market Structure and Technical Efficiency. Few if any studies have been able to measure technical efficiency directly and relate this to market structure elements. The relationship has generally been approached indirectly by estimating the minimum efficient plant size (through the survivor technique, the average plant size of leading firms, or economic engineering studies) and determining the level of concentration necessary for firms to operate plants of this size. These studies indicate that high concentration is warranted on technical efficiency grounds in very few U.S. manufacturing industries.

Technical efficiency studies have concentrated on production economies at the plant level. Firm (multiplant) economies that may exist in procurement, finance, information systems, product development, and marketing activities are generally ignored. Large-firm economies in product development, advertising, and distribution (whether real or pecuniary) may explain the trend toward increasing concentration in consumer goods industries, while concentration is stable or declining in producer goods industries. However, studies have found that while firm economies often are substantially larger than plant economies, most industries are more concentrated than can be explained by firm economies alone (Scherer 1974).

Industry Structure and Economic Stability. The unfolding of events, rather than a priori theories, fostered the belief that market power creates an inflationary bias and excessive unemployment in a market economy. Orthodox economic doctrine that all inflation was of the demand-pull variety (too much money chasing too few goods) was first seriously challenged in the Great Depression when orthodox theory failed to account for economic happenings. As a result of this experience, Means (1935) developed the concept of "administrative" prices (generally referred to as administered prices) in an influential memorandum to the secretary of agriculture. The concept of cost-push, sellers', or administered inflation received additional support from the congressional hearings on pricing behavior in various industries during the 1950s and from the pricing behavior of the steel and other industries during 1969–71.

A number of cross-sectional studies have attempted to test the hypotheses that the market power of corporations and/or unions influences wage and price behavior (Mueller 1974). Although the results are somewhat mixed, they generally indicate that the economic power of unions has enabled them to obtain greater wage increases than unorganized workers; further, when powerful unions bargain with employers who have market power (i.e., successive monopoly), the unions fare even better. The results also indicate that prices in industries with market power tend to be less flexible, respond to changing market conditions with a lag, and are relatively unresponsive to drops in demand. Prices therefore follow a ratchet effect in concentrated industries, where they seldom go down, and have a delayed response to inflationary periods. This results in situations where cost-push inflation follows a period of demand-pull inflation as prices and administered wages play "catch-up."

There is little evidence, however, that price increases in the long run are greater in oligopolistic than in competitive industries. Given the widespread evidence that profits and prices are positively related to concentration, we would expect price increases to be greater in industries that are increasing in concentration vis-à-vis those that are constant or declining in concentration. The structural-change effect on inflation has not been empirically examined, however.

The market power–inflation phenomenon poses an important public policy problem: corporate and labor decision makers that possess considerable discretion in making price and wage decisions may impede effectiveness of macroeconomic policies, which essentially assume that the economy is sufficiently competitive so that prices and wages respond rather quickly and "properly" to contractions in aggregate demand. This assumption simply is not met in much (and perhaps a growing part) of the contemporary industrial economy. The result may be a built-in inflationary bias that prevents attainment of full employment without inflation.

Economists do not know the precise nature and magnitude of the problem, and some believe it does not exist. But there is growing evidence that at any given time inflationary pressures may involve a complicated admixture of seller-push and demand-pull forces that are difficult if not impossible to combat with the conventional Keynesian tools of fiscal policies or Friedmanesque monetary policies. The problem is especially critical to agriculture: if the market power–inflationary bias hypothesis is correct, reliance on monetary and fiscal policies alone forces competitive sectors such as agriculture to bear an especially heavy burden in the fight against inflation. Ironically, agricultural economists have largely ignored this problem despite the fact that the administered prices–inflation hypothesis was germinated in Means's work with the USDA.

Trends and Determinants of Industry Structure. Although industrial organization economists have tended to concentrate on relationships of industry structure to various aspects of performance, several studies have examined the trends in industry structure and causes of structural change. Studies that have examined market concentration trends in producer and consumer goods industries have found the latter to be steadily increasing, while producer goods industries are stable or declining in concentration. In a series of studies during 1966-80, Mueller has found this pattern consistently; further, concentration is increasing more rapidly in highly differentiated than in low and moderately differentiated consumer goods industries (Mueller and Hamm 1974; Mueller and Rogers 1980). For 167 four-digit manufacturing industries from 1947-72, Mueller and Rogers have found that change in concentration was positively and significantly related to television advertising. Rogers (1980) has found similar results in examining 86 food and tobacco industries. Given the massive expenditures for television advertising by many consumer product firms, this finding raises serious questions about the potential for distorting the sovereign role consumers are presumed to play in a market economy. The results also carry insights into the importance of economies of scale in determining industry concentration. Since there is no reason to expect that production scale economies are related to the level of product differentiation, these findings provide no support to the notion that industries are becoming more concentrated because of technological imperatives.

Industry structure is also influenced by mergers and antitrust actions toward mergers. In one of a series of studies of the structural impact of mergers and merger law enforcement, Mueller (1978b) has found that the antitrust agencies have primarily challenged horizontal mergers. Since the Celler-Kefauver Amendment was passed in 1950, only 20 percent of the mergers challenged were conglomerate (mostly product extension), although over 70 percent of all large mergers during the 27-year period were of that type. He concludes that the merger law, as enforced, has been relatively effective in dealing with noncompetitive horizontal acquisitions but ineffective in dealing with conglomerate acquisitions. The latter have steadily increased since 1950.

Studies of the impact of mergers and antitrust enforcement generally have used the industry case study approach rather than econometric analysis. Merger enforcement in the dairy processing and food retailing industries from the mid-1960s to the mid-1970s virtually stopped horizontal and market extension mergers by leading companies (Mueller 1978b). The total number of mergers did not decline but were channeled to small- and medium-sized firms. In the mid-1970s, the antitrust agencies relaxed their posture on food retailing mergers, thereby triggering a wave of acquisition activity by large chains.

Although much is still not known, research has indicated several factors affecting industry concentration; product differentiation activities and mergers are two of these. Although firm economies of scale seldom justify highly concentrated industries, they undoubtedly are a factor influencing the structure of some industries. Unfortunately, research to date has examined the effects of only a few of the factors that could influence industry structure.

Social and Political Effects of Large Firms and Monopolistic Industries. Although this subject has an abundant number of untested hypotheses, research efforts have been meager. Shepherd (1967) has found a positive relationship between industry concentration and measures of racial discrimination in employment of white-collar workers. Siegfried (1975) is the only economist to use the IO paradigm and cross-sectional econometric analysis to examine the relationship between industrial organization and political power as a dimension of performance. Working on the premise that political influence would be reflected by the tax laws, he found a significant negative relationship between absolute firm size and effective tax rate. Political scientists and sociologists have been more active in examining the aggregate concentration of economic and political power than economists. Several semipopular books such as *The Power Elite* (Mills 1956), *America, Inc.* (Mintz 1971), and *Economics and the Public Purpose* (Galbraith 1973) provide some interesting facts and provocative hypotheses. With a few notable exceptions, especially

Galbraith, economists have been noticeably silent on this topic, even though the issues involved are fundamental to a market economy and democratic political processes.

OVERVIEW OF PAST RESEARCH

The many attempts to verify or refute the IO paradigm have added much to the theoretical richness and practical understanding of imperfectly competitive markets. We are certainly far more knowledgeable than we were in the 1960s. However, data, measurement, and methodological and theoretical problems have detracted significantly from the body of research available and have undoubtedly discouraged research on such pressing issues as inflation, employment, conglomerate power, industry productivity, and aggregate economic power. Perhaps the most serious problem has been the difficulty of obtaining data that are consistent with theoretically relevant variables and definitions. For example, secondary data are frequently unavailable for theoretically relevant markets or industries, economic profits, or prices. Measures of market concentration have relied largely on Bureau of Census four- and eight-firm concentration data, which have widely recognized definitional problems. As diversification has increased, profit data for relevant product or geographic markets have become more difficult to obtain from secondary sources. Primary data are often either inaccessible or too costly to obtain.

Adequate quantification of theoretically important variables has also been a continual problem in IO research. Barriers to entry and product differentiation (important elements of market structure) must be estimated largely via surrogate variables, as must the performance dimensions of progressiveness and technical efficiency. For performance dimensions such as equity and product desirability, no suitable surrogate measures have been developed. The difficulty of quantifying conduct and the virtual absence of secondary data on conduct undoubtedly account for the scant attention that has been paid to this part of the S-C-P trilogy (see Chap. 1).

Empirical results provide rather solid ground for limiting horizontal concentration, dealing with horizontal mergers, and taking action to deconcentrate highly concentrated industries. However, our understanding of the economic effects of large conglomerate firms, conglomerate and vertical mergers, government regulations, and dynamics involved in industry restructuring is much more restricted because of theoretical and empirical limitations. Obviously, much remains to be done.

Promising Areas for Industrial Organization Research in the U.S. Food System. Since World War II, the National Commission on Food Marketing (1966b) has identified six major changes that carried important implications for competition in food manufacturing industries. It noted a decline in company numbers, an increase in concentration, a substantial increase in the conglomerate nature of leading food manufacturing firms, an increase in the number of large acquisitions, substantial increases in product differentiation expenditures by large food manufacturers, and a growing differential between the profitability of large food manufacturers and smaller competitors. Connor (1980) has examined each of these changes, using 1963–72 data, and concluded that in every instance except the last the trends have continued since the mid-1960s. Large firms (over $100 million in assets) have remained more profitable in every year since 1960, but the differential has not widened since the early 1960s.

There are important differences in the various food manufacturing industries. Between 1958 and 1972, producer goods industries actually declined in four-firm concentration from 43.8 to 42.6 percent. Concentration in consumer goods food manufacturing industries with low product differentiation increased slightly from 36.8 to 37.4 during the same period. However, food manufacturing industries with high levels of advertising increased in concentration from 53.8 to 61.5. Clearly, the latter industries pose the most serious competitive problems.

Parker and Connor (1979) have estimated "consumer overcharges" in food manufacturing industries as a result of monopoly power. Estimated overcharges ranged from none in meat packing to 29.5 percent in breakfast cereals. Using three alternative estimation procedures, X-inefficiency and monopoly profits for all food manufacturing ranged from $12 to $15 billion in 1975. Although there may be some public benefits from the factors that allow this overcharge (e.g., national brand products, high advertising), they obviously come at a high public cost.

A study of food retailing indicated steady increases in national and local market concentration (Marion et al. 1979). Nationally, the largest 20 grocery chains expanded their share of grocery store sales from 27 percent in 1948 to 37 percent in 1977. Concentration of sales among leading grocery wholesalers is similar to concentration in retailing. The largest 20 wholesalers accounted for 39 percent of wholesale grocery sales in 1979. Together, the largest 20 grocery chains and the largest 20 grocery wholesalers accounted for about 53 percent of the volume sold through retail grocery stores in 1977. Thus concentration of wholesale-retail procurement has increased since the National Commission on Food Marketing (1966a) expressed concern on this subject in the mid-1960s.

Although many local markets are still relatively competitive in structure, the proportion of standard metropolitan statistical areas in which the four largest retailers account for 60 percent or more of sales increased from 5 percent in 1954 to 25 percent in 1972. Based upon the significant positive relationship found between the concentration of grocery retailing in metropolitan markets and grocery prices, Marion et al. (1979) have estimated monopoly overcharges of $662 million in food retailing in 1974.

While the competitive characteristics of producer-first handler markets are less well documented, available evidence indicates high and increasing buyer concentration in many commodities. For example, packer and stockyard data show that in the 10 leading hog slaughter states, where 69 percent of the nation's hogs were slaughtered in 1977, the four largest buyers in each state purchased 82 percent of the hogs, on average. In the 10 largest fed-cattle slaughtering states, which accounted for 80 percent of fed-cattle slaughter in

1977, the average percent slaughtered by the four largest packers increased from 52.5 percent in 1970 to 61.4 percent in 1977. The market share held by the largest packer in each state increased from 18.2 to 24.1 percent during this period (Williams 1979). In fruits and vegetables, Lang (1978) has found that in three-fourths of the markets studied the largest four processors purchased 70 percent or more of the supply in the market area. All in all, there is considerable evidence of high and increasing buyer concentration in many producer-first handler markets. In one of the few studies that has examined the impact of buyer concentration on prices received by farmers, Jesse and Johnson (1970) have found a significant positive relation between contract prices for sweet corn and the number of competing processing plants available to farmers.

Available evidence indicates that there may be important competitive problems and a variety of vertical control arrangements in the farm input industries. For example, the four largest tractor manufacturers did about 80 percent of the U.S. tractor business in 1978; the four leading brands of corn insecticide accounted for about 84 percent of the market (Hamm 1979). Other farm input industries such as feed are more competitively structured but employ a variety of vertical arrangements with agricultural producers. Unfortunately, there have been very few IO studies of the organization and performance of farm input industries.

Given this brief review of the competitive characteristics and trends in the U.S. food system, there are several areas of research that we would place at the "frontier."

IMPACT OF CONGLOMERATES IN THE FOOD INDUSTRY. By conglomerates, we mean firms that operate in two or more distinct product / geographic markets. By this definition, most large food chains and food manufacturing companies are conglomerates. Conglomerate theory holds that such firms are able to engage in competitive strategies unavailable to specialized single-market firms (Mueller 1977). These include cross-subsidization from one business enterprise to another, reciprocal dealing with suppliers and customers, and mutual forbearance between conglomerates (don't rattle my cage and I won't rattle yours). This theory is not universally accepted and has not been adequately tested. More evidence is available to support the cross-subsidization hypothesis than the remaining two. Cross-subsidization may be particularly important in entering or restructuring markets and may have either procompetitive or anticompetitive effects, depending on the structure of the markets involved.

Conglomerates also are hypothesized as substitutes for imperfect capital markets, vehicles for balancing risk, and the natural growth strategy for firms legally blocked from horizontal acquisitions and unwilling to rely on internal growth.

Research is needed to test the above hypotheses in the food industries, especially the practice of cross-subsidization to support what some have characterized as predatory marketing (Mueller 1978a,b). Case studies are useful and often the only method of conducting some aspects of such research.

However, cross-sectional analysis has been used with some success (Cotterill and Mueller 1979; Mather 1979) and holds obvious advantages for generalizing.

The impact of multinational companies (a particular type of conglomerate) on both domestic and foreign markets also warrants research to determine their effect on structure and competitive behavior of industries, industry performance, aggregate economic performance (employment, balance of payments, inflation) and domestic and foreign governments (Connor and Mueller 1977). This is a large and complex area of research that is limited by the inaccessibility of much of the needed data. Case studies of individual companies and industries may be more feasible than cross-section analysis.

FIRM ECONOMIES OF SCALE. Considerable disagreement still exists concerning the extent of firm economies of scale. The pecuniary and real economies associated with procurement, finance, planning, product development, advertising, and distribution functions are particularly in need of research. Unfortunately, the necessary data may not be obtainable, since they are only available from individual companies. Problems of comparability and the possibilities of X-inefficiency also make this research more difficult. However, research on firm economies of scale and the causes and magnitude of X-inefficiency are essential for evaluating the trade-offs that may exist with allocative efficiency under various structural configurations of an industry.

EFFECTS OF VERTICAL AND CONTRACTUAL INTEGRATION ON COMPETITIVE BEHAVIOR AND PERFORMANCE. Numerous theoretical pieces have been written about the reasons for and consequences of vertical integration and contracting (e.g., Stigler 1951; Machlup and Taber 1960; Mighell and Jones 1963; Williamson 1971). However, there have been very few efforts to verify empirically the alternative and in some cases conflicting theories. As a result, public policy toward vertical mergers and various types of vertical coordination arrangements are adrift in a sea of uncertainty. Are such arrangements as harmless as University of Chicago economists contend, or do they allow market foreclosure, squeezing, and fortification of oligopoly power? In-depth case studies of individual firms and industries may be the most appropriate first step to understanding the effect of vertical linkages on such dimensions as technical efficiency, coordination, equity of rights and returns, competitive behavior, and system restructuring. Cross-sectional analysis using data on individual firms within an industry may be possible if the necessary data can be obtained.

SOCIOECONOMIC IMPACT OF ADVERTISING IN THE FOOD INDUSTRY. Although there is rather strong evidence that high levels of advertising are related to high profits and some evidence that television advertising is an important cause of structural change, the total impact of advertising in the food system is not understood. Important differences also remain among scholars concerning the direction of causality. Does high advertising lead to high profits or vice versa?

The interrelationship between advertising, concentration, and profits warrant additional analysis and appears to be researchable.

Advertising may be essential for introduction of new products or new firms, and informational advertising can stimulate competition (Benham 1972). However, extensive advertising may cause subtle changes in consumer values and tastes, making the concept of consumer sovereignty little more than a charade. The influence of advertising on obesity and consumption of foods of questionable nutritional value may be of particular concern.

The impact on consumer values (and ultimately on buying behavior) is an extremely important research topic that underlies much of the social concern about advertising. However, agricultural economists are generally ill equipped to research this topic. An interdisciplinary effort involving scholars from social psychology and mass communication may be useful. Even then there is a serious question of whether the effects of advertising on consumer values and behavior can be unraveled from other cultural influences, which in turn may have been influenced by past advertising.

It seems obvious that advertising influences consumer food purchasing behavior. Otherwise, why would firms advertise? It is not clear, however, how much manipulation of consumer demand occurs through advertising. One important research challenge is to determine when advertising is no longer in the public interest. Blind taste tests or scientific analysis along with comparative price information and firm advertising expenditures would provide useful clues. If two products are physically indistinguishable but one sells at a substantial premium because of high advertising, we would conclude that manipulation of preferences has occurred. Although some will argue that a product gives greater utility to consumers if they think it is superior and therefore warrants a higher price, we cannot accept this rationale. We have seen no evidence that consumers prefer to be deceived.

Phillip Morris–Miller Brewing Company's introduction of its Lite brand of beer is the ultimate achievement of advertising-created product differentiation—being able to sell a lower cost product at a higher price. The Lite story is only a play within a larger play, the consistent theme of which is to persuade consumers to live in worlds of illusion and switch up to premium and super-premium beer brands. The leading brewers' success in increasing the share of premium and superpremium beers from about 30 percent of the market in 1970 to 60 percent in 1978 cost beer drinkers about $400 million in 1978 (Mueller 1978a). There is no evidence from blind taste tests that consumers can detect differences among major U.S. beers.

RELATIONSHIP BETWEEN INDUSTRY ORGANIZATION AND QUALITY AND NUTRITIONAL VALUE OF PRODUCTS SOLD. Preliminary analysis by the Senate Select Committee on Nutrition and Human Needs indicated a negative relationship between industry concentration and nutritional value of products produced. Although this analysis raises more questions than it answers, the relationships among firm size, industry structure, and food product characteristics are worthy of examination. Do firms in concentrated industries tend to emphasize fabricated and highly processed food products that are easier to differentiate through advertising? Are fabricated and highly processed foods inferior from a quality and nutritional standpoint? Is product quality lowered when a large merchandising-oriented firm acquires a smaller firm that has been a quality leader in the industry?

Research on this topic is related to that suggested above on advertising, since its role in selling products of varying nutritional qualities is one of the key issues needing study. The biggest impediment to this research may be the classifying or ranking of food products by their nutritional value and quality. This transcends the training of agricultural economists and suggests a collaborative effort with nutritionists and food scientists.

Corporate research and development personnel might provide insights if their cooperation can be obtained. For example, in a specialized firm that emphasizes quality and uses low to moderate levels of advertising, what happens to the level and emphasis of R & D when it is acquired by a large advertising-merchandising–oriented firm? Several such mergers probably can be identified. Previous employees may be more willing to discuss these matters than present employees.

INFORMATION IN THE FOOD SYSTEM. Information programs provided by the public sector have a long history in agriculture. Their justification relies heavily on the concept that information allows markets to work more perfectly and that individual farmers lack the resources to develop such information.

Research suggests that information provided to consumers on food prices, product characteristics, and nutritional values may hold the potential for restoring some degree of consumer sovereignty and reducing the market power of food manufacturers and retailers that has stemmed in part from consumer ignorance (Devine and Marion 1979). The long-run impact of public information programs on the competitive behavior of industries with different structural and product characteristics has received little attention and is an important frontier for future research. Research on the need for and effect of consumer information lends itself to experimental designs. Experimental programs have provided comparative retail price information; others provide computerized meal planning guides considering nutritional balance, alternative food costs, and consumer food preferences. There is considerable opportunity for imaginative interdisciplinary research in this area and it appears manageable.

Research is also needed on the role of public information and private intelligence systems on the competitive organization and behavior of the food marketing system. Do large firms have superior intelligence systems that give them a competitive advantage over smaller firms that rely mainly on public information sources? To what extent can large firms and concentrated industries manipulate public information on market conditions and prices? Has the type of information provided by government programs kept pace with changes in organization of the food system so that it is relevant for private decision makers? The first of these questions may not be researchable. The last two should be, however, and call for in-depth interdisciplinary analyses of information sources, distribution, and needs in different commodity subsectors.

PRICE LEVELS AND CHANGES IN FOOD INDUSTRY. Consumer unrest and political concern about food prices has been highly correlated with rapid increases in food prices, particularly fresh beef prices. Ironically, commodities where competition among food manufacturers/processors is the healthiest also exhibit the greatest price variability. Conversely, prices are relatively stable (though they may rise substantially over long periods) for such highly differentiated products as soft drinks, detergents, and prepared cereals where there is evidence of substantial monopoly power. However, no systematic analysis has been made of price changes for different products during different stages of the business cycle.

There is a strong need for developing testable models of dynamic pricing. Given changes in demand and/or costs, which firms change prices, at what speed, and based upon what decision rules? Does this behavior vary between differentiated and undifferentiated products, concentrated and unconcentrated markets, and manufacturing and retailing companies? The development of theories of the dynamics of pricing will probably be easier than empirical testing of such theories, which may require tracking pricing behavior over business cycles. For difficult issues such as this, theory development and empirical analysis often need to proceed in an iterative process.

PUBLIC POLICIES, INFLATION, AND AGRICULTURE. Research on pricing behavior in the food system (as discussed above) provides opportunities for gaining insights into the forces affecting inflation. Research should examine the direct effect of food price increases or decreases and the indirect effects on wages, retirement payments, and the like that in some cases are indexed to the cost of living. The multiplier effect of food price changes on inflationary forces throughout the economy needs to be understood. Also, do price increases and decreases have an equal effect, or do institutional arrangements act as ratchets, responding more to price increases than to decreases? A better understanding of the impact of food price changes on the total economy would be invaluable in determining whether a cheap food policy with stable farm prices is socially more desirable than the free-market policies emphasized during much of the 1970s.

Research is also needed on the effect of market organization and institutional arrangements on the incomes of wage earners and executives, which in turn affects aggregate demand. To the extent that concentrated markets (for labor or products) lead to more rapid increases in wages or executive salaries, they may have both a cost-push and demand-pull effect on inflation.

A paramount research need is to unravel the impact of alternative mixes of monetary and fiscal policies on agriculture. Moreover, given the likelihood that wage and price controls of some kind will be applied from time to time, agricultural economists should develop more reliable knowledge of the impact of alternative kinds of wage and price control programs.

STRUCTURE-CONDUCT-PERFORMANCE RELATIONSHIPS IN PRODUCER-FIRST HANDLER MARKETS. The majority of industrial organization studies in the food and fiber system have focused on food processors and manufacturers, food retailers, and to a limited extent farm input suppliers. Very few empirical efforts have examined producer–first handler markets to assess the extent, trends, and effects of monopsonistic buying power; the competitive impact of marketing cooperatives; and the effect of vertical linkages on market behavior. Since most producer–first handler markets are local in nature, cross-sectional analysis is possible where the necessary data can be obtained. In some cases, data on the quantity/quality purchased and prices paid by buyers may best be obtained directly from farmers. Given sufficient resources, however, such research should be feasible.

A related yet somewhat distinct issue is the effect of original and countervailing power of agricultural bargaining and marketing cooperatives on system performance. Although theory and scattered evidence suggest that where a substantial share of the supply is represented in collective action by farmers, higher farm prices result, we have little understanding of the total effects of such action on prices, profits, and other performance dimensions in the food system. The theorem of second best tells us that given the imperfectly competitive conditions at some stages in the food system, social welfare would not necessarily be enhanced by pursuing pure competition in other stages. Theoretically and empirically, there is little basis for determining the optimum level of farmer collective action. This is a difficult area of research but one with a significant bearing on public policy toward cooperatives.

ORGANIZATION AND EFFECTS OF LABOR UNIONS IN THE FOOD SYSTEM. Market power, conduct, and performance of labor unions in the food system has rarely been studied, even though labor costs make up over half the direct cost of marketing food in the United States. How does the market power of unions affect settlement terms over time? What are the interrelationships between increases in wage rates, rising food prices, and inflation in the economy? Do union wage rates lead or lag price increases in the food system, or are the two related on an iterative basis that creates an upward spiral tendency? Do wage rates increase more rapidly in food manufacturing industries with market power and excess profits? What are the institutional factors (pattern setting, bargaining, cost-of-living adjustment clauses) that influence the above relationships? How do wages in the food system compare with those received elsewhere in the economy?

Research on labor unions is not familiar ground for agricultural economists. Collaboration with labor economists would be useful in becoming familiar with data sources and the institutional factors that are important to understand.

IMPACT OF REGULATIONS AND PUBLIC POLICIES ON ORGANIZATION AND PERFORMANCE OF THE FOOD SYSTEM. Through the years numerous federal and state laws have been enacted that impact on the food system. The basic premises of many of these laws are being questioned. Research priorities in this area include analysis of how antitrust laws influence industrial structure, conduct, and performance; benefits and costs of alternative regulations relating to health, safety, and nutrition; and the significance for farmers and others of antitrust immunities provided by the Capper-Volstead Act. Some research by general and agricultural economists has been undertaken in these areas, but there is a growing demand for such work. Both the historical case study and econometric cross-sectional approaches lend themselves to analyzing problems in this area.

OTHER TOPICS

The foregoing discussion is not all inclusive but covers topics that should be assigned a high priority for future research and that are at the frontier of IO research in the food system. Additional research also is needed on topics not listed because in our opinion they are no longer at the frontier (e.g., structure-profit studies) or may be more appropriately done by scholars other than agricultural economists (relationship between industrial organization theory and behavioral theory of the firm).

ORGANIZING FOR INDUSTRIAL ORGANIZATION RESEARCH

Many of the research efforts discussed in this chapter are rather large-scale undertakings. They often require the ability to gain access to the necessary data or to finance its purchase. The special census tabulation of food manufacturing for 1967 and 1972, purchased by NC-117 and the USDA, cost approximately $125,000. This suggests the great expense that can be incurred for certain types of data.

Several of the research efforts are also sensitive and somewhat controversial by nature. If the results are contrary to the interest of the special groups being examined (whether they are farmers, food manufacturers or distributors, or labor unions), they undoubtedly will be challenged. If the results are favorable to the parties involved, they as well as the objectivity of the researchers may well be challenged by consumer and public interest organizations. One must be prepared to withstand the "heat in the kitchen." The ability to do so is what makes academicians the main source of credible work in these areas.

Given these characteristics of the research, collaborative efforts involving several workers are particularly appropriate. An individual researcher who is willing to make a long-term commitment to certain issues and is well supported can make a significant contribution. However, such efforts can probably best be undertaken by a team at one or more universities who together represent a sufficient mass of resources that they can systematically and substantially address some of the central issues and not merely nibble around the edges.

Data are so critical to most IO research that having someone involved who knows the sources and limitations of various data sets is highly desirable. In some cases, the needed data can only be obtained through investigation by congressional committees, special commissions, or regulatory agencies. Previous experience in working with such groups obviously helps but is not essential.

Most of the research topics discussed earlier are also medium to long term in time requirements, depending on the number of people involved. A three-year commitment is probably the minimum one should consider. If a person has had neither training or previous experience in IO research, it hardly makes sense to invest the time required to "tool-up" unless he or she is willing to make a five-year or longer commitment. A ten-year horizon probably makes even more sense; such a commitment permits the researcher to undertake and complete several related projects that are additive, rather than several ad hoc essentially unrelated efforts.

Several of the important research issues in IO call for venturing into relatively unknown territory without a well-developed conceptual map. Like all explorers, one never knows what one will find, if anything. Potential contributions from some of these efforts are high, but so are the risks. We suggest a portfolio approach to planning research so that some low-risk efforts with tested methods and data are balanced against the more innovative and risky efforts. Although younger scholars may be philosophically better suited for pioneering efforts, the tenure and promotion systems at most universities make this hazardous. Because of the policy importance of some of the unanswered questions, however, successful research efforts can quickly thrust one into the national policy arena.

If we were to select the ideal organization for IO research on the food system, it would be a center at a university, established with a ten-year life and funded to support three or four senior economists and several junior economists and specialists, with the ability to purchase data necessary for its tasks. The center should have an advisory and review committee from other universities and government agencies, which would advise the center on research priorities and protect it from becoming ingrown and provincial. A regional research effort with at least a small-core research staff and effective coordination would be our second choice. Either design would be strongly preferred to reliance on individual researchers "doing their own thing" and hoping that the pieces fit together. In our opinion, this is likely to result in a number of piecemeal efforts that fail to answer the critical questions. Although the USDA has the resources to establish a critical mass of IO researchers, it may be vulnerable to adverse reactions to such activity. Thus, although some meaningful IO research can be done at the USDA, a separately organized center or regional project would ensure continuity and greater independence from the pressures of special interest groups.

REFERENCES

Adelman, Morris A. 1959. *A & P: A Study in Price-Cost Behavior and Public Policy.* Cambridge, Mass.: Harvard University Press.

Bain, Joe. 1945. *The Economics of the Pacific Coast Petroleum Industry,* parts 1, 2, 3. Berkeley: University of California Press.

_____. 1956. *Barriers to New Competition.* Cambridge, Mass.: Harvard University Press.

_____. 1968. *Industrial Organization,* 2nd ed. New York: Wiley.

Benham, Lee. 1972. The effect of advertising on the price of eyeglasses. *J. Law Econ.* 15:337–52.

Clodius, Robert L., and Willard F. Mueller. 1961. Market structure analysis as an orientation for research in agricultural economics. *J. Farm Econ.* 43:3, 512–53.

Comanor, William S., and Thomas A. Wilson. 1974. *Advertising and Market Power.* Cambridge, Mass.: Harvard University Press.

Connor, John M. 1980. The U.S. Food and Tobacco Manufacturing Industries. USDA, Agric. Econ. Rep. 451.

Connor, John M., and Willard F. Mueller. 1977. Market Power and Profitability of Multinational Corporations in Brazil and Mexico. Report to the Subcommittee on Foreign Economic Policy of the Committee on Foreign Relations, U.S. Senate. Washington, D.C.: USGPO.

Cotterill, R. W., and W. F. Mueller. 1979. The Impact of Firm Conglomeration on Market Structure: Evidence from the U.S. Food Retailing Industry. NC-117, Work. Pap. 33, University of Wisconsin, Madison.

Cyert, R. M., and J. G. March. 1963. *A Behavioral Theory of the Firm*. Englewood Cliffs, N.J.: Prentice-Hall.

Devine, D. Grant, and Bruce W. Marion. 1979. The influence of consumer price information on retail pricing and consumer behavior. *Am. J. Agric. Econ.* 61:3, 420-33.

Galbraith, J. K. 1973. *Economics and the Public Purpose*. Boston: Houghton Mifflin.

Goldschmid, H. J., H. M. Mann, and J. F. Weston, eds. 1974. *Industrial Concentration: The New Learning*. Boston: Little, Brown.

Hamm, Larry. 1979. Farm input industries and farm structure. In Structure Issues of American Agriculture. USDA, Agric. Econ. Rep. 438.

Hester, Donald. 1979. Customer relationships and terms of loans—Evidence from a pilot survey: A note. *J. Money, Credit and Banking* 11:349-57.

Hoffman, A. C. 1940. Large Scale Organization in the Food Industries. Temporary National Economic Committee, Monogr. 35, Washington, D.C.

Imel, Blake, and Peter Helmberger. 1971. Estimation of structure-profits relationships with application to the food processing sector. *Am. Econ. Rev.* 61:4, 614-27.

Imel, Blake, M. R. Behr, and Peter G. Helmberger. 1972. *Market Structure and Performance: The U.S. Food Processing Industry*. Lexington, Mass.: Heath.

Jesse, Edward V., and Aaron C. Johnson, Jr. 1970. An analysis of vegetable contracts. *Am. J. Agric. Econ.* 52:4, 545-54.

Kelly, W. H. 1969. On the Influence of Market Structure on the Profit Performance of Food Manufacturing Companies. Economic Report to the Federal Trade Commission. Washington, D.C.: USGPO.

Kessel, Reuben. 1971. A study of the effects of competition in the tax exempt bond market. *J. Polit. Econ.* 79:706-38.

Lang, Mahlon G. 1978. Structure, conduct, and performance in agricultural product markets characterized by collective bargaining. In Agricultural Cooperatives and the Public Interest, pp. 118-34. NC-117, Monogr. 4, University of Wisconsin, Madison.

Leibenstein, Harvey. 1966. Allocative efficiency vs. "X-efficiency." *Am. Econ. Rev.* 56:392-415.

Machlup, F., and M. Taber. 1960. Bilateral monopoly, successive monopoly, and vertical integration. *Economica* 27:101-17.

Mann, H. Michael. 1966. Seller concentration, barriers to entry, and rates of return in thirty industries. *Rev. Econ. Stat.* 48:296-307.

Mansfield, Edwin, J. Rapoport, A. Romeo, E. Villani, S. Wagner, and F. Husie. 1977. *The Production and Application of New Industrial Technology*. New York: Norton.

Marion, Bruce W., Willard F. Mueller, Ronald W. Cotterill, Frederick E. Geithman, and John R. Schmelzer. 1979. *The Food Retailing Industry: Market Structure, Profits and Prices*. New York: Praeger.

Marvel, Howard P. 1978. Competition and price levels in the retail gasoline market. *Rev. Econ. Stat.* 60:252-58.

Mason, Edward S. 1939. Price and production policies of large-scale enterprise. *Am. Econ. Rev.*, suppl. 29:61-74.

Masson, Alison, Robert T. Masson, and Barry C. Harris. 1978. Cooperatives and marketing orders. In Agricultural Cooperatives and the Public Interest, p. 209. NC-117, Monogr. 4, University of Wisconsin Madison.

Mather, Loys. 1979. Advertising and Mergers in the Food Manufacturing Industries. NC-117, Work. Pap. 36, University of Wisconsin, Madison.

Means, Gardiner. 1935. Industrial Prices and Their Relative Inflexibility. Senate Document 13, 74th Congress. Washington, D.C.: USGPO.

Mighell, R., and L. Jones. 1963. Vertical Coordination in Agriculture. USDA, Agric. Econ. Rep. 19.

Mills, C. Wright. 1956. *The Power Elite.* London: Oxford University Press.

Mintz, Morton, and Jerry S. Cohen. 1971. *America, Inc.* New York: Dial.

Mueller, W. F. 1974. Industrial concentration: An important inflationary force? In H. J. Goldschmid, H. M. Mann, and J. F. Weston, eds., *Industrial Concentration: The New Learning.* Boston: Little, Brown.

————. 1977. Conglomerates: A nonindustry. In W. Adams, ed., *The Structure of American Industry,* 5th ed. New York: Macmillan.

————. 1978a. Recent Structural Changes in the Beer Industry. Hearings on Conglomerate Mergers, Subcommittee on Antitrust and Monopoly, Committee on Judiciary, U.S. Senate. Washington, D.C.: USGPO.

————. 1978b. The Celler-Kefauver Act: The First 27 Years. Report Prepared for Subcommittee on Monopolies and Commercial Law of the Committee on the Judiciary, U.S. House of Representatives. Washington, D.C.: USGPO.

Mueller, W. F., and L. G. Hamm. 1974. Trends in industrial market concentration, 1947 to 1970. *Rev. Econ. Stat.* 56:511–20.

Mueller, W. F., and Richard T. Rogers. 1980. The role of advertising in changing concentration of manufacturing industries. *Rev. Econ. Stat.* 52:89–96.

Mueller, W. F., J. D. Culbertson, and B. Peckham. 1982. Market Structure and Technological Performance in the Food Manufacturing Industries. NC-117, Mongr. 11, University of Wisconsin, Madison.

National Commission on Food Marketing. 1966a. Organization and Competition in Food Retailing. Tech. Study 7. Washington, D.C.: USGPO.

————. 1966b. The Structure of Food Manufacturing. Tech. Study 8. Washington D.C.: USGPO.

Nichols, Wm. H. 1941. *Imperfect Competition within Agricultural Markets.* Ames: Iowa State College Press.

————. 1948. Reorientation of agricultural marketing and price research. *J. Farm Econ.* 30:43–54.

————. 1951. *Price Policies in the Cigarette Industry.* Nashville, Tenn.: Vanderbilt University Press.

Parker, Russell C., and John M. Connor. 1979. Estimates of consumer loss due to monopoly in the U.S. food manufacturing industries. *Am. J. Agric. Econ.* 61:626–39.

Rogers, R. T. 1978. Structure-profit relationship in food manufacturing firms, App. D. In W. F. Mueller, The Celler-Kefauver Act: The First 27 Years. Report to the Subcommittee on Monopoly, Committee on the Judiciary, U.S. House of Representatives. Washington D.C.: USGPO.

————. 1980. Advertising and Concentration Change in U.S. Food and Tobacco Product Classes, 1958–1972. Paper presented at American Agricultural Economics Association meetings, Urbana, Ill.

Salamon, L. M., and J. J. Siegfried. 1975. The relationship between economic structures and political power: The energy industry. In T. D. Duchesneau, ed., *Competition in the U.S. Energy Industry.* New York: Ballinger.

Scherer, F. M. 1974. Economies of scale and industrial concentration. In H. J. Goldschmid, H. M. Mann, and J. F. Weston, eds., *Industrial Concentration: The New Learning.* Boston: Little, Brown.

————. 1980. *Industrial Market Structure and Economic Performance,* 2nd ed. Chicago: Rand McNally.

Schumpeter, Joseph. 1950. *Capitalism, Socialism, and Democracy,* 3rd ed. New York: Harper & Row.

Shepherd, William G. 1969. Market power and racial discrimination in white-collar employment. *Antitrust Bull.* 14:14–61.

Siegfried, J. J. 1975. Market structure and the effect of political influence. *Ind. Organ. Rev.* 3:1–7.

Stigler, George. 1951. The division of labor is limited by the extent of the market. *J. Polit. Econ.* 59:185–93.

Temporary National Economic Committee. 1940. Large Scale Organization in the Food Industries. Monogr. 35. Washington, D.C.: USGPO.

Weiss, Leonard. 1971. Quantitative studies of industrial organization. In M. D. Intriligator, ed., *Frontiers of Quantitative Economics*. New York: North-Holland.

_____. 1974. The concentration-profits relationship and antitrust. In H. J. Goldschmid, H. M. Mann, and J. F. Weston, eds., *Industrial Concentration: The New Learning*. Boston: Little, Brown.

Williams, Willard. 1979. The changing structure of the beef packing industry. In Small Business Problems in the Marketing of Meat and Other Commodities, part 4. Hearings of the Committee on Small Business, U.S. House of Representatives. Washington, D.C.: USGPO.

Williamson, Oliver. 1971. The vertical integration of production: Market failure considerations. *Am. Econ. Rev.* 61:112–23.

EMERSON M. BABB
MAHLON G. LANG

3

INTRAFIRM DECISION MAKING: PRIVATE AND PUBLIC CONSEQUENCES

INCREASINGLY, resource use in the food system is determined by intrafirm decisions. As firms integrate both horizontally and vertically, fewer resource allocation decisions are disciplined directly by the market. The consequences of this trend are the subject of this chapter.

The problem area is defined by looking beyond the assumptions of the neoclassical firm. The large firm is a unique bureaucracy. It differs in many respects from the monolithic decision-making structure assumed in neoclassical theory. While disciplined at its boundaries by market forces, the modern firm is characterized by what Cyert and March (1963) have called "organizational slack." Slack affords the firm an option to select alternative operational goals and associated strategies, while passing the test of economic survival. The option to pursue alternative goals implies alternative ways of allocating scarce resources and a variety of possible responses to external stimuli. Slack also implies the possibility of X-inefficiency as defined by Leibenstein (1966). X-inefficiency occurs when, for example, the absence of appropriate incentives prevents an otherwise (allocatively) efficient number of workers from working as hard as they might.

This image of the firm contrasts sharply with the neoclassical firm operating with zero slack and responding predictably to policy changes and other external stimuli. The entrepreneur has perfect control over the firm. Changes in policy and market environments will lead the entrepreneur to behave in determinate ways to achieve the single goal, profit maximization.

EMERSON M. BABB is Professor of Agricultural Economics, Purdue University, West Lafayette, Indiana. MAHLON G. LANG is Assistant Professor of Agricultural Economics, Purdue University, West Lafayette, Indiana. The authors are indebted to P. L. Farris, B. C. French, A. C. Hoffman, B. W. Marion, D. I. Padberg, J. D. Shaffer, and E. F. Stone for constructive suggestions on earlier drafts.

In the modern firm, internal resource allocation is shaped not only by market forces but also by intrafirm decision-making options. Resource allocation within the modern firm is not determinate. The existence of slack permits it to pursue alternative goals with respect to sales, market shares, firm growth, stability, and other areas, none of which are necessarily consistent with profit maximization. The firm may focus on a variety of related strategies in pursuit of its goals. These may include technology development, acquisitions, diversifications, foreign investment, vertical integration, and others.

When the market structure is purely competitive, there is little corresondence between models of the firm proposed by economists and organizational theorists. In fact, organizational theorists would have no interest in such an environment. Conversely, economists have developed a variety of models of firms in an environment characterized by imperfect competition. These vary with respect to goals, behavioral assumptions, risk preferences, information availability, and the like. While economists and organizational theorists have fundamental differences in their views, their models of imperfectly competitive firms tend to merge or become more consistent.

The importance of the problem area derives from the fact that most resource allocation decisions are made within large, horizontally and/or vertically integrated firms. Marketing economists are logically concerned with their inability to explain the goal and strategy selection process and to predict the effects of market and policy changes on firm behavior and therefore resource allocation. The executives of large firms are also concerned. They face the challenge of harmonizing activities of many individuals and groups with the goals of the firm.

Thus the subject of intrafirm decision making is of importance to private as well as public interests. The public sector has an interest in learning how intrafirm decisions are made so that public policy decisions can be informed with respect to their expected consequences. Managers have an interest in controlling the vast resources of the firm to achieve private goals. Huge shares of our food producing, processing, and distributing resources are affected by public and private decisions concerning intrafirm resource use.

The issue at hand concerns what kinds of food system organization and control are consistent with public and private interests. More specific sets of issues are discussed later in the chapter.

The objectives of the chapter are to provide a brief review of past research in the field of intrafirm decision making; to identify the purposes of such research as well as its principal uses, to identify needed research and major barriers to it; and to offer a list of priorities and approaches with respect to research in the problem area.

OVERVIEW OF PAST RESEARCH

The literature relevant to intrafirm decision making is massive. Three major handbooks have numerous chapters that deal specifically with the topic (March 1965; Lindzey 1968; Dunnette 1976). At least a dozen journals publish papers on the subject, and numerous texts address the field. Theoretical and empirical contributions have come from many disciplines, including social psychology, sociology, management, and economics. Clearly, even a superficial review and assessment of past work is impossible here.

The early work in organizations and their decision making viewed them as closed systems that make rational choices in achieving well-defined goals. This work emphasized processes and transactions internal to the firm such as specification of tasks, standardized role performance, centralized decision making, uniformity of practice, and avoidance of duplication. Contributions were made by Max Weber (1948) on bureaucracy, Luther Gulick (1937) on public administration, Fredrick Taylor (1911) on scientific management, Chester Barnard (1938) on executive functions, and Alfred Marshall (1922) on economic goals and behavior of firms.

The pioneering work of Simon (1958) marks the beginning of modern organization theory. His most important contribution was development of the concept of bounded rationality. Under this concept, optimizing is replaced by satisficing, choices are discovered through imperfect and limited search processes, and repertories of action programs are developed. This concept has been greatly refined and is important to the study of the firm's responses to external forces, including public policy (March and Simon 1958).

A major thrust of the work on organizations has been directed toward the linkage between them and their environments (Lawrence and Lorsch 1967; Chamberlain 1968; Hall 1972; Katz and Kahn 1978). External forces of a technical, legal, political, economic, demographic, ecological, and cultural nature impinge on firm decision making. The firm may recognize the importance of these external forces and attempt to modify and shape them. Internal environmental factors relating to interactions of individuals and subunits, communications, authority, and the like also shape decision making.

In classical theory, all the energy of the firm is used for internal activities that achieve well-defined goals. Current theory recognizes that some energy may be diverted from internal activities to the functions of reacting to and influencing the external and internal environment. Under extreme pressure, an organization may divert most of its energy to dealing with these forces. Goals may be altered by changes in the external environment. An example of such behavior is the reaction of the Navel Orange Administrative Council (NOAC) to proposals by the Cost of Living Council (CLC) that would have increased prorates for shipments of fresh oranges (Nelson and Robinson 1978). NOAC may have subverted the goal of maximizing grower income in its efforts to diminish the influence of the CLC, which it viewed as a threat.

Organizational theorists now view the firm in the context of an open behavioral system (Katz and Kahn 1978). The firm has overall objectives but expends resources in reacting to and influencing its environment as a part of its effort to achieve those objectives. Many firm goals, including survival, are related to profit but do not imply profit maximization, at least in the short run.

Differentiation (specialization) is necessary in large organizations. Subunits are established with specialized missions essential to overall firm goals, but such subunits may in turn develop their own goals or place weights on objectives that are different from those that would best achieve overall firm goals. Further, individuals within subunits have personal goals that may conflict with those of the firm. Thus integration to achieve unity of purpose among diverse subunits and individuals is an important determinant of firm success. There is a continuing bargaining process among subunits, and the integrator is responsible for coordination and conflict resolution. The integrator's challenge is that of creating incentive structures to bring the actions of the individual into harmony with the goals of the firm. Hierarchical position may not dominate in the bargaining process but may give way to personalities, committees on long-range planning, and the like. Integrators have more lateral than vertical contacts, and many of their decisions affecting resource use may turn on interpersonal relations as well as firm goals.

One reason for greater differentiation is to broaden the interface between the firm and its external environment. Many subunits such as the sales, legal, and public relations departments are established to deal with the external environment (Adams 1976). These subunits work at the boundary between the firm and its external environment in an attempt to reconcile the two. They are more distant from the firm than other subunits and may view their roles as those of agents of the firm. Boundary positions are thus under pressure from the firm and the external environment.

Much has been written about the discretion firms have in using resources to pursue various objectives. Larger firms are thought to have greater discretion. Firms value this for safety and other reasons. While firms desire greater discretion, society attempts to narrow the range they have in decision making. There is thus continuing conflict between society and firms as each tries to achieve its objectives by controlling the environment. In general, firms seek to minimize their exposure to the discipline of markets, while public policy is directed at obtaining greater exposure.

These summary comments do not begin to address the complexity and diversity of intrafirm decision making but should help in understanding why the consequences of public policies may deviate from expectations. Unfortunately, in spite of extensive research, our understanding of firm decision making is still at a rather abstract level. Reasons for this are discussed later. A wealth of hypotheses have been advanced by organizational theorists (Thompson 1967), but it has been difficult to muster data that lead to conclusive acceptance or rejection of hypotheses. This is in part due to diversity of firms and their environment. Each firm seems to have many unique features, and no single organizational structure is optimal.

The bulk of empirical work on intrafirm decision making has used field studies (March 1965; Lindzey 1968; Dunnette; 1976). These have ranged from case studies of a single firm or subunit to cross-cultural studies of firms. Most of the work has involved studies of firms in one industry or cross sections of firms in different industries. Simulation has been used to examine how firm goals are affected by variations in environmental factors, to determine how firm decisions affect their performance, and to model firm decision-making processes to gain a better understanding of them (Bonini 1963; Cyert and March 1963; Cohen and Cyert 1965; March 1965; Patrick and Eisgruber 1968). Laboratory studies have been used to gain a better understanding of decision-making processes, determine how firms respond to external environments, and find ways to improve organizational processes and efficiency (Siegel and Fouraker 1960; Lowin and Craig 1968; Graen et al. 1971; Jones and Babb 1975; Perreault and Miles 1978). While field, simulation, and laboratory studies have produced useful results, there are disagreements about their strengths and weaknesses. This is partly due to the fact that the contributions made using each of the approaches have been quite different.

PURPOSES AND USES OF RESEARCH

The vast area of past and potential research has three possible applications: to understand the intrafirm decision-making process, enhance intrafirm allocative efficiency, and improve firm performance relative to public goals. In the following pages we discuss these applications and, where possible, provide illustrations.

Understanding the Intrafirm Decision-making Process. Studies conducted to learn how an organization functions have been primarily in the domain of social psychology and administrative sciences. Relatively few economists have had more than fleeting interest in decision making internal to the firm.

A classic set of studies by Cyert, Dill, March, and Starbuck (Cyert and March 1963) provides an example of such work by economists. They studied four decisions by three well-established, successful firms by detailed analysis of memoranda, letters, and other file material; intensive interviews with participants in the decisions; and direct observation of the decision process. Their aim was to provide a start for empirical research.

Other studies have focused on many areas, including organizational structures, goal selection and modification, information search processes, communication networks, decision rules used for routine and strategic decisions, leadership styles, differentiation and integration, interpersonal relations (power, influence, authority, interactions of participants), individual motivation, and organizational development.

The potential application of such studies lies in their use as a base for development of a theory of intrafirm decision making and as benchmarks for studies to improve performance with respect to firm and public goals.

Enhancing Firm Efficiency. Agricultural economists have conducted substantial research aimed at improving decision making relative to firm goals. Some research has focused on decisions per se. For example, much effort has been expended in development and application of decision-making techniques. Recent interest has been on incorporating risk and uncertainty in decision models and on optimization where the organization has a multiple objective function (Anderson et al. 1977; Candler and Norton 1977; Charnes and Cooper 1977). Other research has been oriented toward the provision of information to be used by the firm. Examples include assessment of new technology and analyses of productive efficiency. These were reviewed by French (1977).

It has been argued that too much marketing research has focused on the private sector and that firms have been the principal beneficiaries of this effort (Shaffer 1968). While this may be true, the delineation between private and public purposes is fuzzy at best. First, the public may capture some of the gains from research, which reduce costs or improve the firm's efficiency, even in an industry characterized by monopoly. More important, traditional firm marketing research provides information that is essential to public decision making regarding marketing policy. To analyze the consequences of alternative marketing policies requires considerable firm level information such as input-output relationships, costs, and other technical coefficients. Since these change, much information from earlier studies becomes obsolete. In the absence of current firm-level information, sophisticated models developed for policy analysis are not likely to assess alternatives accurately or to be useful in designing a better functioning marketing system. Further, public research may still be influenced in part by the flow of new information. There also remains a broader public interest in viability of small firms.

Reasons for deemphasizing firm-level research are that firm size has grown to the point that many have the capability of conducting research internally, and numerous private research firms can meet private needs. Unfortunately, information developed by firm researchers or acquired by consulting firms is proprietary and usually not available to the public. It is thus argued that there is a continuing need for public research on firm-level goals, behavior, and production processes. The focus of such research should be to acquire information needed for public decision making, but it is clear that it will also be potentially useful for firm decision making. The private value of public research is a vital key to cooperation of private firms with public researchers.

Improving Industry Performance Relative to Public Goals. We have stressed the importance to policymakers of research to determine how firms will respond to changes in their external environments. The basic assumption of such research is that both the firm and the public attempt to shape the external environment to achieve sometimes conflicting goals. If the public is to be effective in stimulating desired industry performance, it must understand how firms interact with their external environments.

This interaction has received increased attention of social psychologists and administrative scientists. But the focus of their work is on improving administrative processes that enable the firm to cope with its environment, especially in its boundary subunits. While these studies may provide some insights on intrafirm decision making that relate to environmental interaction, they are too general to provide sufficient information for the design of public policy. For example, one study attempted to determine the type of organizational structure that could best deal with varying economic and market conditions. Lawrence and Lorsch (1967) analyzed competing firms in several industries, including one sector of the food industry. They found that successful organizations had different structures depending on the industry, degree of automation of production processes, extent of market uncertainties, and stability of the external environment. Degree of differentiation in firm subunits was related to external factors and to firm success. It was concluded that no single organizational structure was optimal. The desirability of a given structure was much affected by the external environment to which the firm was exposed. Many of their findings may be useful to the firm in developing an organizational structure that can function most effectively in achieving firm goals, given a specified external environment. But their findings are of limited use for policy analysis that includes firm response to alternatives being analyzed.

Given the interest of social psychologists and administrative scientists in the firm–external environment interaction, it may be possible to induce them to collaborate with economists on research to provide more specific information about firm response to policy alternatives. Their experience in conducting research on firm decision-making processes would be most valuable.

Economists' interest in firm behavior goes back to Adam Smith, but the first empirical work is of more recent vintage. Hall and Hitch (1939) have found that managers of firms used full cost rather than marginal cost as the basis for price determination. The perspective of economists is quite different from that of other social scientists studying the firm. On a theoretical level, greatest attention has been focused on the goals of the firm and implications of different goals for price and output decisions (Penrose 1952, 1959; Baumol 1959; McGuire 1964; Morris 1964). Even here, there are few empirical studies of firm goals and the behavior implied by these goals (Hathaway et al. 1966; Ladd 1967). Perhaps the greatest number of empirical studies in the agricultural sector relate to bargaining between buyers and sellers, but this represents only one dimension of the external environment of the firm (Ladd and Hallberg 1965; Babb et al. 1969; Knutson 1971; Zusman and Amiad 1977).

There are more voluntary organizations in agriculture than in most other industries. The firm–external environment interactions for voluntary organizations may be different from those for proprietary firms. There may also be differences as a result of functions performed by the firm, size of the firm, and commodities handled. Numerous studies have assessed the consequences of public policies and regulation (Dobson and Buxton 1977; MacAvoy 1977; Kilmer and Hahn 1978; Nelson and Robinson 1978). While such research is useful, it has usually given little indication of why policies did or did not produce desired results. Such after-the-fact studies at an aggregate level have limited value in designing an environment that will produce desired outcomes. There is great need to augment the theory of the firm and to conduct empirical research on firm–external environment interactions needed for policy analysis and design of public policy for agricultural marketing.

FUTURE RESEARCH NEEDS

While much research has addressed the subject of intrafirm decision making, many questions remain unanswered. This is especially true in the U.S. food system. A series of research questions, some general and some specifically applicable to the food system are listed below:

1. What are the areas of conflict and harmony in objective functions of the firm and the public? This involves clarifying specific objectives and the weights attached to them. Public goals and expectations for firm performance may be beyond the scope of traditional market performance measures, especially for firms less disciplined by market forces. Such expectations are usually placed in the genre "social responsibility," which is not well-defined or easily measured. Information about goals would be useful to policymakers in that where goals are in harmony, the self-interest of firms may dictate behavior that will more likely conform with social norms. Where there is harmony, there may be opportunities for cooperation and joint benefits from research. Also, policymakers could focus more attention on areas where there are conflicts in goals. For example, the recent controversy related to the Federal Trade Commission's case against three major breakfast cereal manufacturers involved the public's desire for innovation and the manufacturer's willingness to emphasize new cereal products. But if product differentiation is used as a barrier to entry, the public is not pleased. There is harmony with respect to the goals of progressiveness and innovation. There is conflict with respect to the use of differentiation to affect competition.

2. What are the firm's goals? Why and to what extent do goals vary from one firm to another and what are the consequences of these variations? The first question requires inductive research, but hypotheses about the consequences of variations in goals can be tested. We can deduce price and/or quantity decisions consistent with such goals as sales maximization, sales maximization subject to a profit constraint, market share maintenance, and the like. Research directed at this question might be useful to firms, as well as to the public, in that it might suggest changes in internal organization and decision making that would improve firm performance relative to both public and private goals.

   Firms may choose various strategies to achieve even the same goal. For example, they may emphasize acquisitions, foreign investment, technology development, or diversification to achieve sales growth. Research to determine why firms focus on one goal or another and why various strategies are selected would be useful to private firms and to policymakers alike.

3. Why do firms diversify and enter or exit sectors of the food industry (change the relative proportions of their business in various sectors or change linkages) and what are the consequences? Hypotheses concerning diversification might be based on risk aversion, growth objectives, unused managerial services (Penrose 1959), external environment including public policies, financial flexibility, market access, and discretion.

4. How is decision making influenced by the competitive environment? For example, how do firms interact with one another and what factors are considered in making decisions about these interactions? Probably the majority of public policies that affect the external environment concern the firm's competitive environment. This area has clearly captured public attention, even though the public goals are less than clear-cut and consist-

ent. Theories of imperfect competition provide the basis for hypotheses concerning price and output decisions. Unfortunately, many conflicting hypotheses could be developed from existing imperfect competition theory, since each is predicated on different firm goals, behavioral assumptions, interdependencies, and the like. If the researcher had access to data, this would be a fruitful area for inductive (descriptive) research, which in turn might improve the state of theory. Firms within an industry clearly have common interests in many areas and conflicting interests in many others. Identification of areas in which they cooperate, as evidenced by subjects addressed in trade journals and trade meetings, may help to identify areas in which they compete.

5. How is decision making influenced by the external environment, excluding compétitive environment, and what factors are considered? More specifically, how is decision making influenced by firm and industry growth and by changes in technology; organization of the economy; level of economic activity; and legal, political, demographic, ecological, and cultural dimensions of the environment? At this time, research is clearly limited to inductive work.

6. How is decision making influenced by public policy and regulation? Regulations are generally directed at specific firm behavior. In adjusting to the regulation, firms may alter their behavior in areas outside the scope of regulation. For example, suppose a dairy processing firm has the goal of maximizing sales subject to a constraint on return to equity. It views fluid milk pricing under federal orders as limiting sales growth and may thus reallocate its resources to dairy products not regulated, other foods, or even nonfood products. Not only investments in facilities and personnel but also sales, promotional, and product development efforts are diverted. Prices charged consumers would be different from those under a profit maximizing goal unless the return on equity constraint was binding at the profit maximizing quantity. If the regulatory policy had the objective of ensuring a specified return to farmers, it might not achieve that objective. Further, the firm's response would have other longer run consequences relating to productive efficiency in marketing, progressiveness, and consumer prices.

The relationship between firm goals and firm response to regulation is one area where deductive research can be done. One might also expect that response to be affected by the number and effectiveness of boundary subunits that deal with regulations. Since regulations may have great impacts on firm performance, one possible source of competitive advantage to large firms over small firms is the capacity to employ boundary subunits. One consequence of regulation may be altering of the size distribution of firms in an industry in a way not desired by the public.

Future work in evaluation of regulations should attempt to analyze how the responses of firms have affected the consequences of regulation. Anticipatory response should be analyzed; that is, are firm decisions influenced by the threat of changes in public policies or regulations? It is possible that perceptions of the public mood and policy changes that might follow induce greater change in firm behavior than promulgated regulations.

7. What are the interactions between firms and legislative, executive, and regulatory bodies and how do these interactions affect achievement of public and firm goals? This question appears to be in the domain of investigative reporting, and perhaps it should remain there. However, objective studies of the question might assist in development of rules for interactions that protect the public interest and resolve conflict more rapidly. Recent moves toward open meetings, freer access to public records, and increased reporting by persons in such interactions may provide considerable data.

8. How can we predict changes in firm decision making that result from changes in the competitive environment, other external environments, and public policy? Whether anything can be predicted depends on the findings for the above questions, especially 4, 5, and 6. It may be difficult to conduct research on firm response to competitive environment and ignore other external environments and public policies. There is a high probability of interaction among these external factors. Assuming that relations between firm response and various external environments can be quantified, simulation might provide a mechanism to systematically explore expected firm behavior under varying settings of external environment.

9. What information should be provided to firms by public institutions? We previously indicated that considerable research to obtain technical coefficients was needed for policy analysis and that such research might also be useful to firms. We would place a higher priority on this type of research than on that directed primarily at firm uses. Intermediate and longer run projections on consumption, production, and external factors would be encouraged and short-run outlook would receive less research effort. There are many sources of short-run outlook information, and longer run projections would be less likely to duplicate efforts of firms and others. Major resource allocation questions are based on longer run projections. Given the asset fixity of many investments, better projections could reduce economic losses from decisions based on poor information. Further development of techniques to improve decision making should be encouraged. It should be noted that the above areas of research (intermediate projections and improved decision techniques) would have significant value for public decision making.

10. What have been the impacts on firm performance and achievement of public goals of research that provided decision-making techniques and information used directly in making decisions? This question serves only as a reminder that some day it might be asked. There is no obvious theoretical framework on which to base this area of research.

11. What generalizations can be made about standard operating procedures, decision rules, policy guidelines, and general policy frameworks used to

control the firm? Because chief executives cannot make every decision
personally, they delegate authority to subordinates, in some cases with
very specific guidelines such as standard operating procedures. In other
cases, less specific directives are extended to personnel who may exercise
some discretion in their decisions. To the extent that research provides
descriptive information that can ultimately be generalized about what
standard operating procedures, decision rules, guidelines, and other control mechanisms are used and under what conditions, our capacity to
predict firm behavior will be enhanced.

12. What are the organizational structures, mandates, and functions of
    boundary subunits in the firm? Boundary subunits such as public rela-
    tions, personnel, legal, sales, and other departments have direct contact
    with the external environment of the firm. It is certain that such units af-
    fect the flow of information into and out of the firm, thereby shaping its
    responses to the policy environment and the policy environment itself.

13. How are internal (transfer) prices determined? Most resource allocation
    choices are not disciplined directly by prices determined in markets. Thus
    the firm must establish its own prices for intermediate products. The way
    this pricing is done may affect how the firm allocates resources internally
    and therefore externally. Allocative efficiency is affected by the transfer
    pricing process. Thus there is public as well as private interest in how it is
    conducted.

    A large meat packing firm, for example, purchases livestock, labor,
    management, energy, and other factors in a market but faces no other ex-
    ternal pricing points until it sells wholesale meat. The same is true of
    many other food packing and processing firms.

BARRIERS TO RESEARCH

There are many barriers to research on the topics listed. We have few suggestions as to how these barriers can be reduced but believe that the importance of these problems warrants expending resources to do so.

Inadequate Theoretical Frameworks. Research to create new theoretical
frameworks is both inductive and deductive. The inadequacy of theoretical
frameworks is to a large degree the reason for inductive work. Such research
identifies and quantifies firm behavioral parameters rather than assuming
them. In many areas, the capacity to proceed to deductive research will probably come only after much inductive work.

There is not a total void in theoretical frameworks. Perhaps the best
established dimension relates to goals of the firm. This may be a key determinant of firm response to its external environment, and the theoretical basis for
hypothesis testing is available. Beyond this, other social scientists have
developed propositions about the functioning of organizations. Their work
needs to be examined for possible modifications that could be made to provide
more adequate theoretical frameworks. High priority should be given to

development of these frameworks, and we believe research of an inductive nature is a vital prerequisite to their development.

Identification and Quantification of Variables. The literature reveals many variables that affect intrafirm decision making. In the absence of a theoretical framework, researchers face problems in identifying key variables for analysis. This situation has the potential of resulting in fragmentation of research and lack of additivity. Until a critical mass of common observations on firm behavior is accumulated so that better theoretical frameworks can be developed, there is no quick solution to the problem. Our suggestion is to restrict deductive research to hypothesis testing about the relations between goals and firm response to environment until a critical mass of descriptive data is accumulated. Inductive work on weights placed on alternative firm goals and the consequences of bounded rationality also seems important.

The theoretical frameworks of other social sciences often contain variables that are difficult to quantify. Key variables such as differentiation and integration are examples. It is likely that theory developed for this research area would also contain variables that are difficult to measure. If so, the range of analytical techniques will be limited. Thus researchers addressing such topics will face a trade-off between precision of analysis and number of possibly important variables considered.

Lack of Generality. Closely related to the lack of theoretical frameworks is the serious risk that we will not readily be able to generalize about firm–external environment interactions for different types of firms (especially with the large number of voluntary organizations in agriculture) in different industries. If this proves to be so, it will reduce the value of research findings for policy analysis and make the needed research more expensive. Clearly, there is lack of generality in much of the research by other social scientists regarding organizational structure and behavior. They have identified considerable uniqueness in firms in particular internal and external environments. Some of this lack of generality may be due to their focus on administrative processes internal to the firm. Future studies may reveal more generality about firm responses to changes in environment.

Access to Data. The most serious problem will probably be that of obtaining data from firms on such things as goals, strategies, decision rules, operating procedures, sources and uses of information, and interactions of personnel. The firm is reluctant to collaborate in this type of research for the following reasons: there is a cost of executive time and other resources, there is a risk that information provided may be beneficial to competitors, there may be legal risks inherent in data provided, and the study findings may adversely affect their external environment (i.e., reduce their discretion in resource allocation and increase exposure to market discipline). A record of useful research findings and objectivity on the part of the researcher may make data more accessible, but it does not solve the fundamental problem of firm security.

Cooperatives may be more willing than proprietary firms to provide needed data and may thus be a good laboratory for both inductive and deductive research. This is important since a large share of the food system is made up of cooperative firms.

Professional Image. It was indicated earlier that the greatest research void in the area of the professional image is at the interface of public and private decision making. It also places the researcher on the boundaries between economics, social psychology, and administrative science. It may be more difficult to develop a reputation at the boundaries of disciplines, and the obvious call for interdisciplinary research does not solve this problem. Further, much of the inductive research will be considered "soft." This will be compounded if variables cannot be quantified and subjected to rigorous analysis. In short, young researchers facing tenure and promotion decisions run considerable risk in conducting soft research outside the mainstream of their discipline. Perhaps this should be left to older workers who are more readily forgiven for doing soft research in murky areas.

Alternatively, the profession may grow to recognize the trade-offs between relevance of research in terms of variables studied and the accuracy of measurement and confidence in generalizations. We may, with relatively high levels of precision, measure prices, selected outputs, and selected resources used. But if we choose not to generalize about decision-making behavior because its variables are hard to measure with confidence, we may forego a greater capacity to predict behavior.

If the profession holds fast to its traditional concern with high confidence in a few measurable variables, progress in the area of intrafirm decision making will indeed be slow. If the profession develops a tolerance (or even respect) for what is pejoratively called soft research, expected returns (as well as expected risk) will increase. The fact that Simon was awarded the Nobel Prize in economics for his work on decision making does much to enhance the respectability of research in the area.

RESEARCH PRIORITIES AND THEIR CHALLENGE

The possible research questions listed earlier is neither exhaustive nor refined; much remains to be done in terms of adding to the list and identifying subsets. While major obstacles exist to research in the general area, there is variation in the degree of difficulty associated with specific questions. In addition, some appear to be more important than others. Both the degree of importance and of difficulty associated with each area should be considered in developing a research agenda in the area of intrafirm decision making.

As a guide to persons interested in conducting such research we offer our assessment of the importance and difficulty associated with each of the listed questions (Table 3.1). A rank of 1 indicates that an area is most important or most difficult and a rank of 3 indicates lesser importance or difficulty.

The table serves to integrate the discussion in the two preceding sections

TABLE 3.1. **Assessment of importance and degree of difficulty of research in intrafirm decision making**

	Rank	
Research question	Importance	Difficulty
1. What are the areas of conflict and harmony in public and private goals?	2	2
2. What are the goals of the firm, why do they vary among firms, and what are the associated consequences?	1	3
3. Why do firms diversify?	2	2
4. How is decision making influenced by the competitive environment?	1	2
5. How is decision making influenced by its external environment (excluding competitive environment)?	2	1
6. How is decision making affected by public policy?	1	2
7. How do firms interact with legislative, executive, and regulatory bodies?	3	2
8. How can we predict changes in firm decision making that result from the external environment and public policy?	1	1
9. What information should be provided to private firms by public institutions?	3	3
10. What is the impact on firm performance resulting from public research on decision making and the informational needs of firms?	3	1
11. What can be generalized about standard operating procedures, decision rules, and policy guidelines used by firms?	2	2
12. What are the structures, mandates, and functions of the firms' boundary subunits?	2	3
13. How are transfer prices determined?	1	2

of this chapter. It is clear that the rankings included in the table are subjective and may well be revised as research proceeds. They are offered as a starting point for those who would consider research in intrafirm decision making.

REFERENCES

Adams, J. S. 1976. The structure and dynamics of behavior in organizational boundary roles. In M. D. Dunnette, ed., *Handbook on Industrial and Organizational Psychology,* pp. 1175–99. Chicago: Rand McNally.

Anderson, J. R., J. L. Dillon, and J. B. Hardaker. 1977. *Agricultural Decision Analysis.* Ames: Iowa State University Press.

Babb, E. M., S. A. Belden, and C. R. Saathoff. 1969. An analysis of cooperative bargaining in the processing tomato industry. *Am. J. Agric. Econ.* 51:13–25.

Barnard, Chester I. 1938. *The Functions of the Executive.* Cambridge, Mass.: Harvard University Press.

Baumol, W. J. 1959. *Business Behavior, Value and Growth.* New York: Macmillan.

Bonini, C. P. 1963. *Stimulation of Information and Decision Systems in the Firm.* Englewood Cliffs, N.J.: Prentice-Hall.

Candler, W., and R. Norton. 1977. Multi-Level Programming and Development Policy. World Bank Staff Work. Pap. 258, IBRD, Washington, D.C.

Chamberlain, N. W. 1968. *Enterprise and Environment: The Firm in Time and Place.* New York: McGraw-Hill.

Charnes, A., and W. W. Cooper. 1977. Goal programming and multiple objective optimizations. *Eur. J. Oper. Res.* 1:39–54.

Cohen, K. J., and R. M. Cyert. 1965. Simulation of organizational behavior. In J. G. March, ed., *Handbook of Organization,* Chap. 7. Chicago: Rand McNally.

Cyert, R. M., and J. G. March. 1963. *A Behavioral Theory of the Firm.* Englewood Cliffs, N.J.: Prentice-Hall.

Dobson, W. D., and B. M. Buxton. 1977. Analyses of the Effects of Federal Milk Orders on the Economic Performance of U.S. Milk Markets. University of Wisconsin Agric. Exp. Sta. Res. Bull. 2897, Madison.

Dunnette, M. D., ed. 1976. *Handbook on Industrial and Organizational Psychology.* Chicago: Rand McNally.

French, B. C. 1977. The analysis of productive efficiency in agricultural marketing: Models, methods, and progress, part 2. In L. R. Martin, ed., *A Survey of Agricultural Economics Literature,* vol. 1. Minneapolis: University of Minnesota Press.

Graen, G., J. B. Orris, and K. M. Alvarez. 1971. Contingency model of leadership effectiveness: Some experimental results. *J. Appl. Psych.* 55:196–201.

Gulick, Luther, and L. Urwick, eds. 1937. *Papers on the Science of Administration.* New York: Institute of Public Administration.

Hall, R. H. 1972. *Organizations: Structure and Progress.* Englewood Cliffs, N.J.: Prentice-Hall.

Hall, R. L., and C. J. Hitch. 1939. Price theory and business behavior. *Oxford Econ. Pap.* 2:12–45.

Hathaway, D. E., R. L. Feltner, J. D. Shaffer, and D. Morrison. 1966. Michigan Farmers in the Mid-Sixties. Michigan State University. Agric. Exp. Sta. Res. Rep. 54, East Lansing.

Jones, L. D., and E. M. Babb. 1975. An analysis of behavior and performance in the food retailing industry using experimental business gaming. *Decision Sci.* 6:541–44.

Katz, D., and R. L. Kahn. 1978. *The Social Psychology of Organizations.* New York: Wiley.

Kilmer, R. L., and D. E. Hahn. 1978. Effects of market-share and antimerger policies on the fluid milk processing industry. *Am. J. Agric. Econ.* 60:385–92.

Knutson, R. D. 1971. Cooperative Bargaining Developments in the Dairy Industry, 1960–70. USDA, Farmer Coop. Serv. Res. Rep. 19.

Ladd, G. W. 1967. Analysis of Ranking of Dairy Bargaining Cooperative Objectives. Iowa State University Agric. Exp. Sta. Res. Bull. 550, Ames.

Ladd, G. W., and M. Hallberg. 1965. An Exploratory Study of Dairy Bargaining Cooperatives. Iowa State University Agric. Exp. Sta. Res. Bull. 542, Ames.

Lawrence, P. R., and J. W. Lorsch. 1967. *Organization and Environment.* Cambridge, Mass.: Harvard University Press.

Leibenstein, H. 1966. Allocative efficiency vs. "X-efficiency." *Am. Econ. Rev.* 56:392–415.

Lindzey, Gardiner, ed. 1968. *Handbook of Social Psychology.* Reading, Mass.: Addison-Wesley.

Lowin, A., and J. R. Craig. 1968. The influence of level of performance on managerial style. *Organ. Behav. Hum. Perform.* 3:440–58.

MacAvoy, P. M., ed. 1977. *Federal Milk Marketing Orders and Price Supports.* Washington, D.C.: American Enterprise Institute.

McGuire, W. 1964. *Theories of Business Behavior.* Englewood Cliffs, N.J.: Prentice-Hall.

March, J. G., ed. 1965. *Handbook of Organizations.* Chicago: Rand McNally.

March, J. G., and H. A. Simon. 1958. *Organizations.* New York: Wiley.

Marshall, Alfred. 1922. *Principles of Economics.* London: Macmillan.

Morris, R. 1964. *The Economic Theory of Managerial Capitalism.* New York: Free Press.

Nelson, G., and T. H. Robinson. 1978. Retail and wholesale demand and marketing order policy for fresh navel oranges. *Am. J. Agric. Econ.* 60:502–9.

Patrick, G. F., and L. M. Eisgruber. 1968. The impact of managerial ability and capital structure on growth of the farm firm. *Am. J. Agric. Econ.* 50:491–506.

Penrose, E. T. 1952. Biological analogies in the theory of the firm. *Am. Econ. Rev.* 42:804–19.

————. 1959. *The Theory of the Growth of Firms.* New York: Wiley.

Perreault, W. D., and R. H. Miles. 1978. Influence strategy mixes in complex organizations. *Behav. Sci.* 23:86–98.

Shaffer, J. D. 1968. A Working Paper Concerning Publicly Supported Economic Research in Agricultural Marketing. USDA, ERS.

Siegel, S., and L. E. Fouraker. 1960. *Bargaining and Group Decision Making.* New York: McGraw-Hill.
Simon, H. A. *Administrative Behavior,* 3rd ed. 1958. New York: Macmillan.
Taylor, Frederick W. 1911. *The Principles of Scientific Management.* New York: Harper.
Thompson, J. D. 1967. *Organizations in Action.* New York: McGraw-Hill.
Weber, Max. 1947. *Theory of Social and Economic Organization.* English translation by A. M. Henderson and Talcott Parsons. New York: Oxford University Press.
Zusman, P., and A. Amiad. 1977. A quantitative investigation of a political economy—The Israeli dairy program. *Am. J. Agric. Econ.* 59:88-98.

BEN C. FRENCH
HOY F. CARMAN

PRODUCTION AGRICULTURE AS A FORCE AFFECTING THE FOOD SYSTEM

THE PROCESSES associated with structural changes in agriculture have been described by Shaffer (1968a) as the scientific industrialization of the food and fiber system. Shaffer argued that this process would require new institutions to deal with technological change and the unemployment, displacement, and large-scale organizations created by this process. He suggested there was also need for new institutions to deal with the associated coordination problems, conflicts, externalities, frustrations, and alienations. He did not, however, offer any suggestions as to specific types of institutions or organizational systems that might be considered.

This chapter, written some 11 years after Shaffer's stimulating observations, is a further examination of the industrialization process and its special implications for food and fiber marketing. It first reviews the major trends in the structure of production agriculture and explores in more specific detail changes that seem likely to be of greatest significance for the marketing system. Then it considers the further implications for various approaches to economic analysis of food and fiber system behavior and suggests some issues for future consideration.

SUMMARY OF STRUCTURAL CHANGES

The analysis of structural change in the agricultural economy has not been a neglected topic among agricultural economists. A conference sponsored by the Center for Agricultural and Economic Development, Iowa State University (1965), stands out as one of the early efforts to describe such changes and

BEN C. FRENCH and HOY F. CARMAN are professors of Agricultural Economics, University of California, Davis.

to evaluate their implications. Structural change was also the subject of National Advisory Commission on Food and Fiber reports (1967), a North Central Regional Research Commitee (NCR-20) seminar (1968), a book edited by Ball and Heady (1972), and a book by Wilcox et al. (1974). It was an especially popular topic in 1978. Gardner et al. (1978) evaluated new census evidence concerning structural change; Carter and Johnston and Veeman and Veeman presented papers at the 1978 American Agricultural Economics Association meeting, which focused on the broad implications of these changes for the future of United States and Canadian agriculture; and the Congressional Budget Office published a report prepared by Emerson (1978), which covers similar material, with special emphasis on current public policy options. Structural change was also an underlying topic in the Perry Foundation seminar (1978) on the survival of the family farm and a collection of essays by the staff of the Economics, Statistics, and Cooperative Service (ESCS) (1979b).

These articles, books, and conferences have covered many different aspects of structural change. For convenience of discussion, we have grouped them into six categories: socioeconomic characteristics of farm operators; number and size distribution of farms and farm income; capital and entry requirements; farm input markets; ownership and control of agricultural production and sales; and specialization, productivity, and organization of production activities.

Socioeconomic Characteristics of Farm Operations. Socioeconomic characteristics of the farm population have been changing slowly. The average age of farmers increased from 48.7 to 51.7 years between 1945 and 1974. The average age of operators for the largest farms is lower, with operators of farms with sales between $40,000 and $499,999 having an average age below 50 years.

The average number of years of schooling of farmers has increased just as it has for the nonfarm population. Tied to this increase in educational level has been increasing managerial sophistication.

The availability of family labor on farms has decreased. The number of farm families has declined and average family size has decreased from 3.92 persons per household in 1960 to 3.62 persons per household in 1975. The decrease in number of farmers and unpaid family labor has been partially offset by an increase in permanent hired labor.

The proportion of farm operators who reside off the farm has been increasing for at least 25 years and reached a level of 18.8 percent in 1974. The proportions are higher in the plains and western states.

Number and Size Distribution of Farms and Farm Income. Commercial farms have continued to decrease in number and increase in size, but at a decreasing rate (Gardner et al. 1978). Farm output as measured by total sales is concentrated in the largest units but with little recent change in distribution. Table 4.1 shows that in 1978 farms with $40,000 or more in gross sales were only 21.6 percent of the farms by number but accounted for 81.3 percent of total sales.

TABLE 4.1. **Distribution of farms, cash receipts, income, and capital gain on real estate by annual gross sales, 1978**

Annual gross sales[a]	Number of farms[b]	Percent of all farms	Percent of total cash receipts from farming[c]	Average income per farm operator family[d]	Percent of average income from farming	Capital gain on real estate per farm[e]
($)	(000)			($)		($)
100,000 or more	187	7.0	56.3	63,187	83	123,123
40,000–99,999	390	14.6	25.0	28,482	76	50,492
20,000–39,999	323	12.1	9.9	19,547	60	30,012
10,000–19,999	296	11.1	4.6	15,985	37	20,470
5,000–9,999	281	10.5	2.2	16,854	19	15,092
2,500–4,999	279	10.4	1.1	18,056	11	13,029
Less than 2,500	916	34.3	0.9	18,943	9	10,252
Total or all farms	2,672	100.0	100.0	22,866	44	28,344

Source: Table updated from Emerson (1978).

[a] Includes total cash receipts from farming, government payments to farmers, and other farm income from sources such as recreation, machine hire, and custom work.

[b] A farm is any place that sells (or normally would sell) $250 or more in agricultural products, or any place of 10 acres or more that sells $50 or more.

[c] Includes gross receipts from commercial market sales of farm products as well as loans (net of redemptions) made or guaranteed by the Commodity Credit Corporation and other purchases under price support programs.

[d] Realized net income from farming plus off-farm income of farm operator family divided by number of farms. For the purpose of these calculations, it is assumed that each farm has one resident farm operator family.

[e] Annual change in the value of farm real estate less the net investment in farm real estate.

Net income from farming is similarly concentrated in the largest units. However, it is important to note that in 1978 only 44 percent of all farmer income came from farming. While farms with sales of $40,000 or more received over 76 percent of their income from farming, the 55 percent of farms with gross annual sales below $10,000 received less than 19 percent of income from farming. And the 45 percent with sales below $5000 per year received less than 11 percent of income from farming.

For policymaking purposes, Emerson (1978) has grouped farms into four classes: family farms, small farms, part-time farms, and corporate farms. While there is no universally accepted definition of family farm, he noted that 90 percent of all farms fall within one commonly used definition: farms using less than 1.5 man-years of hired labor and not operated by a farm manager. Such farms accounted for about 60 percent of cash receipts from farming in 1977.

Farms not classed as family farms fall in two groups—industrialized farms (including most corporate-type farms) and nonindustrialized larger-than-family farms. Only about 2 percent of farms fall in the industrialized category, but they account for 15–20 percent of all cash receipts from farming.

Emerson has defined small farms as those with annual gross sales of less than $20,000 in 1976. They accounted for 71 percent of all farms in that year, but only 11 percent of cash receipts were from farming. Part-time farms are defined as those for which the operator is employed off the farm 200 days or more. In 1974 they accounted for 20 percent of all farms and 20 percent of total cash receipts. About 44 percent of farm operators worked off the farm and 40 percent of small-scale farmers fell in the part-time category.

Corporate farms are commonly thought of as large nonfamily types and would include a major portion of the so-called industrialized farms. However, the term "corporate farm" encompasses many types and sizes, including family farm corporations. While the number of corporate farms has increased substantially since the 1950s, it is estimated that they account for between 1 and 2 percent of all farms but about 15 percent of cash receipts (Emerson 1978; Gardner et al. 1978). Some of the attraction for outside capital into farming may have been slowed by the Tax Reform Act of 1976.

Participants in the Perry Foundation seminar (1978) have expressed some doubt concerning survival of traditional family farms. Small farms were thought to have substantial staying power because of their larger off-farm income, but at the same time they may be vulnerable to loss of markets. The larger-than-family farms, which have been gobbling up family farms, were viewed as the kind that big corporations want. Thus many may be in an intermediate stage. However, the corporate invasion of agriculture remains small in numbers of farms although significant in volume of product.

Capital and Entry Requirements. Rapidly increasing capital requirements have made entry into farming more difficult. For example, between the census years of 1969 and 1974, the average value of land, buildings, machinery, and equipment increased from $84,996 to $168,755 per farm. Increases since 1974 have also been dramatic. Entry into particular enterprises, however, remains more flexible because of the possibility of shifting resources already in farming. Increasing reliance on custom operations also appears to be a facilitating factor. Farm real estate debt has increased in absolute value but has declined as a percent of real estate assets since 1970.

Farm Input Markets. Farm input markets have affected farm operations through changes in the nature and availability of major inputs. Changes in the availability of seasonal labor and developments in unionization have influenced farmers' choice of enterprises. The changing labor market has also speeded the rate of adoption of harvest mechanization in a number of crops that were very labor intensive. The size, complexity, and cost of farm machinery have increased dramatically.

Land values have increased more rapidly than realized net farm income, partly because of inflation and entry of outside (nonfarm) capital. It is difficult to assess the reasons for entry of nonfarm capital. Carter and Johnston (1978) questioned whether investors' expectations have exceeded some bounds of rationality.

Large irrigation developments have also had a substantial impact on agricultural production, particularly in the western United States. New chemicals and insecticides, which have been instrumental in increasing productivity, have also created some environmental problems. Use of commercial fertilizers, especially nitrogen, has increased as a result of increased knowledge, irrigation, and new crop varieties.

Production agriculture has become very dependent on purchased energy. The changeover from horses and mules to petroleum as the major source of power was an important factor in rapid increases in agricultural productivity.

Because of the bulky and perishable nature of many agricultural products, energy use in assembly, processing, and distribution is significant. Energy costs are also a major component of the costs of fertilizer, agricultural chemicals, and irrigation water. While major changes are not yet evident, higher energy costs may ultimately have profound effects on the organization, location, and structure of agricultural production and marketing.

Ownership and Control of Agricultural Production and Sales. During the period 1959–74 the percentage of full-owner farms increased, the percentage of tenant farms decreased, and the percentage of part-owner farms increased slightly (Gardner et al. 1978). Much of the increase in full ownership has been in smaller farms. Many farmers with larger sales have been expanding by renting land to supplement what they own.

Personal and corporate income tax provisions are also associated with changes in ownership, control, and the legal structure of farming. Special farm income tax provisions, especially cash accounting, deductibility of some expenses of a capital nature, and capital gains treatment for assets whose costs may have been deducted as a current expense, have encouraged nonfarm investment in agriculture. Crop and livestock enterprises offering tax shelter advantages have undoubtedly expanded over what they would otherwise be. Nonfarm investors use hired management and farm management companies. Thus, while ownership may be dispersed, management may be quite concentrated. Management companies may control marketing of a significant volume of produce and gain some degree of market power.

Tax laws can have other important impacts. Progressive tax rates can influence the choice of cropping systems, enterprise combinations, and optimum farm size. The differential between ordinary income and capital gains tax rates can also influence enterprise choices and management decisions, while the difference between individual and corporate tax rate structures may encourage farm incorporation and growth of the farm firm.

TABLE 4.2. **Methods of coordinating production of selected agricultural commodities, 1970 estimates**

| | | Corporate | | | |
| | | Contracts | | | |
Commodity	Vertical integration	Individual producers	Producer bargaining associations	Producer cooperatives	Open markets
			(% of production)		
Sugar beets	2	. . .[a]	98	. . .	. . .
Sugarcane	60	23	. . .	17	. . .
Fluid grade milk	3	15	. . .	80[b]	2
Broilers	7	85	. . .	5	3
Processing vegetables	10	69	9	7	5
Citrus fruits	30	14	3	38	15
Turkeys	12	42	. . .	17	29
Potatoes	25	24	13	8	30
Deciduous fruits and nuts	20	. . .	8	30	42
Eggs	20	20	. . .	15	45
Fresh market vegetables	30	21	. . .	5	44

Source: USDA (1975).
[a] Indicates data not available.
[b] Includes producer bargaining associations.

Another aspect of ownership and control is the extent of integration of farm production with marketing and processing operations. Table 4.2 summarizes the 1970 estimates of the percentages of production of several major commodities sold through the various types of coordination methods. These data suggest that, with the inclusion of producer cooperatives, farmer-first handler exchange through a pure open market accounts for less than half the production of these commodities.

Table 4.3 provides an indication of the changes in the percentages of output produced under contracts or vertical integration from 1960 to 1970, as compiled by Mighell and Hoofnagle (1972). (For a description and analysis of the legal-economic aspects of contract farming, see Harris and Massey 1968.) The data suggest that most of the structural arrangements in effect in 1970 were established more than 10 years earlier. Exceptions may be noted in the increase in contracting and vertical integration for turkeys, eggs, and fed cattle, but these changes have not approached the dramatic adjustment that occurred earlier in the broiler industry.

TABLE 4.3. Estimated percentage of agricultural output produced under production contracts and vertical integration, United States, 1960 and 1970

Product	Production contracts		Vertical integration	
	1960	1970	1960	1970
		(%)		
Crop				
Feed grains	0.1	0.1	0.4	0.5
Hay and forage	0.3	0.3	. . .[a]	. . .
Food grains	1.0	2.0	0.3	0.5
Vegetables for fresh market	20.0	21.0	25.0	30.0
Vegetables for processing	67.0	85.0	8.0	10.0
Dry beans and peas	35.0	1.0	1.0	1.0
Potatoes	40.0	45.0	30.0	25.0
Citrus fruits	60.0	55.0	20.0	30.0
Other fruits and nuts	20.0	20.0	15.0	20.0
Sugar beets	98.0	98.0	2.0	2.0
Sugarcane	40.0	40.0	60.0	60.0
Other sugar crops	5.0	5.0	2.0	2.0
Cotton	5.0	11.0	3.0	1.0
Tobacco	2.0	2.0	2.0	2.0
Oil-bearing crops	1.0	1.0	0.4	0.5
Seed crops	80.0	80.0	0.3	0.5
Miscellaneous crops	5.0	5.0	1.0	1.0
Total crops[b]	8.6	9.5	4.3	4.8
Livestock or livestock product				
Fed cattle	10.0	18.0	3.0	4.0
Sheep and lambs	2.0	7.0	2.0	3.0
Hogs	0.7	1.0	0.7	1.0
Fluid-grade milk	95.0	95.0	3.0	3.0
Manufacturing-grade milk	25.0	25.0	2.0	1.0
Eggs	5.0	20.0	10.0	20.0
Broilers	93.0	90.0	5.0	7.0
Turkeys	30.0	42.0	4.0	12.0
Miscellaneous	3.0	3.0	1.0	1.0
Total livestock items[b]	27.2	31.4	3.2	4.8
Total Crop and Livestock[a]	15.1	17.2	3.9	4.8

Source: Mighell and Hoofnagle (1972).

[a] Indicates data not available.

[b] Estimates for individual items are based on the informed judgments of a number of production and marketing specialists in the USDA. Totals were obtained by weighting individual items by the relative weights used in computing the ERS index of total farm output.

A USDA (1979a) report on the status of the family farm has indicated that the major change in contract farming since 1970 has been a sharp increase in use of forward sales contracts in marketing cash grains, oilseeds, and cotton. About 8 percent of wheat, 11 percent of soybeans, and 20 percent of cotton were shown to be under contractual arrangements in 1974. Other changes include a growth in feeder-pig contracting and a decline in fed-cattle contracting as a result of a temporary decrease in custom feeding of cattle. It was estimated that use of contracts in production or marketing increased from 17 percent of all agricultural commodities in 1970 to 21 percent in 1974.

Overall, the percentage of farm output under production contracts and vertical integration increased from 19 to 22 percent from 1960 to 1970, with contract production accounting for about 2 percent and vertical integration (ownership within a single firm) about 1 percent. Although these figures are not suggestive of highly dramatic changes, the continual search for better ways to market (e.g., see Forker and Rhodes 1976; Knutson et al. 1978) seems likely to lead to continued alterations of the forms of coordination and integration for at least some commodities. Such changes would not necessarily be revealed clearly by statistics of the type compiled by Mighell and Hoofnagle (1972).

The means by which farmers have organized to sell their products seems to be another important aspect of the structure of production agriculture although more often discussed as a part of marketing. The collective actions stemming from the sanctions of the Capper-Volstead Act of 1922, the marketing agreement and order legislation of the 1930s, and the more recent state bargaining acts (e.g., the Michigan Agricultural Marketing and Bargaining Act of 1972) have been important components of the economic structure. Although the number of farmer cooperatives and their total membership has declined in the postwar period, the volume of agricultural products marketed by cooperatives increased from 20 percent in 1950–51 to 30 percent in 1975–76 (Swanson and Click 1977).

Marketing order programs, although increasingly criticized, have continued to play major roles in the sale of fruits and vegetables and milk; nearly all fluid grade milk sold in the United States is subject to either marketing order or state marketing controls. In California, which is by far the main user of marketing orders for fruits and vegetables, 83 percent of the 1975 value of fruit and nut crops and 65 percent of the value of vegetable crops were subject to some type of marketing order program (French et al. 1978). Some of the marketing orders provide only for collection of funds for research or advertising and have little direct influence on farmer choice. However, about three-quarters of the value of California fruit and nut crops was subject to quality control and more than half to volume or rate-of-flow controls in 1975, although the relative importance of these types of marketing programs has clearly declined in recent years.

Although it seems clear that bargaining organizations have expanded in number, size, and influence and much has been written about bargaining, up-to-date statistics on the scope of these activities are sketchy. A 1973 Farmer Cooperative Service report by Biggs and Samuels includes information on 11 fruit bargaining cooperatives with 1 million tons of fruit valued at $91 million in 1971, 8 vegetable bargainers with 4870 growers and 2.5 million tons of vegetables valued at $105 million, and 5 regional sugar beet associations that bargained for 28,515 growers whose 23 million tons of sugar beets were valued at $415 million. Also reported were 29 American Marketing Association affiliates. Twenty-one of these multicommodity affiliates had a total volume of $114 million, averaging 37,000 tons of fruit and vegetables valued at $2.5 million; 37,000 head of livestock valued at $1.3 million; and 11 million pounds of poultry valued at $1.5 million. In addition, the National Farmers Organization had 53 area offices from coast to coast, including 440 collection and dispatch points for livestock, 270 for grain, 45 for cotton, and 28 for milk. To this may be added the exclusive agency cooperatives of the type established in Michigan. Thus it is evident that collective bargaining is well established as a significant structural dimension of agricultural production. It seems likely that we will see further developments in this area, possibly with particular impetus where marketing order programs have been under fire.

Specialization, Productivity, and Organization of Production Activities. Shaffer (1968a) noted three inseparable developments in the organization of production activities. The first was the transfer of work from the farm to specialized nonfarm firms where the activity can be performed more efficiently (e.g., development of specialized feed manufacturing). Other transfers may be found in development of services for application of fertilizers and chemicals, harvesting, and computerized accounting. It should be noted, however, that not all transfers have been away from the farm. Lettuce packing, for example, which was once done mostly in packing houses, is now done almost entirely in the field. More stringent environmental requirements may also dictate that some partial processing be done in the field where residues can be disposed of more easily.

Shaffer's second development was the substitution of external sources of power for that originating on the farm, in particular, the substitution of mechanical power for labor (the substitution for animal power now seems virtually complete). His third development was the specialization in production knowledge and substitution of knowledge for other inputs. Such knowledge is often combined with other inputs (e.g., optimal pesticide applications, feed rations, fertilizer use).

The developments noted above are reflected in the index of total farm productivity, which has increased at an average annual rate of 1.9 percent per year since 1950. Carter and Johnston (1978) note that there is some evidence that the rate of increase may be slowing. These productivity measures encompass movements along and shifts in the production function associated with technological change.

Many of these technological changes have had rather profound effects on the optimal size of farm firms and organization of agricultural production and on regional specialization for particular commodities. Examples of the latter include shifts in cattle feeding in the Southwest, poultry production in the

Southeast, and processed tomato production in California. Production of some crops has increased in other countries at the expense of U.S. producers (e.g., vegetables in Mexico and asparagus in Taiwan).

IMPLICATIONS FOR THE MARKETING SYSTEM

The dividing line for agricultural production and marketing has traditionally been the farm gate. This division, while convenient, is becoming increasingly difficult to employ. Improved coordination of production and marketing extends traditional marketing activities into production and vice versa. The coordination and specification of planting schedules, cultural practices, application of major inputs, and harvest schedules by vegetable processors is a well-known example. Some traditional farming activities such as cattle feeding have also moved off the farm and taken on the characteristics of large-scale marketing firms.

Because of the numerous interactions, changes in the structure of production agriculture have implications for the marketing system. The marketing factors most often affected include spatial product distribution, pricing and exchange, advertising and promotion, channels of distribution, product quality, market risks, and integration.

A matrix of the major production agriculture structural dimensions and major marketing factors with which they most often interact is shown in Table 4.4. Entries in the row-column intersections designate the major areas of interaction. No direction of causation is implied by the entries because the interactions may very well be simultaneous in nature. Changes in structure may affect marketing, which in turn affects structure.

Not all structural changes in agriculture have significant implications for marketing. Changes in socioeconomic factors and capital requirements, for example, probably have only minor interactions with the listed marketing factors (Table 4.4). The major sources of interaction are found in the implications of farm size for farm-to-consumer marketing, technological change and specialization, changes in availability and character of farm inputs, and

TABLE 4.4. **Major interactions between production agriculture structural dimensions and marketing factors**

Production agriculture structural dimensions	Marketing factors						
	Spatial product distribution	Pricing and exchange	Advertising and promotion	Channels of distribution	Product quality	Market risks	Integration
Socioeconomic factors				x			x
Number and size of farms	x	x		x		x	x
Capital requirements		x				x	
Inputs	x				x	x	
Ownership and control		x	x	x	x	x	x
Specialization	x	x				x	

changes in ownership and control. There is also interaction among these categories. For example, there may be close relationships between farm size and technology, inputs and capital requirements, technology and inputs, farm size and specialization. These interactions and their implications are discussed in more detail below.

Socioeconomic Factors. Interactions between socioeconomic and marketing factors are not well defined, but some aspects of socioeconomic change may have implications for marketing research. For example, changes in land ownership-tenure arrangements may affect the method of marketing agricultural products. Alternatives include agricultural production cooperatives, land purchase associations, and community land trusts. Vertical integration may affect returns, risk, and managerial status of participating farmers.

Differences in land productivity between full-time and part-time farms suggest that there are variations in land quality and/or that there is inefficient land use by part-time farmers. Statistical data may be partially to blame, since they tend to ignore some production such as home-grown food. Increased home gardening and community gardening programs may have minor seasonal marketing impacts in some areas.

While meriting concern, research pertaining to the above topics is not likely to be a major part of future frontiers of agricultural marketing research.

Farm Size Distribution. Interactions between changes in farm size and marketing factors are concentrated in pricing and exchange, channels of distribution, and integration. These marketing factors will often differ for farmers marketing small volumes of product and those marketing large volumes. The differences are best illustrated in a historical sense. In the era of self-sufficient, small-scale agriculture, the channels of distribution were very short, with most sales being directly between the farmer and consumer. The movement toward large-scale specialization in production was accompanied by growth in specialized marketing middlemen and a lengthening of channels. Now producers with large volumes of product can bypass these agents through various forms of integration. (We will discuss these efforts in a later section.) However, this process may leave many small farmers with limited access to markets, thus contributing to their further economic disadvantage (see Rhodes, Perry Foundation seminar, 1978).

Public concern with the plight of the small farmer resulted in the enactment of Public Law 94-463, the Farmer-to-Consumer Direct Marketing Act of 1976. The purpose of this law is to promote, through appropriate means and on an economically sustainable basis, the development and expansion of direct marketing of agricultural commodities from farmers to consumers. Direct marketing includes but is not limited to farmers' markets, roadside stands, and pick-your-own operations. A U.S. Senate committee report (1978) on extension of the act describes some potential benefits from a direct marketing program. The report states, "A program of direct marketing contains the promise of substantial economic benefits to the Nation. The program aids smaller farmers, whose farms are in the more populated areas of the country, to stay in business. The program also makes it possible for more consumers to purchase fresh, field-ripened produce often at lower prices than are otherwise available.''

There is an abundance of descriptive and extension-oriented information concerning direct marketing. Rather extensive bibliographies are available in reports by Lindstrom (1978) and Roy et al. (1978). Information on the current status, economic importance, and economic impact of direct marketing, however, is fragmentary. Lindstrom presents estimates of the number of direct marketing outlets by type for 41 states based on responses from a survey of state department of agriculture and extension marketing specialists. The 1976 act directed the secretary of agriculture (through the Economic Research Service of the USDA) to conduct continuing surveys of existing methods of direct marketing in each state, with the initial survey to be completed within one year of passage of the act. In addition to ascertaining the volume of products moving through various methods of direct marketing, the act calls for measurement of the impact of direct marketing on financial returns to farmers (including the impact of those returns upon improving the economic viability of small farmers) and on food quality and costs to consumers. This survey, however, has not been completed because of its cost and the small initial appropriation to the USDA.

Viewed within this rather narrow framework, it appears unlikely that direct marketing will ever be important to more than a small proportion of ideally located farms. Although a variety of items are offered through direct marketing outlets, seasonal sales of fruits and vegetables appear to account for most of the dollar value. The amount is a small proportion of total food sales. However, a broader view of direct marketing might suggest potentially greater importance. Possibilities include new systems such as direct distribution from farmer cooperatives to consumer cooperatives, large-scale direct delivery to consumers (e.g., parking lot sales of frozen locker beef), contracts for direct delivery between consumer groups and individual farmers, and consumer group purchase of growing crops. Such changes could interact with farms of various sizes.

Developments other than direct marketing, which may improve small farmers' access to markets, include electronic commodity markets and formation of cooperatives to construct their own marketing facilities (see Rhodes, Perry Foundation seminar, 1978). The effectiveness of such developments will be subjects for future research.

Technological Change and Specialization. Technological change in production agriculture may have an impact on spatial distribution patterns, product quality and characteristics, transportation costs, and pricing patterns. The adoption of production innovations is usually based on reduced costs, and the impact is thus on the supply curve. Early adopters capture profits, but under free entry, expanded output will reduce prices until profits disappear. This process has been repeated for many innovations including planting and harvesting mechanization, new crop varieties and hybrids, new handling equipment and methods, and new production practices. Research indicates that the rate at which an innovation is adopted is a function of profitability and the investment required (Griliches 1957; Mansfield 1961).

Technological change may or may not be associated with regional specialization. In addition, regional specialization may be as much a function of marketing factors as production factors. A brief discussion of innovations and structural change in production of several major farm products and product groups helps to illustrate some of the interrelationships.

BEEF CATTLE. Major participants in beef cattle production include cow-calf producers, cattle feeders, and cattle slaughter firms. There has been little innovation or structural change in cow-calf production since the 1960s. The major changes in beef production have been construction of large-scale feedlots concentrated in the High Plains area, and location of specialized beef slaughter plants away from terminal markets and near commercial feedlots. A combination of factors contributed to these changes, including improved truck transportation, efforts to realize economies of scale in both feeding and packing, income tax laws, and location factors (i.e., climate favorable for rapid gains, nearness to feed and feeder cattle, and minimization of total costs of assembly and distribution).

Efficient operation of large-scale beef slaughter plants requires large uniform supplies of fed cattle. An innovative marketing arrangement between Iowa Beef Processors, Inc., and a group of northwestern cattle feeders has attracted the attention of cattle producers, feeders, packers, and government agencies. The arrangement, consisting of a five-year profit sharing contract between Iowa Beef and North West Feeders, Inc., a cooperative of six member feedlots, calls for the cooperative to supply about 350,000 head of fed cattle annually to Iowa Beef. This arrangement, if successful, is likely to be adopted in other areas.

HOGS. Innovations in pork production include the move to meat-type hogs and total confinement operations. The average scale of production has increased substantially and has become much more capital intensive. Regional specialization in production, which has changed little through time, remains in the midwestern states. Slaughter plants have moved away from terminal markets and into production areas.

Many people believe that confinement hog production is a candidate for broiler-type vertical integration between feed processors, producers, and packing firms. Initial efforts, however, have met with very limited success. While vertical integration has been slow, it is possible that the capital-intensive nature of confinement production will lead to more stable hog production. Candler and Manchester (1974) report the results of a substantial research effort documenting the interrelationships between hog production and marketing and the impacts of changing production and marketing systems.

DAIRY. Major innovations in milk production include mechanized milking systems and bulk handling of milk. These are related to structural changes that include a decrease in the number of dairy farms and milk processors, with significant increases in average size. The existence of economies of scale has favored movement toward large-scale dairies and specialized milk processors. Bulk handling and improved transportation have expanded production areas serving processing plants and have allowed expansion of fluid milk markets. The Northeast, Great Lakes states, and California continue to be the leading milk producers. Shifts in production include a decrease in the Corn Belt and increase in the Pacific Coast states. Major marketing changes, with a weak relationship to production structure changes, include a consolidation in producer cooperatives and the movement of large retail chains into fluid milk processing.

EGGS AND BROILERS. There has been a steady stream of new technology in the breeding and feeding of broilers and handling of eggs, which has increased productive efficiency. Since 1950, feed use per unit of product has been reduced more than 25 percent for eggs and nearly 30 percent for broilers. Adoption of these technologies, with their scale economies, has led to a decrease in the number of producers, handlers, and processors, with significant increases in average size of each.

In the case of eggs, the large-scale producers have been able to assume marketing functions formerly performed by middlemen. The relatively small number of large-scale producers in some areas has facilitated group action. The high degree of vertical integration in broiler production and marketing is due more to the efforts of feed manufacturers and processors than to changes in production technology.

The regional location of poultry production is influenced by a number of factors. The most important include efficiency of production, assembly, packing, and distribution; prices paid for major production inputs (feed, chicks, labor); and degree of coordination of production, input supply, and marketing functions. Egg production is concentrated in the South Atlantic, South Central, and Pacific regions, while broiler production remains centralized in the southern states. The density of production for both eggs and broilers, which has cost advantages, has been increasing in the South Atlantic and South Central regions.

VEGETABLE CROPS. There have been significant technological advances for individual vegetable crops. Included are improved varieties; new production systems; and mechanization of planting, cultivation, and harvesting. Innovations have occurred for many crops, but the most dramatic in terms of structural change and marketing implications is the mechanical harvesting system for processing tomatoes. A brief discussion of this illustrates changes that can take place.

Development of the tomato harvester dates back to the 1940s, but commercially produced machines were first used in 1961. In a review of the history of harvester development, Rasmussen (1968) states, "The tomato harvester resulted from the 'system approach.' A team made up of an engineering group and a horticultural group, with advice and assistance from agronomists and irrigation specialists, developed suitable plants and an efficient harvester at the same time. The necessary changes in planting, cultivation, and irrigating were developed concurrently.''

Rapid acceptance of the harvester began in 1964 as a result of demonstrated cost savings and a feared shortage of harvest labor. Parsons (1966), in a comparative cost study of machine versus hand harvesting, has found that the output rate per worker for mechanical harvesting was almost double that of hand harvesting and that machines offered substantial cost savings. Termination of Public Law 78 (the bracero program) in 1964 led to heightened grower interest in mechanical harvesting and undoubtedly speeded its adoption. Mechanization was rapid. The proportion of the California processing-tomato crop that was mechanically harvested increased from less than 2 percent in 1963 to 100 percent in 1970. The technology was not easily transferred outside California, however, primarily because of weather. The economic superiority of the mechanical harvest system permitted California to continue to gain production at the expense of midwestern and eastern states. California's share of U.S. processed-tomato production increased from 60 percent in 1963 to 85 percent in 1975, while acreage increased from 129,000 to 305,000 during the same period (Brandt et al. 1978).

There are economies of size associated with the mechanical harvest system. From 1964 to 1974 the number of California farms producing tomatoes decreased from 1883 to 1493, while average acres per farm increased from 85 to 174 (Brandt et al. 1978). Improved harvester capacity and electronic sorters will probably lead to further increases in acreage of tomatoes per farm. There has also been a reduction in the number of California tomato processing firms from 44 to 28, a tripling of output per processor, and an increase in average hauling distance from 31 to 100 miles. Deflated farm-to-processed-product price spreads, although fluctuating considerably from year to year, remained about level overall.

FIELD CROPS. The major innovations in field crop production have been associated with new varieties (including hybrids) and adaptation of crops to new geographic areas. Examples include successful adaptation of soybeans to the western United States, large-scale production of sunflowers as an oil crop, and introduction of new higher yielding hybrid wheat. Expanded or new output in an area requires new facilities for handling, storage, and processing.

Processing innovations and new crop uses can have dramatic marketing impacts, e.g., use of corn to produce high-fructose corn syrup, which substitutes directly for sugar in many processed food products and costs less to produce than domestic cane or beet sugar. The escalating cost of energy may result in new uses for existing crops or introduction of new crops that substitute for petroleum products. Production of alcohol from wheat, corn, potatoes, or sugar for use as a fuel (gasohol) has potential and is being tested. New crops such as jojoba beans are being investigated. While these crops are not currently a significant factor, continued pressures on energy sources plus some biological breakthroughs could drastically alter the situation. Competition for land (and other factors) between food and energy crops could impact dramatically on the location of production and organization of the marketing system.

TREE CROPS. Innovations in production of tree crops include mechanical harvesting for nuts and some processing fruit, bulk handling, and new varieties and planting techniques (dense plantings, semidwarf rootstocks, diease-resistant rootstocks). Climate limits the location of tree crops. There have been some changes in location as a result of irrigation development and urbanization. Associated with these changes in production, without implying causation, have been decreases in the number of packing and processing plants and increases in the average size of remaining firms. The marketing implications of new crop development are well illustrated by the growth in California acreage of pistachio nuts from less than 400 acres in 1966 to over 31,000 acres in 1976. As these trees begin to bear, a complete marketing system will have to be established.

Availability and Character of Farm Inputs. Farm production inputs interact with several aspects of marketing, especially spatial product distribution, product quality, and market risks. This interaction occurs in two dimensions, the marketing of the inputs to farmers and the marketing of farm products. Our comments are restricted primarily to effects that changes in input markets have on the marketing of farm products.

Combinations of several inputs are necessary for agricultural production, and changes in the character of use of any input will have an impact on production. Changes in production must be comparatively large in aggregate or regionally concentrated to have a significant effect on the marketing systems. Inputs that have the greatest potential for interaction with marketing include land and water development, agricultural chemicals, and capital-labor market changes. Implications of changes in these inputs are varied.

The impact of land and water development is primarily in the area of spatial product distribution. For example, large irrigation developments have brought thousands of acres of land under irrigation in the Columbia Basin and the west side of the San Joaquin Valley in California and have resulted in production of new high-value crops, which are necessary to pay the high water costs. The concentration of production has required an expanded marketing system. As the crops come into production, they require handling, packing, processing, and storage facilities, which are constructed as output reaches necessary levels. Large volumes of new product may alter established spatial and seasonal marketing patterns as the product moves into distribution channels.

Expanded output, as a result of land and water development, affects price impact, which is felt by all growers and especially by established tree crop producers with older rootstocks and varieties, wider spacing, and the like. The impact of new development on cost structures can also be significant, since the most recent technology will be incorporated in the crops grown and the handling and processing facilities.

Changes in the availability and cost of water as a result of such things as drought and energy costs affect regional cropping patterns and spatial marketing patterns. Future large-scale irrigation developments will probably be limited; in fact, there are some indications of surface water and groundwater shortages. Desalinization of seawater, once viewed as offering real promise for many areas, may be of restricted use because of its energy-intensive nature.

Expanded use of agricultural chemicals (pesticides and herbicides) has had a significant impact on the cost structure and supply functions for affected crops. Chemicals can have an impact on product quality (e.g., freedom from blemishes) and on costs of achieving a given quality level. Growing concern over environmental and health aspects of heavy chemical use may lead to further restrictions. There is also the additional problem of pests developing resistance to pesticides. Although research on integrated pest management programs that make use of natural predators is progressing, restricted use of some chemicals can have product quality and price implications. These developments present important topics for fruitful economic research.

The changing labor market structure, including unionization of farm workers, has influenced the rate of mechanization in agriculture and has affected farmer perceptions of risks of growing various crops. This in turn has influenced producers' decisions on crops grown, with consequent regional changes in patterns of production. Union organizing activities and contract disputes have also affected demand and retailing activities through consumer boycotts.

The adoption of mechanical harvesting for processing tomatoes provides an illustration of how changes in the labor market may interact with technological developments. While the harvester would have been adopted in any event because of cost advantages, grower concern over labor availability was certainly a factor in the rate of adoption.

The asparagus industry provides another example of how grower concern about labor availability and unionization may affect production and, eventually, organization and behavior of the food system. The California asparagus industry, like the tomato industry before mechanization, is heavily dependent on large numbers of seasonal hired harvest laborers. Unlike the tomato industry, however, the asparagus industry was not able to develop mechanical harvesting technology that was economical under California conditions. Consequently, with the termination of the bracero program there was a significant downward shift in the asparagus supply level that is not accounted for by cost or price changes (French and Matthews 1971).

Developments in the tomato and asparagus industries since the mid-1960s provide starkly contrasting examples of what can happen to commodity industries as a result of interacting effects of input market changes, technological innovation (or failure to innovate), and changing product demands. The tomato industry combined expanded demand and lower costs (with less labor uncertainty) to more than double output, while the California asparagus industry, faced with declining domestic demand and increased foreign competition and unable to mechanize, has dropped to about half its 1963 size.

These changes have in turn had significant impacts on the operations of processing and marketing firms, product prices, and regional distribution of production. It seems likely that other industries may have comparable experiences in the future.

Ownership and Control. Changes in the structure of ownership and control of farm resources and farm production affect (and are affected by) the system of exchange between farmers and first handlers. The traditional view of farmers as atomistic price takers whose production decisions are independent of marketing decisions has been increasingly modified by contractual arrangements, joint (vertical) ownership of agriculture production and marketing facilities, and collective actions of farmers through bargaining associations and marketing order programs. These changes have stemmed from the continual search for means to obtain control over markets and sales outlets to reduce uncertainty and increase returns.

Many of the changes in the structure of ownership and control have provided both public and private benefits. Contractual arrangements between vegetable producers and processors, for example, have permitted growers and processors to plan their operations to achieve greater efficiency and stability. Integration through producer cooperatives has clearly been a major factor in providing farmers with assured outlets and services and, with possibly a few exceptions, has been a factor in maintaining a competitive marketing environment and lower prices to consumers. Collective actions may also provide for more equitable treatment of farmers and add to the stability of product flow to markets.

Some changes in ownership and control may have less desirable social effects. Contractual arrangements may in some cases alter the social status of farmers and certainly may restrict their freedom of decision making. Collective actions that may be viewed by a majority of farmers as a means to increase their returns may be viewed by others as an interference with their basic freedom of choice. Changes in vertical organization that concentrate economic power may have potentially undesirable competitive effects.

The research task facing economists is to evaluate the probable impacts of various modifications of the structure of ownership and control on the status and returns to producers; costs to consumers; and distribution of benefits, both positive and negative. Implications for product innovation and quality and growth and development are also of concern. We need to go beyond mere description of how the system works. Such research needs to focus on the public policy options that may influence the nature of integrative and control arrangements.

IMPLICATIONS FOR ECONOMIC ANALYSIS

Effective analysis of the vertical interdependence between agricultural production and marketing requires systems-oriented research programs, which are often referred to as subsector studies. French (1974) divides approaches to subsector studies into four qualitative and four quantitative classes. In the qualitative or descriptive class he lists (1) base studies that attempt to identify problems, describe subsector activites, and pull together what is known about the parameters of the system; (2) analogous experience, which involves attempts to generalize for one subsector from an in-depth appraisal of another; (3) the Delphi approach, which involves bringing a range of expert opinion to bear on issues of concern; and (4) systems analytic description, which may be similar to base studies but attempts to examine the subsector in depth within a systems taxonomy.

The quantitative class is separated into (1) design models, which show how the subsector system (or some significant part) might be reorganized to reduce costs and increase profits; (2) comparative static models, which involve more complete supply and demand components to permit evaluation of equilibrium positions before and after varying instruments of change; (3) dynamic economic models, which add lagged variables and time rates of adjustments to static models; and (4) general systems simulation, which is distinguished from econometric modeling by its greater detail and flexible approach to specification of equation forms and estimating the parameters of the equations of the system. In the discussion that follows we shall attempt to show how the influence of structural changes in farming are incorporated into the various types of subsector studies and to appraise the role and importance of these models as research approaches to the identified issues.

Qualitative or Descriptive Analysis. Structural changes in farming enter into models of food system behavior primarily as exogenous influences. However, they may also enter as endogenous factors in the sense of being induced first by changes in the marketing system. The nonfarm effects of changes in farming structure frequently are so complex or so subtle as to defy, at least to this point, analysis of a more quantitative nature. In these cases qualitative or descriptive analysis may provide some useful insights into the nature of the interactions, identification of problem areas, and formulation of behavioral hypotheses, although the latter may remain untested. Some descriptive analysis is essential as a first step in more quantitative studies.

The analysis of changes in the structure of ownership and control and coordination of exchange arrangements has been particularly difficult to quantify, judging by the large volume of descriptive literature and the almost nonexistent quantitative studies in this area. Much of what has been done or is known pertaining to vertical interdependence is referenced or summarized in North Central Public Policy Education Committee papers (1972) and Marion (1976). While these papers provide a valuable service in outlining our current state of knowledge, the elusiveness of the subject is revealed in a concluding comment by Marion in his NC-117 monograph: "At this point in time, we know relatively little about the influence of vertical organization, exchange arrangements and other coordinating mechanisms on vertical coordination and other performance dimensions. . . . We are confronted with a great deal of rhetoric about vertical coordination and subsector analysis, but limited abilities to test its validity." Whether it will ever be possible to achieve much success in this area with quantitative models seems doubtful. However, some possibilities are discussed in a following section.

Another area of agricultural production and marketing interdependence, which has been studied almost entirely by descriptive analysis, is direct farm-to-consumer marketing. Viewed in the limited context of institutions such as roadside stands, farmers' markets, and consumer cooperatives, it is easy to see why this has been the case. The economic issues are not particularly complex and may in fact be approached more appropriately through technical assistance programs than any type of quantitative economic analysis. A broader view, however, might explore new channels and methods of distribution and evaluate their implications within the context of changing life-styles, possible energy shortages, and environmental concerns. System design models of the type described in the next section may offer substantial promise for dealing with such research questions.

Quantitative Analysis. Quantitative models that deal with various aspects of vertical interdependence of agricultural production and marketing may be distinguished by their objectives and empirical methodology.

The design models mentioned previously are optimizing models that focus on the efficiency of subsector organization. When applied within market areas, they provide a means of showing how changes in production location, hauling technology and costs, and scheduling of harvesting and processing activities may affect the optimal number, size, and location of processing facilities. They may also suggest means by which actual systems may be altered to move closer to the optimally designed system, although that aspect has often been neglected. Many such studies are summarized and referenced by French (1977).

When these design models are applied across areas, they may show how changes in farm technology (through cost effects) and resource availability influence the "optimal" location of production and regional distribution flows. The well-known regional programming models of Heady and others fall in this category (see Day and Sparling 1977). Linear programming has been the primary methodological tool for these kinds of models.

If we add estimates of regional supply and demand functions to interregional models of optimal production location, they become spatial price equilibrium models that typically fall within the category of comparative static models (see Day and Sparling 1977). In these models the impact of changes in farm structure is felt mainly through the supply functions. A key element of analysis thus is to find ways to measure the effects of changes in farm size, farm technology, and resource availability on supply functions. In commodity subsectors characterized by group actions in product sales and small numbers of first handlers, some modifications of the competitive model specifications may be required. Efforts and success with this aspect appear to be very limited. For an interesting and potentially important modeling development for oligopolistic structures see Chern and Just (1978).

Dynamic econometric models, in contrast to design models, attempt to model actual price and production behavior within commodity subsectors rather than optimal organization and flows. They are also usually more aggregative and, as the name suggests, rely more on statistical analysis of behavioral relationships (see French 1974; Tomek and Robinson 1977). They differ from comparative static models by their consideration of dynamic elements such as partial adjustment rates, learning, and adoption rates of technological change. The impacts of structural changes in farming again enter through their effects on the supply functions that interact with the derived demand functions to determine prices, production, and marketing spreads. Increasingly, there is need for specification of supply-demand structures to take account of changes in the system of coordination and exchange. There is need in particular for better models to handle bargaining, long-term contracting, and changes in input markets.

General systems simulation includes a variety of quantitative modeling approaches. The country models such as developed at Michigan State University for Nigeria and Korea are illustrative of a modeling approach that may take account of a wide variety of interactions and is extensive in scope. (See Rossmiller 1978; for a more general review of systems analysis and simulation models, see Johnson and Rausser 1977). While not excluding econometric estimates of supply and demand functions, the models may rely on a variety of sources or methods to quantify the parameters of the system. At this stage it appears that access to data required for parameter estimation (beyond that of the usual econometric estimation) remains a major stumbling block.

Another modeling approach within the general systems simulation class focuses on the behavior of individual participants, rather than the outcomes of their aggregate behavior only (e.g., Desai 1968; Duewer and Maki 1966. Application of the latter model was limited by inability to develop good estimates of some of the important parameters.) Although not much used, this approach may well have great potential for dealing with the issues discussed here if properly developed. For example, suppose we wished to develop a model of the production-marketing system for a commodity such as prunes. On the production side, we would classify prune producers by their socioeconomic characteristics and include representative firms from each class in the model. Through surveys and other types of analysis we would develop behavioral rules with respect to planting, removal of trees, and other key decision variables for each producer class. These rules would relate to profit and cost experiences, tax provisions, exchange arrangements, and other variables affecting grower choices. In a similar manner, a set of representative processing and marketing firms would be specified. With appropriate scaling, the model based on the behavior of individual decision units would generate industrywide production and prices. The advantage of this approach over the more aggregative models is that it would permit us to consider entry and exit in farming and marketing, to evaluate impacts of factors such as changes in tax laws, and potentially to deal more effectively with issues concerned with coordination and control. It would offer a possible approach to analyzing the impacts of many structural changes that tend to be submerged in the data aggregation of other kinds of models.

This type of modeling approach would not be easy nor would it be inexpensive in the early stages. Initially, such models would of practical necessity focus rather narrowly on particular industries or limited subsectors of the agricultural economy. They would not replace other research approaches but would simply add to our kit of research tools and economic research capability. However, the possibilities seem endless. We have the computer technology to handle this kind of study. What we need now is the imagination and skill to develop it.

ISSUES FOR FUTURE CONSIDERATION

We turn now to our major conclusions with respect to significant areas for future research, to some methodological points of concern, and to suggestions for research organization to better deal with analysis pertaining to the interdependence of the structure of agricultural production and the organization and behavior of the food system.

Although there has been much discussion of direct farm-to-consumer marketing, it is not likely to be a very important area for economic research as treated within the Direct Marketing Act of 1976. It represents only a small component of the total system, and the issues involved seem more suitable for technical assistance and description than sophisticated economic analysis. Analysis of the impact of possible new direct distribution symptoms and channels may offer more challenging research possibilities.

Research on vertical coordination and control seems likely to remain elusive for some time. The most significant issues pertain to the implications of structural change on the efficiency of exchange and the structure of competition. We will certainly need to continue to monitor changes in the system, to understand them, and to deal as well as possible with public policy implications. Because of the two-way interaction, it is difficult to isolate the separate influence of structural changes in farming. Our ability to do much quantitatively in this area will remain limited as long as our data and access to information remain limited. (For an example of a quantitative approach to the public policy issues relating to control under a marketing order program see Minami et al. 1979.) We shall comment further on this aspect. We feel that microsimulation may have significant possibilities in this area, but it remains an undeveloped approach at this time.

It seems to us that models designed to predict the impacts of technological change and input availability (water, energy) will offer the most fruitful opportunities for quantitative research in the near future. We have had some success in this area and there is much more that can be done. While most technological changes occur gradually, we need to give some attention to the implications of more radical types of changes; e.g., one of our engineering colleagues has indicated that it is technically feasible to grow grapes somewhat artificially as an annual crop. He suggested that grape canes could be inserted in long tubes that would contain all the nutrients needed in liquid form. The grapes would be harvested mechanically. Although probably not cost effective at this time, it takes little imagination to note that if such a process were to be developed, it could have a substantial impact on production location and organization, processing, and marketing. We need to anticipate such developments and formulate models to predict their consequences.

The primary deterrent to effective research in all these areas is not in our capacity to conceptualize but in the limitations of our data and ability to test adequately the models that are formulated. We still lack many kinds of price and production series required for quantitative analysis; e.g., with the exception of a few price series compiled by the Bureau of Labor Statistics, we need good price and production series for commodities beyond the farm level. We also lack continuous series on transportation charge data, which are needed for studies involving interregional analysis. Access to marketing and processing-firm data on costs and firm decision procedures are also inadequate.

There are three kinds of organizational developments that we feel would improve our ability to deal with the research issues pertaining to structural interdependence. First, we would like to reactivate the National Commission on Food Marketing (or something like it) on a continuing basis. The NC-117 project is a move in that direction, but we need a national focus, a broader charge, and a more permanent structure. We visualize a small core staff supplemented by visiting professors and contract studies. Part of the focus would be on maintaining and extending data such as developed in the National Commission on Food Marketing technical studies. The agency would also be empowered to work on ways to obtain cost and return data for food system firms as required to monitor and analyze the operations of the food system. Second, we see a need for better information systems pertaining to research in commodity subsectors. This might be tied in with the reactivated food commission or could take the form of commodity "desks" in ESCS as first suggested by Shaffer (1968b) and further elaborated by French (1974). In either case, we need improvements in the efficiency with which we compile, integrate, and transmit information pertaining to research and analysis on commodity subsectors. Finally, we would like to see a regionally stratified national consumer panel established. This may seem a bit removed from analysis of the effects of structural changes in farming on the marketing system. However, the data that would be generated by such a panel would provide information essential to estimation of regional demand and consumption behavior, which is needed for formulating spatial equilibrium and related types of models. Such models represent an important approach to analysis of the effects of technological changes in farming, changes in farm input markets, and changes in the methods of coordination and exchange.

REFERENCES

Arthur, H. B., R. A. Goldberg, and K. M. Bird. 1967. The United States Food and Fiber System in a Changing World Environment, vol. 4. Technical Papers, National Advisory Commission on Food and Fiber, Washington, D.C.

Ball, A. Gordon, and Earl O. Heady, eds. 1972. *Size, Structure, and Future of Farms.* Ames: Iowa State University Press.

Biggs, Gilbert W., and J. Kenneth Samuels. 1972. Where bargaining cooperatives are today. *News for Farmer Cooperatives,* ACS, USDA, December, 39:4-5.

Brandt, Jon A., Ben C. French, and Edward V. Jesse. 1978. Economic Performance of the Processing Tomato Industry. Giannini Foundation Inf. Ser. 78-1, University of California, Davis.

Candler, Wilfred V., and Alden C. Manchester. 1974. Hog-pork subsector study in review: Research and administrative experience. *Am. J. Agric. Econ.* 56:1023-29.

Carter, Harold O., and Warren E. Johnston. 1978. Some forces affecting the changing structure, organization and control of American agriculture. *Am. J. Agric. Econ.* 60:738-48.

Chern, Wen S., and Richard E. Just. 1978. Econometric Analysis of Supply Response and Demand for Processing Tomatoes in California. Giannini Foundation Monogr. 7, University of California, Davis.

Day, Richard H., and Edward Sparling. 1977. Optimizing models in agricultural and resource economics. In L. R. Martin, ed., *A Survey of Agricultural Economics Literature,* vol. 2. Minneapolis: University of Minnesota Press.

Desai, M. J. 1968. The Computer Simulation of the California Dairy Industry. Giannini Foundation Special Report, University of California, Davis.

Duewer, L. A., and W. R. Maki. 1966. A study of the meat products industry through systems analysis and simulation of decision units. *Agric. Econ. Res.* 18:79-83.

Emerson, Peter M. 1978. Public Policy and the Changing Structure of Agriculture. U.S. Congress, Congressional Budget Office. Washington, D.C.: USGPO.

Forker, Alan D., and V. James Rhodes, eds. 1976. *Marketing Alternatives for Agriculture.* New York: Cornell University Press.

French, Ben C. 1974. The subsector as a conceptual framework for guiding and conducting research. *Am. J. Agric. Econ.* 56:1014-22.

_____. 1977. The analysis of productive efficiency in agricultural marketing: Models, methods and progress. In L. R. Martin, ed., *A Survey of Agricultural Economics Literature,* vol. 1. pp. 93-206. Minneapolis: University of Minnesota Press.

French, Ben C., and Jim L. Matthews. 1971. A supply response model for perennial crops. *Am. J. Agric. Econ.* 53: 478-90.

French, Ben C., Niniv Tamimi, and Carole Nuckton. 1978. Marketing Order Program Alternatives: Use and Importance in California, 1949-1975, Giannini Foundation Inf. Ser. 78-2, University of California, Davis.

Gardner, B., Delworth Pope, and Rulon D. Pope. 1978. How is scale and structure determined in agriculture? *Am. J. Agric. Econ.* 60:295-302.

Griliches, Zvi. 1957. Hybrid corn: An exploration in the economics of technological change. *Econometrica* 25:501-22.

Harris, Marshall, and Dean T. Massey. 1968. Vertical Coordination via Contract Farming. USDA, ERS Misc. Publ. 1073.

Iowa State University Center for Agricultural and Economic Development. 1965. Structural Changes in Commercial Agriculture. CAED Rep. 24, Ames.

Johnson, S. R., and Gordon C. Rausser. 1977. Systems analysis and simulation in agricultural and resource economics. In L. R. Martin, ed., *A Survey of Agricultural Economics Literature,* vol. 2. Minneapolis: University of Minnesota Press.

Knutson, Ronald D., Wallace Barr, and William E. Black. 1978. Who Will Market Your Products? Texas A & M University Agric. Ext. Serv. D-1053. College Station.

Lindstrom, H. R. 1978. Farmer-to-Consumer Marketing. USDA, ESCS-01.

Mansfield, Edwin. 1961. Technical change and the rate of imitation. *Econometrica* 29:741-66.

Marion, Bruce W., ed. 1976. Coordination and Exchange in Agricultural Subsectors. NC-117, Monogr. 2, University of Wisconsin, Madison.

Mighell, Ronald L., and William S. Hoofnagle. 1972. Contract Production and Vertical Integration in Farming, 1960 and 1970. USDA, ERS-479.

Minami, Dwight D., Ben C. French, and Gordon A. King. 1979. An Econometric Analysis of Market Control in the California Cling Peach Industry. Giannini Foundation Monogr. 39, University of California, Davis.

National Advisory Commission on Food and Fiber. 1967. Food and Fiber for the Future. Superintendent of Documents, Washington, D.C.

North Central Public Policy Education Committee. 1972. Who Will Control U.S. Agriculture? NCR Ext. Publ. 32, University of Illinois, Coll. Agric. Spec. Publ. 27, Urbana.

North Central Regional Research Committee on Economics of Marketing. 1968. Agricultural Organization in the Modern Industrial Economy. Ohio State University, Dep. Agric. Econ. NCR-20-68, Columbus.

Parsons, Philip S. 1966. Costs of Mechanical Tomato Harvesting Compared to Hand Harvesting. University of California, Agric. Ext. Serv. Publ. AXT-224, Davis.

Perry Foundation and University of Missouri. 1978. Can the Family Farm Survive? Report of seminar sponsored by M. A. and Johnnye D. Perry Foundation and University of Missouri. Agric. Exp. Sta. Spec. Rep. 219, Columbia.

Rasmussen, Wayne D. 1968. Advances in American agriculture: The mechanical tomato harvester as a case study. *Technology and Culture* 9:531–43.

Rossmiller, George E., ed. 1978. Agricultural Sector Planning, A General Simulation Approach. Michigan State University, Dep. Agric. Econ. Agricultural Sector Analysis and Simulation Projects, East Lansing.

Roy, Ewell P., Don Leary, and Jerry M. Law. 1978. Organization and Operation of Louisiana Farmers' Markets. USDA, DAE Res. Rep. 532, Louisiana State University, Baton Rouge.

Shaffer, James D. 1968a. The scientific industrialization of the U.S. food and fiber sector: Background for market policy. In Agricultural Organization in the Modern Industrial Economy. Ohio State University, Dep. Agric. Econ. NCR-20-68, Columbus.

_____. 1968b. A Working Paper Concerning Publicly Supported Economic Research in Agricultural Marketing. USDA, ERS.

Swanson, Bruce L., and Jane H. Click. 1977. Statistics of Farmer Cooperatives, 1972–73, 1973–74 and 1974–75. USDA, FCS Res. Rep. 39.

Tomek, William G., and Kenneth L. Robinson. 1977. Agricultural price analysis and outlook. In L. R. Martin, ed., *A Survey of Agricultural Economics Literature,* vol. 1, pp. 329–409. Minneapolis: University of Minnesota Press.

U.S. Department of Agriculture. 1975. The Food and Fiber System—How It Works. ERS Agric. Inf. Bull. 383.

_____. 1979a. The Status of the Family Farm. Second Annual Report to the Congress, ESCS Agric. Econ. Rep. 434.

_____. 1979b. Structure Issues of American Agriculture. ESCS Agric. Econ. Rep. 438.

U.S. Department of Commerce. 1977. 1974 Census of Agriculture, United States Summary and State Data, vol. 1, part 51.

U.S. Senate. 1978. Extension of the Farmer-to-Consumer Direct Marketing Act of 1976. Rep. 95-854. Washington D.C.: USGPO.

Veeman, Terrence S., and Michele M. Veeman. 1978. The changing organization, structure and control of Canadian agriculture. *Am. J. Agric. Econ.* 60:759–68.

Wilcox, Walter W., Willard W. Cochrane, and Robert W. Herdt. 1974. *Economics of American Agriculture,* 3rd ed. Englewood Cliffs, N.J.: Prentice-Hall.

5

MARKETS FOR AGRICULTURAL INPUTS: CURRENT STATUS AND NEEDED RESEARCH

ECONOMISTS tend to overlook the importance of factor markets until some shock brings one or more to our attention. However, both classical economic theory and historical patterns of economic development underscore the importance of factor markets and production processes. Understanding these relationships is essential to understanding economic behavior and structure at both micro and macro levels.

Dramatic changes in relative prices of some factors of production and the likelihood of their continuance make it imperative that economists give increased attention to factor markets, implications of major changes in these markets, substitutability among factors of production, and how economic systems and markets adjust to such changes.

An understanding of factor markets is important because:

1. Relative prices and supplies of factors determine resource use relative to given technology in the short run and provide incentives for research and development, which has an impact on production technology, in the long run. In turn, technology influences the economic organization of production processes and, ultimately, coordination and exchange mechanisms in product markets.
2. Relative prices and supplies of factors are themselves determined and conditioned by market structures and institutions.
3. The present economic structure of the U.S. food and fiber industry has been heavily influenced by factor supplies and prices. Understanding the sources of structural change, therefore, requires an understanding of past and present factor markets.

JOHN E. LEE, JR., is Administrator, Economic Research Service, U.S. Department of Agriculture.

4. Recent changes and prospects for further changes in factor availability and prices, as well as major changes in organization of factor markets, pose the potential for significant adjustments in organization and performance of the food system, including production and marketing processes.

The technology issue is basic and deserves further elaboration. The organization of industry and economic institutions (and ultimately social institutions) is built on the prevailing (or recently prevailing) organization of production processes, which largely is a function of the technology of production. This in turn is a function of many factors but certainly includes relative prices and supplies of factors of production that encourage development and adoption of new technology. Finally, factor prices and supplies depend on the endowments of nature and the markets for those endowments and their direct and indirect services.

The importance of this set of relationships has been amply demonstrated by the histories of civilizations, cultures, and economic systems. In modern times the differing patterns of development of various regions of the world can be partly traced to the relative scarcity or abundance of various factors of production (as determined by either natural endowments or structure of markets for factor services).

The pattern of development of the agricultural and food system in the United States is further testimony to the fundamental role of factor supplies, prices, and markets. Changes in technology and production practices are related to changes in relative availabilities and prices of land, labor, and various forms of capital (Cochrane 1979). These historical patterns are generally well known but tend to be taken for granted.

During the 1800s, land was generally plentiful, and the technology and production practices of that era tended to be land extensive. But later, when the frontiers were gone, land came to be viewed as a fixed and scarce resource. Subsequently, land and labor became increasingly expensive and capital relatively cheap. Thus modern agriculture in the United States was built on a technology that intensified the substitution of capital for land and especially for labor. The dramatic increases in yields were based on technology and production practices that exploited relatively cheap capital goods, energy, fertilizer, and other chemicals. The extent to which these price relationships and relative input availabilities came to be taken for granted is reflected in the common acceptance of output per unit of labor input as a valid measure of industry productivity.

The relative cheapness of energy and capital in its various forms also influenced the technology of the rest of the food system (and the general economy). Labor saving became the criterion for technological developments in processing, warehousing, and retailing systems and design of food products for households. It was more economic to save labor than to save energy, plastics, packaging materials, and capital-intensive equipment. Perhaps the most dramatic expression of labor saving (alias time saving) and indirectly the relative prices of inputs was the "throw-away" technology, designed to eliminate certain labor functions altogether.

Business organization and management of the food system also have evolved to exploit the possibilities generated by the underlying technology. Even the coordination and exchange mechanisms, including price discovery processes, have adapted to evolving business practices and business organization that were built around the prevailing technology. One could even make the case that life-style options and ultimately societal values trace back to the underlying technological possibilities, which are heavily influenced by factor markets.

The above is a cursory treatment of the role of factor markets and the technology, industrial organization, and economic and societal infrastructure built on or influenced by them. While the references listed at the end of this chapter clearly demonstrate that agricultural economists have not ignored the fundamental role of factor markets, much more attention has been given product markets, and appropriately so. However, dramatic developments in input markets, which threaten major changes in the price relationships on which the modern food system is built, make improved knowledge of factor markets and the potential impacts of changing factor price relationships perhaps the most urgent target of marketing research in the 1980s.

RECENT DEVELOPMENTS

A number of developments during the 1970s have increased public awareness that resources once considered plentiful and cheap can become increasingly dear. The reason, in part, is simply physical exhaustion of supplies available at low cost. In addition, owners of scarce resources have sometimes managed to organize to exploit market demand and improve their terms of trade.

Still another development has been the rapid evolution of institutional constraints on use of resources, which have the effect of reducing availability or making them more expensive. Moreover, these forces interact with each other to compound the overall impacts. For example, a raw material considered plentiful and inexpensive but which embodies large amounts of petroleum-based energy to give it time, place, and form utility may become relatively more expensive as the relative price of petroleum rises sharply.

Examples of developments in several important factor markets serve to illustrate that important and sometimes dramatic changes are taking place. These have important implications for the structure, organization, and performance of the food system and further underscore the need for improved research to better understand these implications.

Energy. The most visible developments in energy markets are related to petroleum. The Arab oil embargo of 1973 was a major shock to the entire economy, including all stages of the food and fiber system. Unknown coefficients that immediately became apparent included price and income elasticities of demand for petroleum and products made from petroleum and elasticities of substitution between petroleum-based inputs and other inputs.

The shortages of petroleum and the periodic shortages of natural gas and other gases have caused higher prices and lower supplies of such manufactured inputs as nitrogen fertilizer and pesticides, which are a critical part of modern high-yield crop technology.

In addition to the obvious impacts of higher petroleum prices, there are more subtle impacts on the utility-creating processes (time, form, location) and secondarily on market exchange and coordination processes in the food system. For example, for some foods the single greatest energy embodiment is that involved in transportation. Cheap energy has permitted a pattern of regional specialization in production and processing. Examples include energy-intensive irrigation systems and wide use of plastics in packaging. A set of marketing arrangements has developed around the regionalization and specialization of production, which have become accepted as a part of the institutional setting for food marketing and distribution. As another example, the coordination and exchange mechanisms in the dairy industry are built around fresh fluid milk and related physical processes (e.g., cooling, processing, refrigerated trucks and stores). If higher relative energy prices lead to major changes in product form (e.g., dried or reconstituted milk or milk treated with ultrahigh temperatures to eliminate need for refrigeration and greatly extend shelf life), there could well be major adjustments required in markets for dairy products and in the organizations that operate in these markets.

We are only beginning to fully appreciate the fundamental role of energy in the national economy and how developments in energy markets in turn impact on other markets. As a result of higher prices and potential shortages of liquid fuel, considerable attention is being given to use of agricultural biomass to produce alcohol. A major alcohol program would have significant impacts on the supply, demand, and pricing of grain; relative input prices for livestock production; markets for soybeans and protein supplements as a result of high-protein distillers' grain generated by grain alcohol production; and markets for land services. Another example is the shift in demand for coal resulting from relatively higher petroleum prices. This increased demand has ramifications for land use, markets for land services, water markets, markets for transportation services, and the structure of rural communities.

Land. Developments in markets for land and land services, while not as sudden and dramatic as those in energy markets, are nonetheless quite significant. Land prices and rental rates have increased sharply in absolute terms and relative to overall rates of inflation. Although great concern has been expressed about the rapid increases in land prices being irrational and speculative, a number of studies (Boehlje 1979; Melichar 1979) have demonstrated that generally the increases have been consistent with growth in returns from landownership, including current earnings from land use and capital appreciation.

Several other developments in land markets are significant. One is the awareness that in concept (and increasingly in practice) the markets for land and land services are separable. Thus one can treat the purchase of land and gaining access to land services as separate though often related decision processes. This concept is not new. We have long had the notion and reality of economic rent. Tenant farming may be almost as old as farming itself. What is new, or at least evolving, is a changing set of terms of trade between landowners and land users and a changing attitude toward rental as a means of acquiring access to land services. This development is one of many interrelated factors including changing patterns of landownership, changing goals and business growth strategies of farm operators, institutional developments and public policies affecting risk and uncertainty, costs and availability of loan funds, and tax policies.

One result is increased competition for land services. In some regions where this competition is intense, intermediaries have emerged to facilitate the flow of land services from owner to user. Nationally, such intermediation processes and agents are more potential than real. But the potential may be important for the future organization of land use, since it increases the entrepreneurial options for marketing and using land services.

Another development of increasing importance in the markets for land services involves a host of economic constraints and regulations. Examples include environmental constraints; health and safety regulations; operating rules of federal commodity programs; acreage limitations in public irrigation projects; public irrigation policies and programs; policies governing access to use of public lands; and, potentially, public policies with respect to research, technology, and structure of farming and landownership (e.g., see USDA 1981; USDI 1981). All these and other developments must be internalized into the supply-demand and pricing mechanisms of land and land service markets.

Finally, there appears to be growing concern regarding the entry of new participants in land markets. Inflation, international instability, tax laws, the growth image of agriculture, the fixed supply of land, and the search for new life-styles are among reasons given for interest in landownership by nonfarm individual investors, investment syndicates (representing pension funds, endowment portfolios, and the like in addition to pooled resources of individual and corporate investors), and alien investors. Concerns expressed by farmers over these developments are that land prices are bid up because the new entrants have tax and other economic advantages that enable them to pay more than farmers can pay, beginning farmers find it difficult to become owner-operators, and operating farmers can find themselves at a disadvantage in obtaining land services from nonfarm landlords.

In addition, these developments are seen by present farmers as a potential threat to the owner-operated family farm as the dominant structural feature of American agriculture. Perhaps the greater issue for efficiency of land use in the future is the large amount of farmland now owned by small farmers and other rural residents who farm part-time or simply wish to live in a rural setting. One result has been to reduce the intensity of land use for farming within commuting ranges of large cities. These developments affect price and supply of and demand for land and land services.

Money. Prior to the 1920s, agriculture was viewed by lenders as a high-risk sector of the economy. Accordingly, the terms of trade in money markets were not favorable for farmers and they paid dearly for borrowed funds. However, a combination of new credit institutions and risk-reducing public policies changed the situation. By the 1960s some economists were arguing that loanable funds were so plentiful and inexpensive to farmers that agriculture was overcapitalized relative to the rest of the economy (Lee 1971).

However, in the 1970s there was increasing concern about prospects of long-term capital shortages relative to the growing demands of a capital-intensive economy. This concern was exacerbated by increasing inflation and outflows of dollars to pay the increasing costs of petroleum imports. Farmers entered the 1980s facing high and volatile (by historical standards) interest rates. With rural banks competing for funds in the central money markets and with constraints on (frequently subsidized) federal farm lending programs, agricultural credit markets are no longer insulated from the volatility and competitive pressures of the large urban money markets.

Continuing high interest rates and scarce loan funds could have substantial impacts on the food system now and in the future. Farm management, growth and resource acquisition strategies, and behavior may be altered. There are signs that financial markets are altering practices and institutional arrangements in agricultural product and factor markets. In grain markets the high cost of funds is causing many farmers to reassess marketing and storage strategies. Periodically, as experienced in the mid-1970s and early 1980s, accounts receivable by some farm-input suppliers declined because higher interest rates forced a reduction in extensions of credit. The rising costs of money may also impact on productivity in the food system. For the past half-century, the technological and managerial revolution in agriculture has been fueled by cheap and abundant credit. The food processing and distribution system has also become highly capital intensive. Sharply higher costs of money and restricted supplies would probably impact on technology and productivity in that system.

Changes in the markets for agricultural funds are more complex than shifts in supply, demand, and prices. Examples include rural banks asking agencies of the Farm Credit System for loan participants; formation of other financial institutions (OFIs), which allow cooperatives, banks, and other establishments to gain access to central money markets through the Federal Intermediate Credit Banks; and the rapid shift of bank depositors' funds from time deposit and transactions accounts to money market certificates and other instruments for capturing competitive returns for savings.

Labor. Labor has long been regarded as scarce and expensive relative to capital and energy. Thus much of our technology has been designed to economize on labor inputs. Output relative to labor input is widely accepted as a measure of performance (although agriculture is one of the few sectors of the economy for which there are ongoing measures of total factor productivity). Several regulatory developments have affected the costs and terms of use of labor sevices. These include elimination of the bracero program, more stringent regulations affecting migrant workers, Occupational Safety and Health Administration standards, increases in minimum wages, and farm labor unionization. Wage rates were not rising as rapidly in the late 1970s and early

1980s as interest rates and energy prices, although some of the regulations mentioned above may have directly influenced labor costs. Thus it is unclear how much relative prices for labor services have changed, and models of farm labor markets, in any case, probably are not adequate for forecasting future developments.

A market has emerged in agriculture for human services other than physical labor. This market is for management services ranging from technical advice on when to apply pesticides to complete management of farms. Little is known about the supply of and demand for these services. This development is part of the continuing industrialization and specialization trend in the farm sector.

Other Factor Markets. Changes in energy and money markets, as already noted, have had major impacts on markets for fertilizer, pesticides, and other chemicals used in farm production. After many years of stable or declining real prices, prices for these yield-augmenting chemicals were rising rapidly as the economy entered the 1980s. The markets for these inputs have also been affected by several environmental and safety-related regulatory constraints.

Farm machinery markets have been impacted by rapid changes in technology and thus in the mix of products, structural and technological changes in farming, and more recently by changes in prices of energy and money. Farm machinery manufacturing is highly concentrated and that concentration has been increasing. A more recent development has been the rapid decline in machinery retailers (from 16,400 in 1963 to 10,000 in 1980) as fewer and larger farmers purchased fewer, larger, more expensive, and more specialized pieces of equipment. Thus there has been rapid growth in concentration of market shares by local retailers (Galbraith 1980).

Finally, water shortages and changes in the conditions under which water can be used have been emerging, especially in the irrigated West. There is growing pressure to have water markets and prices reflect the true social costs of water development and opportunity costs for alternative uses. The economy of many western irrigated areas was built on the assumption of plentiful supplies of relatively cheap water. Major threats to that assumption carry significant implications for land and water use and product mix in irrigated regions and for location of production nationally.

Summary of Recent Developments. The net implications of recent developments in agricultural factor markets are not clear. Dramatic increases in real prices of inputs that have provided the basis for "the miracle of modern agriculture" pose a major dilemma at a time of strong and growing world demand for farm products. The result could be accelerated changes in technology, sharply higher farm product prices, or both.

An emerging characteristic of factor markets that may be as significant as higher price levels is the increased volatility of input prices. Farm-produced inputs such as feeder calves have typically had variable prices. As the share of U.S. agricultural production going to world markets has increased, the volatility of prices for key farm commodities used as input to further production has increased. We now also experience greater volatility in other input markets, including interest rates. Since price variability apparently shifts product supply functions, there would seem reason to believe that factor price variability should influence the demand for factors. But the literature reveals little if any empirical work done in this area.

Another general characteristic of forces affecting agricultural factor markets is the increasing tendency to impose institutional and regulatory constraints on markets to achieve a wide variety of social and political objectives. These constraints simply represent changing the rules under which the factor markets operate. The net effect is to internalize in the workings of private markets those social costs previously treated as externalities. This development may be more significant in the long run than the price increases for energy and money.

An emerging concern with structure of the food system and the attendant concern with who gets access to resources and under what terms also must be noted. These concerns may be translated into additional constraints or rules within which factor markets operate. Their direction is not at all clear, but potential implications provide a rich field of inquiry.

COMMENTARY ON PAST RESEARCH

An exhaustive review of the literature and past research on factor markets is beyond the scope of this chapter. However, some general observations on the focus and tendencies may be more germane to the objective of this book, which is to suggest priorities for future research. A limited list of references is provided at the end of this chapter for the reader seeking more specific information.

The literature reporting on agricultural factor markets is not as extensive as one might suppose, given the fundamental importance of the subject. (Dahl et al. 1971 give more than 500 references on input markets; Tomek 1980 also provides a very useful list.) Most marketing research since World War II has dealt with farm product markets. The review of postwar literature commissioned by the American Agricultural Economics Association and published in three volumes contains few headings or sections treating input or factor markets as such. In volume 1, Tomek and Robinson (1977) provide a brief review of literature on demand and supply of farm inputs. Another section is devoted to agricultural finance and capital markets, and the subject of input markets comes up indirectly in sections reviewing research on production functions and technical change.

The limited research by agricultural economists on factor markets may be both understandable and reasonable. The United States has been blessed with an abundance of low-cost resources. Excess supplies of cropland were held in reserve by public farm policies. Energy, chemicals, water, machinery, and equipment, and (since the 1930s) loan funds have been plentiful and cheap.

Furthermore, for fifty years agricultural economists were preoccupied with product markets in an environment characterized by chronic excess supplies and depressed prices and incomes. Even with depressed commodity prices, farm management analyses tended to suggest that most individual farmers would profit from applying more fertilizers, pesticides, and other capital goods to their land because doing so increased output and revenues more than costs. Clearly, until the late 1970s, product prices, not input prices, were perceived as the more critical obstacle to improving returns to farming.

Generally, research on agricultural inputs has fallen into one of two categories: structure and organization of input industries and markets and the economics of input use within the context of the theory and behavior of farm firms. The first category is defined to include supply and demand relations, although these have not often been the primary focus of studies of the structure of input markets.

A number of studies have described the economic structure of the fertilizer, pesticide, manufactured feed, and farm machinery industries (see References). They have tended to focus more on structural characteristics (such as number and size of firms, concentration of market shares, production processes, degree of horizontal and vertical integration, and marketing channels) than on performance characteristics. There are remarkably few studies of markets for manufactured inputs (as opposed to industry structure) and of price discovery and behavior. Few studies have dealt with the relationship between industry structure and market performance, expressed in terms of pricing practices, monopolistic power, and pricing efficiency.

Perhaps more than any other manufactured input, fertilizer has been the subject of studies of the economics of use in farm production processes. Perhaps this occurred because crop-yield response to successive increments of fertilizer lends itself to controlled experiment. Moreover, the input-output relationships observed tend to conform to theoretical relationships. Generally, these studies tended to identify optimum fertilizer applications, other inputs held constant. Except for the substitution relationships implied between land and fertilizer, little of the fertilizer research dealt with elasticity of substitution between fertilizer and other yield-augmenting technologies. Fertilizer-yield relationships imply demand functions for fertilizer, but few attempts have been made to empirically explain or forecast fertilizer prices by estimating derived demand from production functions and supply functions, consistent with industry structure and costs.

In contrast to fertilizer, little research has been done on markets for other chemical inputs such as pesticides and antibiotics. The use of these in production has been subjected to extensive regulation. Research has been directed to economic impacts of regulation on their use in production, but little is known about how these regulations affect supply, demand, and prices of the inputs themselves.

The Canadian government sponsored one of the more exhaustive studies of the farm machinery industry (Information Canada 1971), which contained detailed analyses of costs of production and estimates of the number of plants required for the most cost-efficient production of different types of farm machinery.

Money markets have been the subject of many studies, few of which focus specifically on pricing mechanisms for money for the farm sector. Much of the work done on financial markets has dealt with structures and problems of credit institutions within those markets. Considerable work has been done on financial intermediation processes. But there has been little research on money market imperfections, how they affect the pricing of money as a farm production input, and how these markets affect the allocation of loan funds within agriculture and among the sectors of the economy.

Because money markets are not perfect, behavior of local (especially rural) markets may be unique. However, there are few studies that examine money supply and demand functions in local markets.

Loanable fund markets are highly regulated. Only limited research has been done on how these regulations affect the structure of money markets and the supply, demand, and price of money.

Again, a derived demand function for money capital is implied by the economics of use of nonmoney production inputs. Little empirical work has been done on such a concept. For much of the past 50 years, interest rates were sufficiently low and stable that little research was done on impacts of limited money supplies and high interest rates on the farm production and marketing system. Exceptions are studies that show the impact of alternative interest rates on land values.

Several studies have revealed the relationship between tax policies and capital flows in the food system, especially in farming (Melichar 1979; Reinsel 1979; Davenport et al. 1982). Through a process of incentives and disincentives, tax laws impact on the supply of equity capital and demand for loan funds. The research done thus far suggests that because the incentives and disincentives are not evenly distributed among farmers, the resulting capital flows tend to be unevenly distributed in a way that contributes to the ability of large, high-equity farmers and wealthy nonfarmers to consolidate farmland into fewer and larger holdings.

Studies of land markets have tended to focus on explanations of land prices. Surprisingly few studies have dealt with markets for land services (rental markets) and on the possibility of improved intermediation processes between landowners and those who desire access to the services of land. To the extent that land use is tied to landownership and that landowners are not the most efficient users of land services, the markets for these services can be improved if they are freed from the constraints of ownership and allowed to flow to the most productive users (highest bidders). Little has been done to either conceptualize or measure the potential impacts of improving markets for land services as distinct from markets for land itself.

Several research efforts have dealt with capitalization of federal commodity program benefits into land values as those benefits tend to shift the demand function for land to the right (Hottel 1971; Reinsel 1972). Conversely, regulations that reduce the discounted benefit stream from landownership could be viewed as shifting the demand function to the left. These are but examples of regulatory developments that become internalized in the workings of land markets and ultimately impact on land prices and use of land services. A review of the literature suggests that work done to date does not scratch the surface of this important area.

Finally, the increased variability of factor prices adds a new element of uncertainty to the food system and especially to farming. Past research on the consequences of price instability has dealt mostly with commodity prices. We know little of how risks arising from factor price instability are transmitted through the food system or of the incidence of costs and benefits.

Three other deficiencies in past research on factor markets should be noted: inadequate treatment of markets for factor services as opposed to factors themselves, inadequate information about elasticities of substitution among factors used in the food system, and lack of understanding of relationships between factor markets and subsequent coordination and exchange mechanisms in the food system.

The first deficiency, noted in the failure of research on land markets to treat adequately the distinction between ownership and access to the services, is relevant for any input not entirely consumed in production but which generates a flow of productive services. In addition to land, examples include capital goods (buildings, machinery, equipment) and labor. Markets for the services from capital goods (hence equipment leasing) exist, but because of ease of ownership of capital goods, leasing of services has not been a highly developed phenomenon in agriculture. Perhaps for this reason it has received limited research attention. However, growing capital requirements for farming along with increasing competition for equity and loan funds suggest the likelihood of increased separation of the capital goods–ownership function and the use of services from those goods. Implied is the need for research to facilitate efficiency and economic equity in those markets.

The literature on markets for labor services is more advanced, since the ownership of the labor good is embodied in the individual laborer. Thus the treatment of markets for labor services can serve as a model for services of other factors of production. However, there is very little literature on the markets for managerial services that also flow from humans.

Another more recently noted deficiency of past research on factor markets is the lack of estimates of elasticities of substitution among factors of production. More precisely, not enough is known about the flexibility of the various production and exchange processes in adjusting to large and rapid changes in factor prices and supplies or about the flexibility of factor markets themselves to adapt to changing supply or demand realities. Such knowledge is critical in understanding how sharply higher energy prices, for example, will impact on production technology and resource use. Moreover, empirical determination of the flexibility for adjustment to sharp changes in input supplies and prices is essential for understanding not only the impacts of such changes in the food system but also the subsequent impacts throughout the economy.

For example, theory suggests that changes in relative factor prices will result in changes in factor mix and will not necessarily be inflationary. But it is not clear that in reality adjustments take place as suggested by theory; e.g., recent sharp increases in energy prices appear to have been inflationary. A plausible explanation is that institutional rigidities in factor markets do not permit rapid or full adjustment to price changes. As one example of such rigidity, increases in energy prices get indexed into higher labor costs through cost-of-living clauses in wage contracts. Also, there are fixed costs in capital inputs and costs of adjustment that may result in lengthy lags. Because so little research has been done in this area of input markets, it is difficult to understand even the qualitative nature of impacts of sharp changes in input prices.

Finally, past research tells us little about the linkage between factor markets and coordination and exchange processes in product markets. The dairy example noted earlier illustrates the point. Exchange processes tend to develop around physical realities. Thus, as changes in input markets lead to changes in production processes, exchange and coordination processes are forced to adjust.

Another illustration from dairy marketing involves the shift out of door-to-door delivery, which could no longer compete with supermarkets where less labor was required per unit of milk sold. When the dairy companies shifted from door-to-door sales to distribution through supermarkets, the coordination and exchange processes also changed. When milk flowed from farmer to dairy company to consumer, the coordination was probably provided by the dairy company. When the flow shifted and went from farmer to dairy company to supermarket chain to consumer, it is likely that the center of market power and coordination shifted to the supermarket chain. This example could be taken further by suggesting that the retail chains exercised their buying power, forcing small dairy firms to grow or consolidate to remain competitive. To increase the competitive pressure, there was some backward integration by chains purchasing or operating their own dairies. The net result was change not only in market channels and exchange processes but also in the structure of the food system itself.

As another example, if sharply rising energy costs force changes in routes, distances, and modes of physical movement as well as physical form of products, it is not hard to imagine that new firms and exchange processes will evolve to serve the new situation or that old firms will be required to make major changes. Past research has contributed little to understanding or quantifying these linkages.

In summary, input markets have tended to be neglected by researchers in agricultural economics. Research on inputs has described the structure of manufactured input industries and analyzed the economics of use of inputs in farm production processes. Less attention has been given to the performance of input markets and to the elasticities of substitution among factors of production. For land and capital goods used in agriculture, emphasis has been on markets for those factors of production rather than on markets for factor services. Little research attention has been given to the internalization into factor markets of regulations and rules changes that shift supply and demand functions. Finally, there have been few empirical attempts to trace and quantify impacts of changes in factor markets on production processes, coordination and exchange processes, and industry structure.

RESEARCH NEEDS: AN AGENDA FOR THE FUTURE

Conceptually there is little if any difference between factor and product markets. A market is a market. Along the continuum from the basic endowments of nature to final consumption (that too is an input), any boundary between factor and product markets is somewhat arbitrary. Nevertheless, important practical reasons exist for studying factor markets separately from product markets. Those reasons were expressed or implied in the preceding pages. Research priorities for the future are also implied in those pages. They can be summarized in two groups: general and input-specific.

General Research

LINKAGE OF FACTOR MARKETS TO STRUCTURE OF THE FOOD SYSTEM. Specific research attention needs to be given to tracing the linkages among factor markets and the subsequent organization and performance of the food system. The first need is to improve the conceptualization of those relationships, i.e., to trace them qualitatively. That contribution alone would serve to improve understanding of structural change processes and of the fundamental importance of relative prices and supplies of major inputs and the institutional arrangements under which they are provided. Much of the present controversy about the structure of agriculture pertains to the distribution of access rights to resources and the conditions of that access.

Documentation of historical relationships between factor markets and food industry structure would be undertaken to provide empirical content to the linkage conceptualizations. There have been a few general treatments of these relationships. But painstaking quantification of how specific input market situations impact on input use, technology, and organization of economic activity in the food system would provide practical insight to one set of forces causing structural change. A fertile period for such study would be the years beginning roughly after World War I and running through the 1960s. For example, what are the forces that created the demand for currently used inputs?

SEPARATION OF MARKETS FOR FACTORS AND FACTOR SERVICES. There probably is significant potential for improving the performance of markets for factor services by developing institutions and intermediation processes to facilitate the flow of services from factor owners to users. Such institutions could further free the flow of services to the most efficient users and eliminate the lumpiness that characterizes their being tied to factor ownership. As an example, a farmer may buy a larger tractor to assure timeliness of operation during short planting and harvesting seasons. During other periods, the tractor may be idle and no services flow from it. Through leasing, the markets for capital goods (e.g., tractors) could be separated from markets for services of capital goods (e.g., use of the tractor). If the factors were scarce (e.g., land) and markets for the factor services were highly competitive, services would flow to the highest bidders (most efficient users).

It would be instructive to compare the efficiency of existing resource use in a geographic area containing all owner-operators with the hypothetical efficiency of resource use if all the land services were released from ownership constraints and the use of land was allowed to flow to the highest bidders (who presumably are the best managers). Moreover, development of improved rental markets for services of land and capital goods may offer the flexibility needed to assure new farmers access to factor services and permit structural alternatives to owner-operated units.

INTERNALIZING REGULATORY REQUIREMENTS INTO THE WORKINGS OF FACTOR MARKETS. All markets operate within a set of rules. Attempts by society to curb undesirable side effects of resource use or to achieve desired objectives through regulations (changing the rules by which markets operate) have the effect of shifting supply or demand functions or both. Members of the broadened constituency of agricultural and food policy seek to achieve environmental, conservation, distributional, and structural objectives through rules or regulations that constrain or in some way modify the workings of markets (including factor markets). This may be one of the more significant economic developments in recent years. Tax laws modify the "real" prices of land, money, and capital goods; restraints on soil runoff impact on markets for land services and water; and acreage limitations on farms served by federal irrigation projects modify markets for land and machinery services in those areas. Research should address the economic impacts on internalizing these regulations into the workings of factor markets, and conversely, research is needed to provide policy guidance on how the objectives of the regulations can be achieved in ways that minimize distortion and inefficiency in factor markets.

IMPEDIMENTS TO ADJUSTMENT IN FACTOR MARKETS. Impediments to adjustment are of increasing concern when relative prices and supplies of factors are changing rapidly and when there is growing concern with structural inflation. This point, discussed in the review of past research, is related to the growing tendency to impose regulatory constraints or institutional rigidities on markets to achieve specific societal objectives. Research is needed on market imperfections and barriers to adjustment to understand how increases in the relative price of a given input will contribute to inflation, to minimize inflationary impacts and reduce distortions in resource use, and simply to measure the probable length of lags in adjustments to new prices—the long-run elasticities. This is a neglected but very important area of research for agricultural economists.

ELASTICITIES OF SUBSTITUTION. Elasticities of substitution among agricultural inputs is a topic about which all too little is known. Yet such knowledge is essential to understanding, for example, how sharply higher petroleum prices will affect the organization and cost of farm production as well as food processing and distribution. Gradual changes in relative factor prices tend to result in substitution among factors through technological change and evolution of new production functions. Factor substitutions in response to short-term rapid changes in relative factor prices are defined by existing technology. Thus research on elasticities of substitution among agricultural and food system inputs must have an empirical orientation and recognize technical realities as well as institutional and structural realities in input markets.

IMPACTS OF NONMARKET ACTIVITY. The transfer of ownership of land, money, and capital goods through gifts and inheritance may result in less efficient use of those factors than would be suggested by the norms of a competitive market. Documentation of the nature and extent of nonmarket activity in factors of production would be useful. If the magnitude of such activity is significant, it is important to understand the incentives for it and the consequences. If the consequences suggest structural or performance problems, public policies to alter incentives for nonmarket transfers of factor ownership and use rights can be analyzed, especially if a causal relationship exists between public policies and incentives for these transfers. For example, what are the implications for structure and performance of the farming sector if the tax and inheritance laws encourage a situation where most of the farmland is held by members of a landed class who inherited it? The implications may depend in part on the performance of secondary markets for use rights to the factors whose ownership was transferred via nonmarket mechanisms. Research on this subject could be an important input to the national dialogue on future policies pertaining to the structure of American agriculture.

Research for Specific Input Markets

ENERGY. Energy is a relatively new research topic but one of obvious importance in the near term; how important it is and for how long depend on how rapidly alternative (to petroleum) energy sources are found and how expensive they are. Knowledge is urgently needed not only on the elasticities of substitution among alternative energy forms but also between energy and other inputs. Use of petroleum energy pervades every facet of the food industry. Thus developments in energy markets will affect various stages of the food system and institutions within those stages differently, depending on how energy intensive they are and their flexibility to substitute among energy forms. The potential for structural change stemming from energy market developments is enormous and begs to be researched. One major energy market development might be the internalizing of subsidy provisions for alcohol from biomass, resulting in a substantial shift (to the right) in the supply function for alcohol fuel. Tracing the impacts of energy market developments could provide the first example of research on quantifying the specific linkages between input markets and industry structure. Needless to say, there is considerable interest in the distribution and incidence of impacts of higher energy prices and shorter supplies of certain energy forms.

LAND. Despite all the past research on land markets, much remains unknown. Moreover, there are no universally accepted explanations of price behavior for land and land services. Some have suggested, for example, that land markets have behaved quite rationally and resemble the behavior of markets and prices for growth common stocks. Better understanding of land markets is also essential because of the close association of concern over farm structure with who has access to land and under what terms and conditions. The possibility of further separation of use and ownership of land has been discussed and is a rich field for research. Also, it should be rewarding to study further the institutional structure of the market for land services (cash, crop-share) and how it affects intensity of land use, risk sharing, and distribution of benefits between landlords and land users.

WATER. Water markets are closely tied to developments in land markets. Use of irrigation water has been affected by rising energy prices. Water pricing deserves much more research because of the importance of policy to the organization of agricultural production and processing in areas served by publicly subsidized surface water supplies. Again, the role of regulation in water markets must be addressed. Research on both land and water markets will become increasingly critical as demand for American farm products continues to grow. Little is known of the supply functions for land and water if incremental increases in cropland are required each year. If those supply functions become highly inelastic with modest increases in food demand, there are major implications for U.S. food and agricultural policies. Without elaborating on those implications, it is safe to say that realistic estimates of supply, demand, and price behavior in water and especially land markets are needed urgently and soon.

MONEY. The importance of further research on money and the needed directions have been suggested earlier in this chapter. As with energy, it appears that strong demand for loan funds relative to supplies will put pressure on markets in the years ahead and will create for money many of the same issues that exist for energy. A number of major regulatory and institutional changes in money markets have occurred and their impacts are largely unknown, especially for agriculture and rural businesses. Moreover, with the structure of agriculture greatly different today from what it was when a number of present credit institutions and policies were established, the workings and impacts (who gets what benefits and how) of the financial markets serving agriculture need to be carefully reexamined.

MARKETS FOR MANUFACTURED INPUTS. Markets for manufactured inputs, including the structure and performance of the industries involved, are important subjects for research. Each year agriculture grows more dependent on manufactured inputs. Typically, market shares are concentrated among a few firms in the fertilizer, chemical, and farm machinery and equipment industries. Past failures to research market structure, performance, and conduct in these industries needs to be rectified. Given the nature of the industries in question, useful data will be difficult to obtain. Perhaps that situation implies that researchers in agencies having subpoena power should take the lead on researching market behavior for manufactured inputs. Alternatively, special study groups could be commissioned, modeled after the Canadian government group that studied tractor and machinery industry structure and performance.

The turbulence of markets for inputs suggests that improved forecasting of input market conditions could contribute importantly to improved public and private decision making and to efficiency of resource use. A prerequisite for improved forecasting is improved understanding of supply, demand, and price relationships. Thus empirical studies of these relationships for major farm inputs should be given increased priority.

   The research needs listed above merely illustrate the importance of factor markets and make clear that they should not be overlooked by researchers. Ideal data will be difficult to obtain for much of the research needed. But important progress can be made on most of the suggested topics with existing data if the topics are given high priority on the profession's research agenda.

REFERENCES

Anderson, Dale G., and P. W. Lytle. 1974. Purchased Farm Supply Markets: Feed, Fertilizer, Farm Machinery. In proceedings of the Farm Supply Seminar WM-61, Denver, Colo.

Baker, C. G. 1968. Credit in production organization of the firm. *Am. J. Agric. Econ.* 50:507-20.

Barse, Joseph R. 1977. Agriculture and energy use in the year 2000: Discussion from a natural resource perspective. *Am. J. Agric. Econ.* 59:1073-74.

Bauer, Larry L. 1969. The effect of technology on the farm labor market. *Am. J. Agric. Econ.* 51:605-18.

Baum, E. L., and S. L. Clement. 1958. The changing structure of the fertilizer industry in the United States. *J. Farm Econ.* 40:1186-98.

Bell, David M., David L. Armstrong, George R. Perkins, and Dennis R. Henderson. 1972. Resource Adjustment in the Fertilizer Industry: With Emphasis on Michigan Marketing. USDA, ERS Marketing Res. Rep. 974.

Binswanger, H. P. 1974. A cost function approach to the measurement of elasticities of factor demand and elasticities of substitution. *Am. J. Agric. Econ.* 56:377-86.

_____. 1978. Measured biases of technical change: The United States. In H. P. Binswanger et al., eds., *Induced Innovation.* Baltimore: Johns Hopkins University Press.

Boehlje, Michael. 1979. Land values, farm income and government policy. In *1980 Agricultural Outlook.* Committee on Agriculture, Nutrition and Forestry, U.S. Senate. Washington, D.C.: USGPO.

Boehlje, Michael, and S. Griffin. 1979. Financial impacts of government price support programs. *Am. J. Agric. Econ.* 61(2):285-96.

Boehlje, Michael, D. Harris, and J. Hoskins. 1979. A modeling approach to flow of funds in localized financial markets. Am. J. Agric. Econ. 61(1):145-50.

Bostwick, Don. 1969. Partitioning Financial Returns: An Application to the Growth of Farm Firms. USDA, ERS-390.

Boxley, Robert F. 1979. Ownership and Land Use Policy. In Structure Issues of American Agriculture. USDA, Agric. Econ. Rep. 438.

Brensike, John V. 1958. The changing structure of markets for commercial feeds. *J. Farm Econ.* 40:1201-10.

Brewer, Michael, and Robert Boxley. 1980. The Potential Supply of Cropland. Paper presented to Resources for the Future conference on the adequacy of agricultural lands. Washington, D.C.

Burton, Robert, and R. G. Kline. 1978. Adjustments in a farm business in response to an energy crisis. *Am. J. Agric. Econ.* 60(2):254-58.

Cochrane, Willard W. 1979. *The Development of American Agriculture—A Historical Analysis.* Minneapolis: University of Minnesota Press.

Congressional Research Service. 1976. Land Use and Energy: A Study of Interrelationships. Committee on Interior and Insular Affairs, U.S. Senate. Washington, D.C.: USGPO.

Cromarty, William. 1959. The farm demand for tractors, machinery and trucks. *J. Farm Econ.* 41:323-31.

Crosson, Pierre. 1979. Agricultural Land Use: A Technological and Energy Perspective. In Max Schnepf, ed., *Farmland, Food and the Future.* Ankeny, Iowa: Soil Conservation Society of America.

Dahl, Dale C. 1980. The U.S. Fertilizer Industry: Analytical Perspectives and Research Issues. In Input Markets in Agriculture—Proceedings of NCR-117 Seminar. NCR-117 Publ. 1, Ohio State University, Columbus.

Dahl, Dale C., J. D. Anderson, and R. D. Peterson. 1971. Purchased Farm Input Markets in the United States 1950-1971: A Bibliography of Economic Studies. University of Minnesota, Agric. Exp. Sta. Misc. Rep. 103, St. Paul.

Davenport, Charles, Michael D. Boehlje, and David B. H. Martin. 1982. The Effects of Tax Policy on American Agriculture. USDA, Agric. Econ. Rep. 480.

Doll, J. P. 1974. On exact multicolinearity and the estimation of the Cobb-Douglas production function. *Am. J. Agric. Econ.* 56:556-63.

Dovring, F. 1975. Macro-constraints on agricultural development in India. *Indian J. Agric. Econ.* 27:46-66.

_____. 1976. Agricultural Technology for Developing Nations. Paper presented at Conference on Farm Mechanization Alternatives for 1-10 Hectare Farms. University of Illinois, Urbana.

Duncan, Marvin, and E. Harshbarger. 1979. Agricultural Productivity: Trends and Implications for the Future. Econ. Rev., Reserve Bank of Kansas City.

Eichers, Theodore R. 1980a. Research Needs on Economics of Pest Control. In Input Markets in Agriculture—Proceedings of NCR-117 Seminar. NCR-117 Publ. 1. Ohio State University, Columbus.

_____. 1980b. Farm Machinery Research Needs. In Input Markets in Agriculture—Proceedings of NCR-117 Seminar. NCR-117 Publ. 1, Ohio State University, Columbus.

Erven, B. L. 1980. Research Opportunities Related to the Hired Farm Labor Input in Production Agriculture. In Input Markets in Agriculture—Proceedings of NCR-117 Seminar. NCR-117 Publ. 1, Ohio State University, Columbus.

Fox, Austin S. 1966. Demand for Tractors in the United States—A Regression Analysis. USDA, Agric. Econ. Rep. 103.

Frederick, Kenneth D. 1980. Irrigation and the Future of American Agriculture. In S. S. Batie and G. Healy, eds., *The Future of American Agriculture as a Strategic Resource.* Washington, D.C.: The Conservation Foundation.

French, B. C., and H. F. Carman. Production agriculture as a force affecting the food system. Chapter 4, this book.

Fuller, Varden, and C. L. Beale. 1967. Input of socio-economic factors on farm labor supply. *J. Farm Econ.* 49:1237-43.

Galbraith, William. 1980. Estimated Changes in Farm Machinery Dealerships. National Farm and Power Equipment Dealers Association, St. Paul, Minnesota.

Griliches, Zvi. 1958. The demand for fertilizer: An economic interpretation of technical change. *J. Farm Econ.* 40:591-606.

_____. 1959. The demand for inputs in agriculture and a derived supply elasticity. *J. Farm Econ.* 41(2):309-22.

Gunjal, Kisan R., R. K. Roberts, and E. O. Heady. 1980. Fertilizer demand functions for five crops in the United States. *South J. Agric. Econ.* 12:111-16.

Hamm, L. G. 1979. Farm Input Industries and Farm Structure. In Structure Issues of American Agriculture. USDA, Agric. Econ. Rep. 438.

Hammonds, T. M., R. Yadav, and C. Vathana. 1973. The elasticity of demand for hired farm labor. *Am. J. Agric. Econ.* 55:242-45.

Heady, Earl O., and J. L. Dillon. 1961. *Agricultural Production Functions.* Ames: Iowa State University Press.

Heady, Earl O., and L. G. Tweeten. 1963. *Resource Demand and Structure of the Agricultural Industry.* Ames: Iowa State University Press.

Heady, Earl O., and M. H. Yeh. 1959. National and regional demand functions for fertilizer. *J. Farm Econ.* 41:332-48.

Headley, J. C. 1968. Estimating the productivity of agricultural pesticides. Am. J. Agric. Econ. 50:13-23.

Hottel, J. Bruce. 1971. Resource pricing of land and mobile rice allotments. *J. Am. Soc. Farm Managers and Rural Appraisers* 35:45-49.

Information Canada. 1971. Report of the Royal Commission of Farm Machinery. Ottawa.

Kurdle, Robert T. 1975. *Agricultural Tractors: A World Industry Study.* Cambridge, Mass.: Ballinger.

Lee, John E., Jr. 1971. The money market—Is it adequate for the needs of today's agriculture? *Agric. Financ. Rev.* 32:1-5.

Lee, John E., Jr., Stephen C. Gabriel, and Michael Boehlje. 1980. Public Policy toward Agricultural Credit. Paper presented at Symposium on Future Sources of Loanable Funds for Agricultural Banks, Kansas City, Mo.

Lippe, J. Chris, and Ryan L. Petty. 1980. Effects of Energy Development and Production on Agricultural Water Requirements. Final report to the U.S. Department of Energy by Radian Corporation, Austin, Tex.

Liu, Yao-Chi. 1979. Technological change and structure. In Structure Issues of American Agriculture. USDA, Econ. Rep. 438.

Melichar, Emanuel. 1979. Capital gains versus current income in the farming sector. *Am. J. Agric. Econ.* 61:1085-92.

Menkhaus, Dale J. 1973. The Effects of Environmental Legislation on the Structure of the Pesticide Industry: A Simulation Study. Ph.D. diss., Purdue University, West Lafayette, Ind.

Minden, Arlo J. 1970. Changing structure of the farm input industry: Organization, scale, ownership. *Am. J. Agric. Econ.* 52:678-86.

Moore, C. V. 1979. External equity capital in production agriculture. *Agric. Financ. Rev.* 39:72-82.

Moore, John R., and R. G. Walsh. 1966. *Market Structure of the Agricultural Industries.* Ames: Iowa State University Press.

National Academy of Sciences, Board of Agriculture and Renewable Resources. 1975. *Agricultural Production Efficiency.* Washington, D.C.

Nelson, Paul E., Jr. 1970. The Farm Machinery and Equipment Industry—Its Changing Structure and Performance. USDA, Mark. Res. Rep. 892.

_____, ed. 1973. Farm/Ranch Input Research Yesterday, Today, and Tomorrow. NCR Res. Publ. 215, Michigan State University, Agric. Exp. Sta. Rep. 208, East Lansing.

Paul, Duane A., Richard L. Kilmer, Marilyn A. Altobello, and David N. Harrington. 1977. The Changing U.S. Fertilizer Industry. USDA, Agric. Econ. Rep. 378.

Paul, Duane, and R. L. Kilmer. 1977. The Manufacturing and Marketing of Nitrogen Fertilizer in the United States. USDA, Agric. Econ. Rep. 390.

Penson, John B. 1972. Demand for financial assets in the farm sector: A portfolio approach. *J. Agric. Econ.* 54:163-74.

Phillips, W. G. 1958. The changing structure of markets for farm machinery. *J. Farm Econ.* 40:1172-82.

Raup, Philip M. 1980. Competition for Land and the Future of American Agriculture. In S. S. Batie and R. G. Healy, eds., *The Future of American Agriculture as a Strategic Resource.* Washington, D.C.: Conservation Foundation.

Reinsel, Robert D. 1973a. The Aggregate Real Estate Market: An Evaluation of Prevailing Hypothesis Explaining the Time Series Trend in United States Average Farm Real Estate Values. Unpubl. Ph.D. diss., Michigan State University, East Lansing.

_____. 1973b. Farm Land Rents, Values and Earnings. Contributed paper, annual meeting, American Agricultural Economics Association, Edmonton-Alberta, Canada.

_____. 1979. Energy and U.S. Agriculture. U.S.-Indian Working Group on Agricultural Inputs. New Delhi, India.

_____. 1980. Economic issues related to energy in the food system. In Input Markets in Agriculture—Proceedings of NCR-117 Seminar. NCR-117 Publ. 1, Ohio State University, Columbus.

Reinsel, Robert D., and R. Krenz. 1972. Capitalization of Program Benefits into Land Values. USDA, ERS-506.

Reinsel, Robert D., and E. I. Reinsel. 1979. The economics of asset values and current income in farming. *Am. J. Agric. Econ.* 61:1093-97.

Robinson, Lindon J. 1980. Researchable issues of importance in agricultural finance. In Input Markets in Agriculture—Proceedings of NCR-117 Seminar. NCR-117 Publ. 1, Ohio State University, Columbus.

Rodewald, Gordon E. 1977. Assessment of the Four-Wheel-Drive Tractor—A Progress Report. USDA, AGERS-31.

Rosine, John, and P. Helmberger. 1974. A neoclassical analysis of the U.S. farm sector, 1948-1970. *Am. J. Agric. Econ.* 56:717-29.

Schuh, Edward G. 1967. The agricultural input markets—A neglected area of agricultural policy and economic research. In Clarence J. Miller, ed., *Marketing and Economic Development,* pp. 367-95. Lincoln: University of Nebraska Press.

Shumway, C. R., H. Talpaz, and B. Beattie. 1979. The factor share approach to production function estimation: Actual or estimated equilibrium shares? *Am. J. Agric. Econ.* 61:561-64.

Smith, Kerry V., and J. V. Krutilla. 1979. Resource and environmental constraints to growth. *Am. J. Agric. Econ.* 61:395-408.

Tomek, William G. 1980. Price analysis for farm inputs: A status report. In Input Markets in Agriculture—Proceedings of NCR-117 Seminar. NCR-117 Publ. 1, Ohio State University, Columbus.

Tomek, William G., and Kenneth L. Robinson. 1977. Agricultural price analysis and outlook. In L. R. Martin, ed., *A Survey of Agricultural Economics Literature,* vol. 1: *Traditional Fields of Agricultural Economics, 1940's to 1970's,* pp. 329-409. Minneapolis: University of Minnesota Press.

Tyner, Fred H., and L. Tweeten. 1965. A methodology for estimating production parameters. *J. Farm Econ.* 47:1462-67.

U.S. Congress. 1979. Pest Management Strategies in Crop Production. Office of Technology Assessment, Washington, D.C.

U.S. Department of Agriculture. 1968. Structure of Six Farm Input Industries—Petroleum, Farm Machinery and Equipment, Fertilizers, Chemical Pesticides, Livestock Feeds, Farm Credit. ERS-357.

_____. 1981. A Time to Choose: Summary Report on the Structure of Agriculture. Publ. 2.

U.S. Department of the Interior. 1981. Acreage Limitation, Draft Environmental Impact Statement. Water and Power Resources Service.

Vosloh, Carl J., Jr. 1980. Structure of the Feed Manufacturing Industry, 1975. USDA, Stat. Bull. 596.

Wipf, Larry J., and D. L. Bawden. 1969. Reliability of supply equations derived from production functions. *Am. J. Agric. Econ.* 51:170-78.

Young, Kenneth B., and Jerry M. Coomer. 1980. Effects of Natural Gas Price Increases on Texas High Plains Irrigation, 1976-2025. USDA, Agric. Econ. Rep. 448.

Zeimetz, Kathryn. 1979. Growing Energy, Land for Biomass Farms. USDA, Agric. Econ. Rep. 425.

KENNETH L. CASAVANT
JAMES K. BINKLEY

6

TRANSPORTATION CHANGES AND AGRICULTURAL MARKETING RESEARCH

FROM THE TIME of Adam Smith, it has been well recognized that trade between specialized producers forms the foundation of economic development. For specialization to be useful, trade must occur; and prior to trade, commodities produced by specialized producers must be transported. Hence transportation, broadly defined as the transfer of goods between specialized producers and consumers, is of vital importance to the economic well-being of society.

This chapter is concerned with the role of transportation in agricultural marketing. In keeping with the purpose of this book, the primary goal is to indicate fruitful avenues along which future research efforts may proceed. Before that can be done, it is necessary to lay some groundwork and to review what has been done in the past. Thus the first part of the chapter will discuss the significance of the role transportation plays in the economy, particularly with respect to agriculture. This will be followed by a discussion of the theory and concepts underlying transportation research and a review of past work in the field.

Of the three basic dimensions of marketing (time, form, and space), transportation primarily involves the spatial dimension. A market is often defined in terms of physical boundaries, which in turn are defined in terms of transportation. One of the proverbs of economics states that the division of labor is limited by the extent of the market, which in turn is limited, not by actual physical distance, but by the cost of overcoming this distance, that is, the cost of transportation. Transportation plays the role of a facilitator of trade,

KENNETH L. CASAVANT is Professor, Department of Agricultural Economics, Washington State University, Pullman. **JAMES K. BINKLEY** is Assistant Professor, Department of Agricultural Economics, Purdue University, West Lafayette, Indiana.

and improvements in transportation have the effect of widening the market for specialized production.

Transportation is of particular importance to the agricultural sector. The location of agricultural production is dependent upon the location of land and water resources, but consumers of agricultural commodities are widely dispersed. The development of agriculture has followed the development of means of overcoming this spatial separation. In the not too distant past, production occurred close to population centers because of the high cost of transportation. Consequently, agriculture was relatively unspecialized. With improvements in transportation, it is now possible for most agricultural commodities to move great distances at relatively low cost. As a consequence, agricultural production has achieved a high degree of specialization.

Agricultural commodities tend to be either highly perishable or to have a very low value-to-bulk ratio. Furthermore, the inherently seasonal nature of agricultural production increases shipping costs because of the uneven demands made upon the transport system. As a result, transportation accounts for a greater share of the delivered cost for agricultural products than is true for other goods. For this reason, the nature and location of agricultural production is particularly sensitive to changes in the transportation system.

The pervasive influence of transportation on economic activity in the agricultural sector ensures that transportation is important in determining the amount and distribution of farm income and the prices paid by consumers for food and fiber products. Thus agricultural interests have played a continuing role in development of transport regulation: a primary impetus for the regulation of railroads was the belief that monopoly rail rates were resulting in depressed farm income. Throughout the history of transport regulation, agriculture has received special treatment in recognition of the fact that the farm sector is particularly dependent on transportation. The exemption of trucking of most agricultural commodities from rate regulation is one example of this. The fact that transport firms providing service to rural areas have often been subsidized is another. Recent attention to the problems brought on by rail line abandonment stems from a realization of the significant effect transportation can have on the economic health of rural regions.

Numerous impacts on the U.S. farm sector have been brought about by institutional and technological changes in transportation. Locational shifts in the wheat milling industry have been induced by rail transit rates. The rise of California and the Southwest as major suppliers of fresh vegetables occurred only after development of refrigerated rail cars; the domination of this market by these producers was a direct result of the availability of high-speed, long-haul refrigerated trucks. Similarly, the shift of the broiler industry from the East to the South reflected development of barge transportation and barge-competitive rail rates. More recently, changes in the structure of the grain marketing industry, as evidenced by reliance upon fewer and larger facilities, have come about primarily because of the transport scale economies made possible by unit trains and large barge flotillas.

Transportation changes are felt at all levels of the food system; e.g., increasing concentration in the food marketing industry has been partially engendered by developments in the transport sector. Centralized distribution of food products to retail outlets and the above-mentioned growth of large grain facilities provide two examples in which transportation efficiencies have contributed to concentration. Such changes in industry structure can be brought about by relatively minor innovations in transportation. The introduction of unit trains, for example, involved no new technology but simply a different way to use that already existing.

Indeed, changes in organization of the transport sector may have a greater potential for increasing the efficiency of the system—the unit train example is only one of many. Discovering and evaluating such potentialities is the role of the transportation economist. At the same time, the sensitivity of the entire agricultural marketing system to transportation changes necessitates consideration and analysis of possible effects of changes in market structure and income distribution that might accompany efficiency gains, regardless of their source. Further, the structure of the transport industry itself (in many cases having characteristics of a natural monopoly), as well as the ensuing regulatory framework in which much of transportation operates, generates a need for information to guide public policy.

There has been an increase in research in the economics of transportation, particularly with reference to the agricultural sector. This reflects a growing recognition of the integral role played by transportation in the nation's economy. Continuing energy problems, the current emphasis on economic deregulation, and the apparent growing difficulty of the transportation industry in meeting the demands placed on it would seem to ensure that this trend will continue.

CONCEPTUAL BACKGROUND

The one concept that explicitly or implicitly underlies all freight transport research derives from transportation's role as a facilitator of trade. As such, it can be viewed purely as an input into the productive process, and production theory and the concepts of derived demand are applicable. From these it follows that transportation improvements in marketing agricultural output allow marketing areas to increase or cause net product prices received in an existing market to increase. With decreased transport costs, inputs delivered into production areas can shift effective supply curves, causing increased production or net returns or both.

But such concepts refer to transportation as a single input, as a method of moving goods from one place to another, a definition which for many research purposes does not apply. When viewed from this perspective, transportation as an input has very simple characteristics (e.g., limited substitutability with other inputs). However, a particular type of transportation may be highly substitutable with another, even though each has markedly different characteristics (e.g., barges and rail in the shipment of grain). But the same types may be totally nonsubstitutable in some other capacity (barges do not carry tomatoes). In other words, there are vast differences in the characteristics of transportation services. The transportation industry is composed of a highly interrelated set of subindustries (modes), which are often quite similar but also often quite disparate. For some purposes, this may not be of much significance; for others, it is often of great significance. In any event, research projects dealing with transportation problems have generally only examined subsectors of the transport industry, have been composed of several disaggregated studies, or have been conducted at a very macro level.

Thus it is possible to divide the conceptual background of much transportation-related marketing research into two categories: that dealing with transportation as an undifferentiated input and that dealing with specific types of transportation inputs. The former includes concepts dealing with interregional trade and the location of economic activity; the latter primarily includes concepts dealing with the behavior of individual firms. A major difference between the two groups is the nature of substitutability implied: the first involves substitution among shipments of raw materials and finished products and possible locations of production and processing; the second primarily deals with substitution among transport modes. There is no abrupt dividing line between these two groups, but they do provide a useful categorization. Indeed, a continuing challenge to transportation economists is to find ways to bring these two sets of concepts together in conducting marketing research.

Location theory provides much of the conceptual background of the first group above. This has been developed over time by Hotelling (1929), Weber (see Friedrich 1929), Losch (1954), Isard (1956), Greenhut (1963), and Von Thunan (see Hall 1966), among others, and refined for agricultural marketing firms in work such as that done by Takayama and Judge (1964) and Bressler and King (1970). Von Thunen's concentric ring theory, explaining the role of transportation costs in determining the location of agricultural production and the resulting site rents, laid the foundation for more refined and analytically useful models. Weber's location triangle specified that location depends upon the weight loss in production and hence explicitly recognized that transportation of final products can substitute for transportation of inputs. This led to his identification of market-oriented, raw material–oriented, and footloose industries. While Von Thunen and Weber considered transport costs in the context of a fixed market with variable production points, Losch and Isard examined fixed production points and variable markets, yielding the conceptual idea of natural marketing areas as determined by cost of transportation. Isard's treatment of transportation as an explicit input into the production process permitted consideration of the interaction between transportation and other inputs. Hotelling's model of spatial duopoly pointed out the interplay between shipping costs and interfirm competition in determining firm location. Greenhut emphasized the role of transport costs in restricting market areas, thus in his view causing oligopoly to be the typical market form.

A large body of transportation research, especially in the area of transportation of agricultural commodities, has been based on the theory of regional trade and concepts of spatial equilibrium. Initially, this research was based on cost minimization or allocation models with fixed production and demand. Samuelson (1952) demonstrated that under the assumption of completely inelastic (fixed) supply and demand, such models yield spatial equilibrium solutions. Work incorporating assembly, processing, and distribution costs, such as that by Stollsteimer (1963) and others, moved research further away from a partial equilibrium framework by incorporating possible substitution between transport and production costs. Based on a framework developed by Takayama and Judge (1964), spatial equilibrium analysis is now possible, incorporating differing resource availability, differing processing costs, and differing demand schedules and even allowing multiproduct analysis at firm, industry, and region levels.

At the micro (firm) level, transportation research has been based on traditional concepts of supply and, to a lesser extent, input demand. The cost and hence supply functions of transport firms are derived from the same theoretical foundations as are those of conventional firms, and analysis of transport firm costs and comparisons of the costs of different modes have been based on these theories. Application of concepts of input demand in studying the demand for transportation at the individual firm level has been somewhat less successful. While the demand for transportation follows from such theory, it is of limited usefulness in dealing with the more interesting question of modal choice. Conventional input substitution models are of little guidance for this purpose, since transport modes in their role purely as movers of products are very nearly perfect substitutes. The only hypothesis generated by such models is that the cheapest mode will be selected. The fact that this is often not the case has generated interest in quality and service differences among modes. While in principle such characteristics can be incorporated into a traditional framework, efforts to do so have met with limited success, primarily because of the difficulty of assigning costs to quality differences. Since the demand for the services of individual modes is a function of the substitutability among them, the service aspect of transport modes is a research area in need of a workable conceptual framework or more creative use of existing concepts. Perhaps the Lancasterian demand theory is appropriate, although the difficulty involved in empirical application of this paradigm has thus far limited its usefulness.

The micro orientation of transport supply forms a component in the broad approach provided by spatial equilibrium analysis, creating a situation where elements from both classes of concepts are joined together. This provides an important framework that includes not only the substitution of transportation between regions and between transportation and other productive inputs but also the substitution between modes. However, although such a framework permits the analysis of transport demand by mode at some rather aggregate level, permitting examination of such issues as cross elasticities between modes, the inability to incorporate much detail and the aforementioned problems associated with quality differences precludes the analysis of modal choice in any meaningful way. This is really less a conceptual difficulty than a practical one, but one of rather large magnitude.

Another characteristic of transportation, largely unrelated to those discussed above, has generated a need for different conceptual approaches. Because many transport facilities involve characteristics of public goods and natural monopoly, regulation of profits and rate of return (including provision of subsidies) has been a dominant factor in transportation. The combination of necessity and monopoly has led to varying degrees of regulation and public intervention into transportation firm decision making, thus constraining firm behavior and changing relative producer efficiencies. Without pure markets, price information has been distorted. Thus evaluation of the benefits and costs of resource use in transportation has come under the purview of economists, who have endeavored to identify the impacts of regulation on consumers, producers, and general social welfare. While this research has made use of some of the concepts outlined above (e.g., in examining the supply and demand characteristics of regulated modes), it has drawn on concepts from imperfect competition theory, public goods and public utility theory, and theories of industrial organization. A major thrust of this type of analysis has been to identify and evaluate secondary and external effects of alternative transport investments and institutions (or the lack thereof).

In conclusion, to say that transportation research has suffered from the lack of guidance provided by a well-developed conceptual framework may not be erroneous. However, this is more to be understood than decried. The transportation industry is made up of diverse yet highly interrelated components. This has generated corresponding micro and macro approaches to researching transportation problems, and a framework appropriate for one is often ill-suited for the other. It is doubtful that dramatically different theoretical and conceptual developments will be forthcoming to eliminate this difficulty. This suggests the need for the imaginative use of existing ideas in researching transportation issues. For example, the marriage of spatial equilibrium theory with more refined modal and firm cost studies provides the basis on which the effects of quality differences among modes can perhaps be analyzed. Of course there are rather large practical difficulties associated with such studies regarding data needs and perhaps computer technology. But this only serves to suggest that the primary constraint on transportation research may not be lack of a conceptual basis. For the most part, the concepts are there; once the mechanical research problems are solved, the concepts can be applied to nearly any transportation problem.

REVIEW OF PAST RESEARCH

Specific transportation studies tend to fall into one of four categories: firm studies, spatial equilibrium or allocation studies, demand studies, and studies of performance under regulation, with considerable overlap among them. This section presents a brief description and review of some of these studies. It is by no means intended to be exhaustive. Rather, the purpose is to indicate in a broad fashion the types of questions that have (and have not) been studied in the past and in so doing present the foundation on which future research will be based.

Many cost studies of transport firms have been conducted. The purpose of many of these has been to evaluate the efficiency of the food marketing sector and provide information to farmers and firms involved with agricultural marketing. Others have been conducted to evaluate transport firm performance with and without regulation.

The bulk of firm cost studies has dealt with commodity movements by truck. Many of these have been conducted by the USDA and include work by Camp (1965) (bulk feed by cooperatives) and Hunter (1963) (general exempt commodites), studies based on examination of actual firms. These have been followed by economic-engineering studies such as those by Kerchner (1967) (milk) and Boles (1973, 1976) (livestock). Casavant and Nelson (1967) combined accounting cost data with synthesized economic-engineering models to evaluate the performance of trucking firms carrying livestock and grain. Many similar studies have been performed by researchers at land-grant universities.

Much of the analysis of transport firm cost has concentrated on scale economies and the question of cost allocation to fixed and variable factors, primarily with an eye toward regulatory implications. This has been of particular importance in studying rail costs. The foundation of most such work was Bort's series of econometric studies of production relations in the railway industry in the late 1950s. Healy (1961) and, later, Griliches (1972) further statistically examined the cost allocation and existence of scale economies in rail transportation. Griliches's study, in which he concluded there were no significant scale economies in railroading, was an attack on methods used to allocate costs in a regulatory context, an extremely important issue, since regulated rates must cover avoidable costs. Harris (1977) made the important distinction between economies of scale and economies of density, the latter referring to cost behavior as traffic increases. He found significant density economies in the railroad industry and concluded that many miles of the railroad system should be abandoned.

The need for cost information for larger studies has generated cost analyses of different modes of hauling bulk agricultural commodities. In the work of Fedeler et al. (1973), Schnake and Franzman (1973), and Tyrchniewicz and Tosterud (1973), techniques for estimating transport cost functions have been developed as input for spatial equilibrium cost minimization models. In addition, a host of studies have developed transport costs as part of assembly and distribution cost functions in optimal plant location models.

The primary contribution of spatial equilibrium and plant location studies, of course, has not been their cost analyses but their ability to model the interaction among components of the transport system and between transportation and other elements of the marketing system. This has permitted researchers to examine the substitutability among transportation, storage, and processing. The original plant location work by Stollsteimer (1963) has been followed by increasingly sophisticated studies dealing with the assembly, processing, and distribution of nearly every major agricultural product. (For a review of these, see French 1977.) Allocative and spatial equilibrium models have directly focused on transportation issues. For example, numerous models examining the costs of the grain transportation system have been constructed and implemented. In addition to those mentioned above, these include studies by Wright (1969), Leath and Blakely (1971), Ladd and Lifferth (1975), and Binkley et al. (1978). (See Judge and Takayama 1973 for many others.)

Many studies have been conducted to evaluate performance under regulation. The general approach has been either to compare regulated and un-regulated firms or to examine rates and costs under regulation as contrasted to what they would presumbly be in its absence. Snitzler and Byrne (1978) have performed a pioneering study of trucking rates on fresh poultry before and after deregulation and found rates to be significantly lower under free competition. Both Farmer (1964) and Miklius (1969) have examined the economic performance of exempt carriers and concluded they provide either higher quality service or lower rates than regulated carriers, a finding corroborated by Cornelius (1977). Miklius and Casavant (1977) have compared the stability of exempt carriers with similarly capitalized industries and found no differences. Separate studies by McLachlan (1972) and Sloss (1970, 1971) have examined trucking in Canada, some of the provinces of which regulate trucking and some not. They found higher rates and more private trucking in regulated provinces but no difference in profitability of trucking firms.

The question of the effects of value of service railroad pricing, in which rates are based on commodity value in addition to costs of movement, has received much attention. It has been contended that such rates lead to resource misallocation by distorting cost-based substitution relationships between rail and other modes and cause cross-subsidization between commodities and regions. In a study of upper Midwest grain shipments, Martin and Dahl (1979) have found that value of service pricing significantly increased truck use relative to rail and led to higher total transport costs to society. Sorenson et al. (1973) have found substantial evidence of cross-subsidization among agricultural commodities in Great Plains rail rates, with a negative correlation between rates and elasticities of commodity demand. In a study of the railroad industry, Friedlaender (1971) has found significant social costs attributed to value of service pricing but much greater costs from excess capacity engendered by regulation. In an earlier study, Harbeson (1969) has estimated much larger social losses from value of service pricing. This was seriously disputed by Boyer (1977), who has contended that estimates of such distortions were grossly overstated as a result of failure to include the effects of quality differences between modes.

Studies of the transportation of agricultural commodities have dealt with other regulatory issues. Felton (1978) has evaluated backhaul restrictions in trucking; both Felton (1971) and Boles (1972) have examined freight car rental and detention policies and have recommended rule changes to increase transport efficiency. Using an optimization model, Shouse and Johnson (1978) have studied seasonal rail rates and found limited incentive for smoothing seasonal shipments resulting from the availability of substitute modes at little increase in cost. However, their study applied only to one market area.

Rail abandonment has been an issue of great concern to agricultural interests, and agricultural economists have responded with several studies, with the prototype conducted by Baumel et al. (1976). These studies have generally concluded that, on the basis of benefits versus costs, many existing branch lines do not warrant upgrading and hence should be abandoned. In an interesting companion study, Miller et al. (1977) found no differences in economic growth of areas that had lost rail service relative to those that had not.

The estimation of demand for transportation of agricultural products represents an area where little work has been done. Indeed, relatively few transport demand studies of any kind have been conducted, a reflection of the difficulties encountered in performing such analysis. Work by Perle (1964), Benishay and Whitaker (1966), Sloss (1971), and Bayliss (1973) has provided a rather meager foundation. The few studies conducted by agricultural economists have been very specific in nature, e.g., those by Limmer (1955) (Florida produce) and Miklius (1967) (California lettuce). A major problem with demand estimation has been the difficulty of measuring quality and service characteristics of the modes involved. Some success in this direction has been achieved by Miklius et al. (1976), who used certain quality variables and proxies for such variables in a logit analysis of the demand for transportation of cherries and apples. More recently, Johnson (1976) has provided a framework for inclusion of quality considerations, which he implemented in a study of the demand for grain transportation in Michigan.

Assessment of Past Research. The above review, which has only presented a sample of the type of studies relevant to issues in transportation and agriculture, illustrates that much work has been accomplished. Analysts have been quite successful in dealing with some questions. Among the notable successes are studies that have analyzed broad efficiency questions within the grain marketing system and studies on the effects (or lack thereof) of rail abandonment on rural areas. But rather than dwell on successes, we feel it is more appropriate here to discuss what needs to be done and point out important specific issues for future analysis. We now wish to indicate some broad areas where transportation research has been deficient.

The first problem, one which will probably haunt transportation economists for some time to come, is determination of costs and benefits associated with quality and service differences between modes. Lack of knowledge of this aspect of transportation has affected the relevance of the entire spectrum of transportation research, from specific demand studies to broad spatial equilibrium analyses. It has become increasingly clear that such nonquantifiable factors as reliability may be as important (and in some cases more important) as rates in shaping many shippers' transportation decisions. Accurate quantification of such characteristics can lead to analyses to determine the true nature of the efficiencies involved in the transport system. For example, if value of service pricing reflects quality differences, rail services may not substitute for trucking as well as is believed. Hence, rate distortions may not generate misallocations as large as some have contended.

In short, transportation economists need to sharpen their measurement capabilities. This is a difficult problem, but if costs and quality characteristics can be properly measured, the existing body of conceptual ideas can be employed to address many important issues that have been of necessity ignored. However, the problem is compounded by the fact that data availability seems to be a particular problem in transportation research. Data are held by firms whose interest in a study may be anything but unbiased and who are often reluctant to make information available. Even when data are accessible, transportation research tends to require very large amounts, and the collection task can be formidable and sometimes overwhelming.

Another general shortcoming of past transportation marketing research is an apparent lack of a systems approach. Many transportation problems can be analyzed in a specific, partial equilibrium framework, with little loss of relevance; but in many cases this is not true. The relationships among modes and among commodities make for a very complex transportation system with complicated substitution relationships. A change in one part of the system can cause effects and countereffects in others. With few exceptions, e.g., work by McDonald (1969) and Conley (1973), most studies of bulk grain shipping have essentially ignored the export sector, even though a large portion of such movements involve export grain. Incorporating interrelationships such as this requires no conceptual or methodological breakthroughs but only an expansion of existing models such as demonstrated by Friedlaender and Simpson (1978), which incorporated linkages between costs and demands for modes and included regional income effects. As methods of measuring transportation parameters are improved, research based on a system approach may be able to realistically address many of the important linkages between components of the transportation industry.

Finally, just about all past research has been concerned with the question of efficiency of the transportation system. But there are other criteria by which to evaluate economic performance, and these are certainly relevant to evaluating transportation. In a previous section, emphasis was placed on the fact that transportation can have significant effects throughout the economy. Such changes can bring efficiency, but they can also bring distributional changes; they can lead to lower transportation costs, but they can have impacts on the economic viability of certain areas. Such nonefficiency changes may be socially desirable, or at least may be a small price to pay for efficiency gains. But clearly, knowledge of such impacts is required before the total effect of any transportation change can be fully evaluated. Research is needed to identify these impacts. To cite just one example, consider motor freight deregulation. The weight of evidence provided by economists (all based on efficiency criteria) is heavily in favor of partial if not full deregulation. But the resistance to this policy from many quarters indicates that there are other issues involved. This may arise solely from unenlightened self-interest. Be that as it may, economists need to consider other than efficiency goals, if for no other reason than that even they must take political realities into account when examining a sector as important to all segments of society as transportation.

FUTURE RESEARCH NEEDS

Specific topics that we believe to be worthy of particular attention in future research are presented here in categories, although there is overlap among specific topics. No specific priorities are assigned, since we feel these categories all represent areas of needed research.

Firm-Level Studies. The efficiency of delivery of transport services should continue as an important topic for research. Gains can be made through cost control and improved management in all modes and modal combinations. Particular emphasis should be given to the interface between modes, between firms within a mode, and between transport firms and other components of the food handling system. There are several questions of particular importance.

How can the different modes be made to operate more as a transport system and what cost reductions are possible? For example, is containerization a feasible alternative in the handling of agricultural commodities?

Given the recent problems with rail car availability, the issue of car utilization is of pressing importance. How can freight car interchange among railroads be improved so that the amount of empty car miles is reduced? Incentive per diem charges have been introduced to increase the cost of holding cars idle. What other rule changes are possible, and how would they influence operation of the rail system? Changes are clearly needed, for in some areas car turn-around time has actually been increasing.

A related question involves the apparent decline in railroad productivity. Much of this has been attributed to changes in the geographic patterns of economic activity. However, at least one study (Allen 1975) has disputed this claim. It is imperative that the causes of decline in railroad viability be identified. In this regard, the role of management, work rules, and railroad financing arrangements warrant attention.

Much attention has been given to efficiencies in the line-haul segment of agricultural commodity movements but relatively little to getting commodities to or from bulk shipping and receiving points. But often the costs of assembly or disassembly outweigh line-haul costs. How can assembly and disassembly of large shipments be improved? One area of particular interest is optimal grain assembly systems (as in the work of Fuller et al. 1976 and Hilger et al. 1977). This presents an example of a problem that should be studied using more of a systems approach than most such work in the past. Hilger et al. have found that the optimal configuration of grain assembly units was very sensitive to availability of unit train rates. This suggests that such considerations as intermodal competition and destination of shipments can be very important. A possible approach to such a problem is the linkage of a detailed regional model (Stollsteimer 1963) with a more aggregated allocative model. Delivery systems for retail food outlets is a case where non–line-haul costs can assume paramount importance in the total food delivery bill. This is an all but neglected topic and one that can legitimately be handled at the micro level, perhaps with the joint efforts of economists and industrial engineers. All such work should endeavor to incorporate interaction and trade-offs between transportation and the rest of the physical distribution system.

Uncertainty. While agricultural shippers have always been faced with uncertainty, it has been causing increasing problems, particularly with respect to grain movements. There have been periods in recent years when transportation was the binding constraint in the grain handling system: markets existed but because of transport bottlenecks and lack of equipment, deliveries simply could not be made. Under these conditions, prices paid for transport services become relatively unimportant in shipper decision making. For highly perishable commodities, this type of uncertainty can completely prohibit market development.

The year-to-year variation in supplies (production plus inventory) and demand generates questions relative to optimal stock and growth of capital. Research is needed to analyze sources of uncertainty and determine whether it is likely to continue to have significant impacts on transportation in the future. If so, knowledge is needed as to how the transportation and marketing system can best adapt to uncertainty. For example, should transport capacity be increased, or should storage and handling facilities and investment in inventories be expanded?

Also, knowledge of the costs imposed on shippers by uncertainty will lead to much more accurate transport demand estimates, since the existence of such costs can be viewed as altering effective prices paid for transport services. Such knowledge can lead to predictions of producer response to rate changes as well as better estimates of inefficiencies introduced by rate regulation.

Energy. Since transportation is obviously an energy-intensive industry, increasing energy costs have significant impacts on shipping costs. There are several important energy-related issues. For example, how will changing energy costs affect competition among modes? What adjustments can be made by shippers, and what implications does this have for movements of inputs and outputs into and out of rural areas? The substitution possibilities available to agricultural shippers will in a large part determine the effects of changing energy costs on prices received by farmers. Perhaps some energy-use minimization models are needed to isolate the energy factor and provide insights into possible ways of alleviating negative energy effects and to determine the most energy-efficient modes and modal combinations in different circumstances.

Another area of inquiry is the effect of institutional factors, such as the freight car interchange rules discussed above, on energy use. During the fuel crisis of 1973–74, the Interstate Commerce Commission relaxed some motor carrier policies deemed energy consuming. There are undoubtedly others (e.g., backhaul restrictions), particularly with respect to rail and truck.

A broader question involves determining possible impacts of increasing energy costs and the seemingly ever-present threat of shortages on general rural development. For example, although many lightly used rail lines may be uneconomic now, may the loss of this (perhaps) energy-efficient mode impair future development for the regions served? Which decisions are irreversible and which are not? It may be easy to restore Amtrak services but quite difficult to reopen abandoned rail lines.

Will higher energy costs reverse the trend toward larger, centralized distribution centers and perhaps lead to a revitalization of rural towns? Similarly, what effects on industrial location can be expected? These are basic questions, and simple analysis based on the concepts of Weber can provide information useful for present and future policy.

International Trade. In the 1970s agricultural exports have become increasingly important to the U.S. economy. The growth of export demand, particularly since it tends to come in surges, has generated serious peak demand problems in the transportation system. This has contributed to the problems of uncertainty and equipment availability mentioned above and has created difficulties relating to port congestion. There is evidence that in some cases port costs may be nearly as important as at-sea costs in international grain shipping (Harrer 1979), suggesting that cost studies of port operations (such as that of Hammond and Salvador 1976) would be useful in suggesting ways to increase efficiency of port handling facilities.

A question of importance is the relationship of the U.S. domestic export transportation system to that of the rest of the world and how this affects the comparative advantage of U.S. agriculture. Given the growing importance of exports in determining U.S. farm income, analysis of the role of transportation (both in the United States and elsewhere) in international trade merits considerable attention. Production of agricultural commodities in many countries has been increasing, suggesting that international commodity markets may become more competitive. Hence transportation advantages (rather than just those involving production) may take on increased importance.

The year-to-year variability in demand and supply for agricultural commodities at various points throughout the world makes the question of optimal location of storage capacity and the effect of this location on the international transportation system important issues. At present, some countries (e.g., Germany) have very little storage. Can the pattern of storage facilities be improved so that some of the pressures on the transport systems of supplying countries are reduced? Given the prevalence of political factors in international grain trade, it is likely that the nature of the world grain handling system is not solely a result of economic forces. Knowledge of benefits of possible changes in the system is needed, even though such changes may be difficult to implement.

Regulation. Total economic deregulation of transportation may become a reality. It is almost certain that some deregulation will occur through either legislative or administrative action. The impacts of this institutional change and its disequilibrating effects on agriculture need to be examined to guide policy and assist in the adjustment process.

Two topics seem to be of special importance. Railroad deregulation presumably would involve variable rates (rates on some fresh produce shipments are variable now). Do agricultural firms need stable rates to minimize risk and uncertainty? What is the trade-off between unstable rates and uncertain equipment availability? Are contract rates between shippers and railroads a solution to this problem? If so, what are the implications of contract rates for the competitiveness of small shippers relative to large ones?

A second issue involves the general effects of transportation deregulation on rural areas. Of special interest here is the question of the availability of general (common carrier) trucking services. As indicated above, this is quite naturally a concern of rural interests and is a major source of support for continued regulation. Previous studies examining motor carrier services to nonmetropolitan areas seem to indicate that they are economically viable and thus not in need of regulatory protection (Banks et al. 1977). However, results are not conclusive. A related issue concerns availability of public passenger transportation to rural areas. It is important to know not only whether such services can survive in a free market setting but also whether their loss would have significant adverse impacts on rural areas. If so, how can they be alleviated?

An issue related to regulation (but perhaps worthy of separate treatment) is the question of subsidies in the transport sector, both direct and de facto. Transport firms receive government subsidies for providing unprofitable services deemed socially desirable. Direct payments have become less prevalent, but if deregulation should become a reality, such subsidies may again assume significance. The question to be addressed is, Given that subsidized service is desired, what is the form of subsidization to choose? A broader question is, Are there alternative measures that can accomplish the purpose of the subsidized service at lower cost (e.g., industrial relocation versus subsidized branch lines)?

De facto subsidies primarily involve government construction and operation of transportation facilities. Clearly, allocation can be distorted by public provision of facilities for some modes and not others. Research is needed to determine the nature of such subsidies (a problem in cost determination) and their allocative effects. A particular question that merits analysis is whether motor carrier user charges underwrite the costs imposed by trucks on the highway system. Without this information, the wisdom of existing motor freight weight restrictions, for example, cannot be evaluated.

A more subtle issue is that of cross-subsidization, both of a geographic and a commodity nature. Primarily because of varying transport demand elasticities, shipments of different commodities from and to different areas move at rates that cover differing proportions to total cost. This leads to situations in which producers in some areas are put in an advantageous position relative to those in another, an advantage not based on different costs. A familiar case outlined by Sorenson (1973) is rail rates on grain from areas with active barge competition vis-à-vis rates from areas without competition. Such discriminatory practices have implications for resource allocation and regional economic growth. A spatial equilibrium approach could be employed to estimate the magnitudes of such effects and determine whether policy to alleviate such discrimination is warranted.

Roads and Highways. The problem of roads and highways is an important research area that has received very little analytical attention. The condition of the nation's roads and highways has been deteriorating, and parts of the system are inadequate to meet user needs. In many areas rural roads are in particularly poor condition, to the point that they have become serious constraints on movement of agricultural inputs and outputs, while such factors as rail-line abandonment and centralized collection and distribution are increasing the use of these roads. Research determining the cost thus imposed on agricultural production relative to the cost of road improvement (and perhaps road abandonment) needs to be conducted. A broader question is, What is an optimal road system and how does this compare to the existing system? This is a difficult problem, for a road system has large external and public goods effects.

Forecasts of future road use in light of continuing energy problems are required to provide a guide to rural planners. Perhaps the most difficult issue is that of finding a workable means to finance rural roads. Local governments are encountering increasing difficulties in maintaining their road systems in even their present substandard state. Baumel (1979) has outlined alternative finance methods; these and others should be explored.

In short, because there has been almost no research conducted to analyze problems associated with the rural road system, there are a host of researchable issues, study of which would make a useful contribution.

Ownership and Control. Earlier, we attempted to underscore the important role transportation plays in the agriculture sector and in the marketing of agricultural commodities. This raises some serious issues regarding how decisions are made for a sector that pervades the entire agricultural economy. Does the market shape the transportation system or does the transportation system shape the market? How are changes in the transportation system implemented? It is very difficult for a single shipper, no matter how large, to introduce transportation innovations, whether they are of a technological or organizational character. It is hardly less difficult for groups of shippers to do so because there are public good characteristics to transport improvements. For example, a simple innovation like standardized containers could lead to lower costs throughout a transport system. But one shipper or even a group of shippers cannot introduce such a change. Are shippers, then, at the mercy of transport firms, and must marketing firms adapt themselves to whatever transportation system happens to exist? How can shipper preferences be articulated, and how responsive is the transportation system to shipper needs?

This issue is of special importance with respect to the railroad industry, for shippers generally cannot operate their own equipment and often are served by a single line. In some cases, these shippers have no viable transportation alternatives (captive shippers). Changes implemented in the past (multiple car rates available in some areas but not others, movement toward exclusive reliance on 100-ton hopper cars) all have had large and varying impacts on different groups of shippers. Since railroads are of such importance to the movement of agricultural commodities, alternative means to increase the responsiveness of the rail system need to be explored. For example, a possibility is to treat the railway as a public utility over which various companies could operate equipment.

A related issue involves the structure of the food marketing system. Scale in transportation has increased efficiency in moving agricultural products. As Greenhut (1963) pointed out, this extends the marketing area of individual firms and hence tends toward an oligopolistic market structure. This has doubtless occurred in some sectors of the food system. This trade-off between efficiency and concentration needs to be examined. Are short-term efficiencies being gained at the expense of long-term monopoly costs? Indeed, given the difficulties involved in articulating shipper preferences, it is not clear what are the true costs of transport changes. The efficiency of any transport arrangement depends upon the alternatives. For example, trucking grain to subterminals and thence to ports via unit train may be cheaper than direct shipment from country elevators in boxcars, but would this be true if smaller, covered hopper cars were available to country elevators? Research to investigate alternative transport structures rather than to merely evaluate the efficiency of existing systems is needed.

Conclusion. The above section has identified many important issues for future research endeavors. The diversity of these issues makes it difficult to suggest any general scheme for organizing this research. Some topics will undoubtedly require joint effort by economists at different universities and government agencies. This is especially true when a similar problem is being examined in different geographic areas. The benefits of coordination among researchers in such cases are obvious. Indeed, increased communication and coordination among transport researchers is probably the best way to achieve more of a systems approach. In addition, the difficulties associated with obtaining data can probably only be overcome if information is collected in a more methodical manner than it has been in the past. Efforts to coordinate data collection among transportation economists and to avoid needless duplication (a problem in past research) are especially warranted.

Finally, it is clear that research should be conducted with an eye to public needs. This includes playing a role in policy formulation as well as policy evaluation. In the context of the interest that has developed in transport deregulation, there is an indication that future transport policy will be designed with a view toward the entire transportation system rather than the mode-specific approach characteristic of the past. This provides yet another reason why a wider perspective must be adopted by transportation researchers. While studies must of necessity continue to deal with rather specific problems, attempts should be made to do so in the wider context of the entire transportation system. If such attempts are even partially successful, the usefulness of transportation research for policy formulation will be considerably enhanced.

REFERENCES

Allen, W. Bruce. 1975. A current view of the eastern U.S. transport market. *J. Transport Econ. Public Policy* 9:50–61.

Banks, R. L., and Associates, Inc. 1977. Service to small communities. In P. W. MacAvoy and J. W. Snow, eds., *Regulation of Entry and Pricing in Truck Transportation.* Washington, D.C.: American Enterprise Institute for Public Policy Research.

Baumel, C. Phillip. 1979. *The Local Road and Bridge Problem and Alternative Solutions.* Washington, D.C.: National Extension Task Force.

Baumel, C. Phillip, John J. Miller, and Thomas P. Drinka. 1976. An Economic Analysis of Upgrading Rail Branch Lines: A Study of 71 Lines in Iowa. Natl. Tech. Inf. Serv. Rep. FRA-OPPD-76-3, Washington, D.C.

Bayliss, B. T. 1973. Demand for Freight Transport—Practical Results of Studies on Market Operations. European Conference on Ministers of Transport, Paris.

Benishay, H., and G. R. Whitaker, Jr. 1966. Demand and supply in freight transportation. *J. Ind. Econ.* 14:243–62.

Binkley, James K., Joseph Havlicek, Jr., and Leonard Shabman. 1978. The Effects of Inland

Navigation User Charges on Barge Transportation of Wheat. Virginia Polytechnic Institute and State University, Res. Div. Bull. 137, Blacksburg.

Boles, Patrick P. 1972. The Freight Car Supply Problem and Car Rental Policies. USDA, ERS, MRR-953.

_____. 1973. Cost of Operating Trucks for Livestock Transportation. USDA, ERS Marketing Res. Rep. 982.

_____. 1976. Operations of For-Hire Livestock Trucking Firms. USDA, Agric. Econ. Rep. 343.

Borts, George. H. 1952. Production relations in the railway industry. *Econometrica* 20:71–79.

_____. 1954. Increasing returns in the railway industry. *J. Polit. Econ.* 52(4)316–33.

_____. 1960. The estimation of rail cost functions. *Econometrica* 28:108–31.

Boyer, Kenneth D. 1977. Minimum rate regulation, modal split sensitivities, and the railroad problem. *J. Polit. Econ.* 85:493–512.

Bressler, Raymond G., Jr., and Richard King. 1970. *Markets, Prices, and Interregional Trade.* New York: Wiley.

Camp, Thomas H. 1965. Costs and Practices of Selected Cooperatives in Operating Bulk-Feed Trucks. USDA, FCS Gen. Rep. 132.

Casavant, Kenneth L., and David Nelson. 1967. An Economic Analysis of the Costs of Operating Grain Trucking Firms in North Dakota. North Dakota State University Agric. Exp. Sta. Agric. Econ. Rep. 54, Fargo.

Conley, Dennis M. 1973. An Analysis of the Domestic and Foreign Distribution of U.S. Heavy Grains under Alternative Assumptions Regarding Possible Future Supply, Demand, and Transport Costs. Unpublished Ph.D. diss., Iowa State University, Ames.

Cornelius, James C. 1977. An assessment of the economic performance of motor carriers of exempt agricultural commodities. In Proceedings of the National Symposium on Transportation for Agriculture and Rural America. U.S. Dep. Transp. Rep. DOT-TST-77-33.

Farmer, Richard N. 1964. The case for unregulated truck transportation. *J. Farm Econ.* 46:398–409.

Fedeler, Jerry A., Earl O. Heady, and Won W. Koo. 1973. Interrelationships of Grain Transportation, Production, and Demand: A Cost Analysis of Grain Shipments within the United States for 1980. CARD, Iowa State University, Ames.

Felton, John Richard. 1971. The utilization and adequacy of the freight car fleet. *Land Econ.* 47:267–73.

_____. 1978. The Impact of Rate Regulation upon Common Carrier Back Hauls. University of Nebraska Agric. Exp. Sta. Rep. 88, Lincoln.

French, Ben C. 1977. The analysis of productive efficiency in agricultural marketing: Models, methods, and progress. In L. R. Martin, ed., *A Survey of Agricultural Economics Literature,* vol. 1. Minneapolis: University of Minnesota Press.

Friedlaender, Ann. 1971. The social costs of regulating the railroads. *Am. Econ. Rev.* 61:226–34.

Friedlaender, Ann, and Robert Simpson. 1978. Alternative Scenarios for Federal Transportation Policy: Freight Policy Models, vol. 1. U.S. Department of Transportation, Washington, D.C.

Friedrich, C. J., ed. and trans. 1929. Alfred Weber's *Theory of the Location of Industries.* Chicago: University of Chicago Press.

Fuller, Stephen W., Paul Randolph, and Darwin Klingman. 1976. Optimizing subindustry marketing organizations: A network analysis approach. *Am. J. Agric. Econ.* 58:425–36.

Greenhut, M. L. 1963. *Microeconomics and the Space Economy.* Chicago: Scott, Foresman.

Griliches, Zvi. 1972. Cost allocation in railroad regulations. *Bell J. Econ. Manage. Sci.* 3:26–41.

Hall, Peter, ed. 1966. *Von Thunen's Isolated State,* trans. Carla M. Wartenberg. New York: Pergamon.

Hammond, Jerome J., and Michael Salvador. 1976. Simulation of Export Grain Flows through Gulf Ports. Paper presented at meetings of the American Agricultural Economics Association, State College, Pa.

Harbeson, Robert W. 1969. Toward better resource allocation in transportation. *J. Law Econ.* 12:321–38.

Harrer, Bruce. 1979. Ocean Freight Rates and International Trade. Unpubl. M.S. thesis, Purdue University, Lafayette, Ind.

Harris, Robert G. 1977. Economies of traffic density in the rail freight industry. *Bell J. Econ. Manage. Sci.* 8:556–64.

Healy, K. 1961. *The Effects of Scale in the Railroad Industry*. New Haven: Yale University Press.

Hilger, Donald A., Bruce A. McCarl, and J. William Uhrig. 1977. Facilities location: The case of grain subterminals. *Am. J. Agric. Econ.* 59:674–82.

Hotelling, Harold. 1929. Stability in competition. *Econ. J.* 39:41–57.

Hunter, John H., Jr. 1963. Costs of Operating Exempt For-Hire Motor Carriers of Agricultural Commodities: A Pilot Study in Delaware, Maryland, and Virginia. USDA, ERS-109.

Isard, W. 1956. *Location and Space Economy*. Cambridge, Mass: MIT Press.

Johnson, Marc A. 1976. Estimating the influence of service quality on transportation demand. *Am. J. Agric. Econ.* 58:496–503.

Judge, George G., and Takashi Takayama, eds. 1973. *Studies in Economic Planning over Space and Time*. Amsterdam: North-Holland.

Kerchner, Orval. 1967. Cost of Transporting Bulk and Packaged Milk by Truck. USDA, ERS, MRR-791.

Ladd, George W., and Dennis Lifferth. 1975. An analysis of alternative grain distribution systems. *Am. J. Agric. Econ.* 57:420–30.

Leath, Mack N., and Leo V. Blakely. 1971. An Interregional Anaysis of the U.S. Grain Marketing Industry, 1966/67. USDA, ERS Tech. Bull. 1444.

Limmer, E. 1955. The elasticity of demand for rail transportation of Florida produce. *J. Farm Econ.* 37:452–60.

Losch, August. 1954. *The Economics of Location*. New Haven: Yale University Press.

McDonald, Hugh J. 1969. A Linear Programming Model to Optimize the Transfer Cost and Facility Requirements for U.S. Grain Eports. Unpubl. Ph.D. diss., Ohio State University, Columbus.

McLachlan, D. L. 1972. Canadian trucking regulations. *Logist. Transp. Rev.* 8:59–81.

Martin, Michael V., and Reynold P. Dahl. 1979. Social Costs of Regulating Railroad Grain Rates in the Upper Midwest. University of Minnesota Agric. Exp. Sta. Tech. Bull. 319-1979, Minneapolis.

Miklius, Walter. 1967. Estimating the demand for truck and rail transportation: A case study of California lettuce. *Agric. Econ. Res.* 19(2):46–50.

_____. 1969. Economic Performance of Motor Carriers Operating under the Agricultural Exemption in Interstate Trucking. USDA, ERS, MRR-838.

Miklius, Walter, and Kenneth L. Casavant. 1977. Stability of motor carriers operating under the agricultural exemption. In P. W. MacAvoy and J. W. Snow, eds., *Regulation of Entry and Pricing in Truck Transportation*. Washington, D.C.: American Enterprise Institute for Public Policy Research.

Miklius, Walter, Kenneth L. Casavant, and Peter V. Garrod. 1976. Estimation of demand for transportation of agricultural commodities. *Am. J. Agric. Econ.* 58:217–23.

Miller, John J., C. Phillip Baumel, and Thomas Drinka. 1977. Impact of rail abandonment upon grain elevator and rural community performance measures. *Am. J. Agric. Econ.* 59:745–49.

Perle, E. D. 1964. The Demand for Transportation. University of Chicago Dep. Geography Res. paper 95.

Samuelson, P. A. 1952. Spatial price equilibrium and linear programming. *Am. Econ. Rev.* 42:283–303.

Schnake, L. D., and John Franzman. 1973. Analysis of the Effect of Cost-of-Service Transportation Rates on the U.S. Grain Marketing System. Oklahoma State University Tech. Bull. 1484, Stillwater.

Shouse, James C., and Marc A. Johnson. 1978. Anticipated Consequences of Seasonal Railroad Rates in the Oklahoma Wheat Transportation Market. Oklahoma State University Agric. Exp. Sta. Res. Rep. P-773, Stillwater.

Sloss, James. 1970. Regulation of motor freight transportation: A quantitative evaluation of policy. *Bell J. Econ. Manage. Sci.* 1:327–66.

_____. 1971. The demand for intercity motor freight transport: A macroeconomic analysis. *J. Bus.* 44:62–68.

Snitzler, James R., and Robert J. Byrne. 1958. Interstate Trucking of Fresh and Frozen Poultry under Agricultural Exemption. USDA, AMS, MRR-224.

Sorenson, L. Orlo. 1973. Rail-barge competition in transporting winter wheat. *Am. J. Agric. Econ.* 55:814–19.

Sorenson, L. Orlo, Dale G. Anderson, and David C. Nelson. 1973. Railroad Rate Discrimination:

Applications to Great Plains Agriculture. Great Plains Agric. Counc. Publ. 62, Manhattan, Kans.

Stollsteimer, J. F. 1963. A working model for plant numbers and location. *J. Farm Econ.* 45:631-45.

Takayama, T., and G. G. Judge. 1964. Equilibrium among spatially separated markets: A reformulation. *Econometrica* 32:510-24.

Tyrchniewicz, Edward W., and Robert J. Tosterud. 1973. A model for rationalizing the Canadian grain transportation and handling system on a regional basis. *Am. J. Agric. Econ.* 55:805-13.

Wright, Bruce H. 1969. Regional and Sectoral Analysis of the Wheat-Flour Economy. USDA, ERS, MRR-858.

HAROLD S. RICKER
DALE L. ANDERSON
MICHAEL J. PHILLIPS

7

TECHNOLOGY ADOPTION
IN THE AGRICULTURAL
MARKETING SYSTEM

AN INTEGRAL PART of any attempt to address and explore the future of agricultural marketing research is the role of technology development and adoption. Technology has been defined as the science or study of practical or applied science. It is important to distinguish between two types of technology, that which improves economic efficiency in the traditional sense and that which improves the environment, human health, and the like, but does not necessarily provide economic incentives for adoption by firms. This chapter emphasizes the traditional technology but recognizes the impacts and interrelationships with environmental technology. While agricultural marketing research covers the .spectrum from basic science through applied research to developmental research, the focus of this chapter is directed at adoption of technology. In this context we address problems in measuring productivity, incentives and disincentives to technology adoption, private and public sector research and development in agricultural marketing, and the role of technology assessment and suggest some priority issues for a research agenda.

A deep concern for the decline in research effort devoted to improving the physical processes of marketing has prompted us to emphasize the need for economic cost-benefit research on the potential adoption of alternative

HAROLD S. RICKER is Deputy Director, Market Research and Development Division, Agricultural Marketing Service, USDA; DALE L. ANDERSON is President, Distribution Productivity, Fort Washington, Maryland, formerly Staff Scientist for Transportation and Facilities, Science and Education, Agricultural Research, USDA; and MICHAEL J. PHILLIPS is Project Director, Food and Renewable Resources Program, Office of Technology Assessment, Congress of the United States.

technologies that arise from research in the physical and biological sciences and could improve efficiency in the assembly, processing, and distribution of agricultural products. While agricultural economists tend to think of marketing in broader economic terms (markets, structure, policy, economics of the business firm), there are definite opportunities to improve economic efficiency. Determination of economically efficient alternative methods and processes for storing and marketing agricultural products should help firms lacking this capability to make better choices for capital investment and directly or indirectly serve to improve industry productivity.

MEASURING PRODUCTIVITY

Productivity statistics are frequently cited as indicators of the extent to which technology adoption is progressing in major segments of our economy. They also are used to identify sources of economic growth, compare economic performance among sectors, justify appropriations for agricultural research, and justify price changes.

The most commonly used definition of productivity is real output per work-hour. In this sense, it is a rough measure of effectiveness with which we use an important productive resource—labor. It is a concept that has significant social and economic implications, since it takes into account not only the chief means of satisfying individual and social wants (i.e., output of goods and services) but also the major real cost of getting the output, namely, work-hours. The trend in output per work-hour has a direct bearing on movement of labor costs, prices, and real earnings.

While this index of productivity serves many analytical and policy purposes, it is a partial index based on only one factor of production. Additional measures are needed to fully evaluate productivity growth. For example, when we measure output per unit of capital and labor combined and adjust for quality change, we have a partial measure of the efficiency of resource use. A number of measures are needed to provide a clear understanding of the impact of technical change. Confusion arises, however, when different ratios are used by related organizations to indicate changes in productivity. Therefore, it is important in selecting a ratio to understand its components and, when making comparisons, to know how and why the numbers differ.

An additional word of caution centers on selection of the base periods. A judicious choice can show either an increase or decrease in productivity growth rates as illustrated by the following example. The USDA multifactor agriculture production productivity index showed a growth rate of 1.8 percent between 1950 and 1970 and 2.1 percent between 1970 and 1978. However, the growth rate could also be shown to be 2.3 percent between 1950 and 1965 and 1.4 percent between 1965 and 1977 with the same index. Or one could show a 1.5 percent annual growth rate in agricultural productivity since 1930.

In evaluating agricultural marketing, a preferred definition of productivity is marketing output per total factor inputs. This definition includes efficiency and effectiveness and enables us to briefly highlight important productivity changes in agricultural marketing. However, most of the ratios still cling to the narrower traditional definition of output per work-hour. This is justified by the fact that 46 percent of the farm food marketing bill represents labor costs, making it a relatively more significant input than in agricultural production where it was estimated to be 13 percent of inputs in 1978. Therefore, our evaluations of productivity growth in the marketing sector are restricted to the partial productivity index of output per labor input.

Productivity in the food industry has tended to be highest in certain food manufacturing categories and lowest in the retail service areas, with the service area experiencing declines (Table 7.1). The malt beverage and wet corn milling industries had increases in output per employee-hour exceeding 6.0 percent between 1972 and 1977, while restaurants had an increase of only 0.2 percent and retail stores had a decline of 1 percent during the same period. The poor productivity rates in food retailing and food service more than offset the food processing increases because of the size of the service industries and the large number of people employed. Food industry sectors having 1.0 percent or better growth since 1972 represent less than 10 percent of total food industry employment.

A number of factors have probably contributed to the decline in productivity in the food and fiber marketing system. These include reduced investment in capital and research and development, an increase in marketing services, more government regulations, labor restrictions that limit the testing or adoption of innovative technology, work disincentives such as unequal system of rewards or incentives, and changing public attitudes concerning technology.

TABLE 7.1. Average annual rate of change in output/employee-hour, selected food and agriculture industries

Standard industrial classification	Industry	Employment	Output/employee-hour, all employees	
			1972–77	1947–77
		(000)	(% change)	
2082	Malt beverage	51	6.8	5.5
2046	Wet corn milling	13	6.2[a]	4.9[b]
2086	Bottled and canned soft drinks	137	5.5	2.4
2047,48	Prepared feeds for fowl and animals	74	4.5[a]	3.9[b]
204	Grain mill products	147	3.2[a]	3.4
203	Canning and freezing	289	2.8[a]	2.9[c]
2111,21,31	Tobacco products	57	2.0	2.9
2061,52,63	Sugar	30	1.2	3.7
2045	Blended prepared flour	10	1.0[a]	2.0[b]
205	Bakery products	240	0.3	2.3
2041	Flour and other grain mill products	28	0.2	3.6
58	Eating and drinking places	4205	0.2	1.0[d]
2044	Rice milling	5	0.1	2.4[b]
2043	Cereal breakfast foods	17	−0.3[a]	1.5[b]
7011	Hotels, motels, tourist courts	969	−0.8	1.9[d]
54	Retail food stores	2379	1.0	2.0[d]
2065	Candy, confectionery products	56	−2.8	3.5
401—Class 1	Railroads (revenue traffic)	520	0.5	4.7
4123 Part	Intercity trucking	652	3.2	2.7[e]

Source: U.S. Department of Labor 1978.
[a] 1972–76. [c] 1947–76. [e] 1954–77.
[b] 1963–76. [d] 1958–77.

In the case of changing public attitudes, a number of social concerns exist in the agricultural sector, such as increased use of pesticides, environmental contamination from chemicals that might enter our food supply, and increased use of drugs and chemicals as additives to feed for livestock. Restrictions on use of such chemicals tend to impact negatively on productivity. In the marketing sector, social concerns relate to the increased use of fabricated foods and food additives, impacts of new packaging technologies, electronic checkout, electronic funds transfer, central meat cutting, and similar technologies that could affect productivity. If these concerns are translated without adequate analysis into legislative or regulatory action or threat of actions, the result could be further reduction of innovation, technology adoption, and productivity.

The linkage between societal concerns and labor productivity in the food sector is indirect for most technologies. Productivity measurement in agricultural marketing is difficult and imprecise and therefore tends to be heavily oriented to labor productivity. It should also be recognized that the apparent decline in productivity in food marketing may reflect, at least in part, an expansion in service activities that, when included, contributes to lower productivity rates. Productivity analysis represents a promising area of research activity.

INCENTIVES AND DISINCENTIVES

Much of our theory about markets is based on competition with freedom of entry and a belief that competing firms will cause any new and more efficient processes or products to be adopted quickly. From the viewpoint of the competitive firm, innovation pays in the short run as per unit costs are lowered while prices remain unchanged. In the long run, as more firms adopt the technology, output is increased and prices decline to the competitive equilibrium.

Rogers's (1962) concepts of how innovations are adopted are explained in the charting of adoption of hybrid seed corn by U.S. farmers. However, the situation described involved almost perfect competition among small family farmers, each relatively independent and almost completely free to adopt a new technology. Also, the costs of adoption of the new technology (hybrid seed) were low. No new equipment was required; the only real change being the higher price of the seed. Of course, there was the risk that the new seed might not be as disease resistant as established varieties or yield as expected. Loss of a crop is a high cost to the producer. Rogers identifies those willing to take the risk as early adopters. Webster (1968) indicates that the firms most likely to be early adopters are those for which the innovations are most likely to be profitable, those that can best incur the risk, and those that have the highest performance aspirations and access to the most information about the invention. Technical innovation constitutes change and often opens the market to entry by other firms. Schumpeter (1954) postulated that the entrepreneurs are a dynamic factor in the economy because they introduce new techniques and methods of production (innovation) that constantly disturb the equilibrium. Innovations cause a boom to start. Profit is associated with expansion but falls to zero under perfect equilibrium as more firms adopt the new technique.

Firms in a purely competitive environment lack the ability to internalize the benefits of innovations. With too many firms competing, each would have difficulty generating investment capital. Schumpeter proposed that innovation rates would be highest with fewer firms because of greater incentives and profits, and Bain (1948) argued that a monopolist partially protected from other firms would be more likely to innovate, since his chances of recovering investment and returning profits are greater.

In a study examining 20 food manufacturing industries (Mueller et al. 1980), the evidence did not support the Schumpeter and Galbraithian hypotheses that great market power and large firm size best promote research and development (R & D) effort. Mueller et al. have determined that, at least in food manufacturing industries, decreasing returns to firm size and market power occur beyond medium firm size and moderate levels of concentration. Industry R & D apparently is maximized at even lower levels of market concentration.

Conventional wisdom suggests that rapid technological progress is greater when market structure is a blend of competition and monopoly. Scherer (1967) and others believed that innovation rates would be higher when there were neither too few nor too many competitors. A few firms would hold monopoly positions and would not need to invest in innovations, and many competitors would limit a firm's ability for adequate returns on research investments. Oligopolists generally are believed to possess the ability to internalize a significant share of the benefits from innovation.

Mueller et al. (1980) have found that the greater part of inventive and innovative activity inducing greater efficiency in food manufacture originates outside these industries. The largest share (55 percent) was made by firms that manufactured food machinery and provided other plant operating services, followed by government laboratories (21 percent) and food processors and ingredient manufacturers (13 percent).

Drucker (1957) indicates that innovation entails three risks: risks of being overtaken by (lack of) innovation (even large companies fail), risk of failure of the innovation attempt (technology that is wrong or too late), and risk of successful innovations (impacts beyond expected changes). The adoption of technology does not necessarily happen automatically. Industry experts have hypothesized that in any industry much of the technology needed to make the changes considered necessary by management is already "on the shelf." Research funding is ineffective if it causes R & D to stop too short of the transfer point for industry application to rely on momentum of the discovery or market forces to bridge the gap (U.S. Department of Commerce 1976). Davis (1977) draws two conclusions. First, there is no justification for assuming that better research results will be translated into action. Second, we can assume that we still have a challenge to develop better marketing methods and better techniques for technology transfer.

A brief examination of the innovation cycle and adoption lags may help

TABLE 7.2.   Average rate of development of selected technological innovations

Factors influencing rate of technological development	Average time interval		
	Incubation period	Commercial development	Total
	(years)		
Time period			
Early twentieth century (1885–1919)	30	7	37
Post-World War I (1920–44)	16	8	24
Post-World War II (1945–64)	9	5	14
Type of market application			
Consumer	13	7	20
Industrial	28	6	34
Source of development			
Private industry	24	7	31
Federal government	12	7	19

Source: Lynn (1966).

to support this observation in the food industry. It also illustrates one of the important disincentives to innovation: the risk of a long time lag before adoption and possible financial returns to the research investment.

Mansfield (1968) breaks down innovation into incubation period (begins with basic discovery and establishment of technological feasibility) and commercial development (begins with recognition of commercial potential and commitment of development funds, ends with commercial introduction of product). Lynn (1966) evaluates rates of development by time period, source of research, and whether consumer or producer goods are involved (Table 7.2).

Anderson (1977) has identified approximately a six-year lag for food marketing productivity after establishment of the research program under the Marketing Act of 1946 and approximately a six-year lag to declining productivity after the research program was largely eliminated. This program involved relatively applied research aimed at solving practical food industry distribution problems. This suggests a more rapid response rate for certain kinds of research and industries.

Evenson et al. (1979) projected a lag time between release of an innovation and its peak adoption of 5½ years for state-supported agricultural research and 8 years for federally supported research. A possible explanation for this difference is the size and nature of the projects undertaken.

Rosen (1976) indicates that the incubation period for 37 new products from concept to realization averaged 22.4 years, although the 8 products conceived after 1940 averaged 8 years to realization. Two agricultural marketing examples averaged 12 years to commercialization.

In food manufacturing, smaller firms (annual sales under $10 million) contributed 44 percent of the innovations. About one-fourth of these innovators were acquired by large corporations, suggesting that large firms used acquisition to acquire innovations (Mueller et al. 1980).

Several studies suggest that the best and most creative inventors or research teams are not necessarily in the largest, best-funded laboratories. Also, there are suggestions that creativity and innovation are related to both human and organizational factors that can be learned or ordered (Steiner 1965; Kivenson 1977).

Organizational impediments to basic research occur when R & D functions are decentralized, operating divisions have a heavy impact on project selection and funding, R & D managers have little or no discretion to initiate work, and formal quantitative procedures are used to select projects (Nason and Steger 1978). The increased difficulty of getting good research data because of company protection policies, fear of antitrust action, and company awareness of a market value for information hampers public researchers and can also distort findings.

Most authors examining cause and effect relationships in technology growth and innovation suggest that there are unexplained factors in addition to the traditional economic inputs and outputs examined. Variations as to the type of research techniques used and type of technology are apparent, but other nonnumerical factors must also be considered. Government control and regulation, including international trade barriers, have become important factors in setting business priorities that influence investments in technology development and/or adoption. Rostow (1962), in describing the process of economic growth, indicates that during the drive to maturity of an economy the industrial process is differentiated, with new leading sectors gathering momentum to supplant older faltering sectors. A firm, or small group of firms, that dominates a given market is likely to resist new technology that seems to threaten their market domination. Such resistance may rely on government regulatory agencies to help create barriers to entry. Scherer (1971) provided a good discussion on the relationship between R & D expenditures and the structural configuration of an industry.

Labor has had a significant impact on innovation through union work rules that perpetuate employment and resist technological changes. Often such resistance is specific to an individual process, such as the former rule forbidding the entry of boxed beef into the Chicago area.

The researcher is rewarded for finding and cataloging new facts, and the innovator is rewarded for applying technology successfully to solve current practical problems, but rarely do they either work together or communicate to develop the best problem solution. An integrated team approach that adequately rewards all participants offers a possible solution to some of these problems. Economic analyses that could identify potential payoffs would help to bring the parties together and get the innovation adopted.

A number of other barriers to technological change result from new social values. Environmental and health concerns have had impacts on the use of chemicals in the agricultural marketing system and point to the need for more careful assessment and testing before use.

Another aspect of our current society seems to be a general resistance to change. In a society where technological and social change has been rapid, more and more people seem to be trying to maintain a status quo, and creativity appears to be either absent or suppressed. No-growth movements are gaining in communities and even in states and regions. The celebrated and feared "technological imperative," the view that technology has a force of its own that demands its continual development and fulfillment, no longer has the vitality it had in 1970. Many areas of technological growth have been stalled while advocates of technology are forced to examine more closely the implications of changes they would make. While this is where technology assessment and futures studies come into economic analysis, Rostow (1962) feels there is also a need for more sociological assessment.

Impediments to technology development and adoption occur in almost all areas and help to explain the decline in technology adoption in food marketing. To counter this trend, more effort is being devoted to development of impact analyses for specific projects and to technology assessments for broader areas of investigation to help determine priorities and direction and to identify disincentives to innovation that might be changed by R & D, legislation, or other government actions.

One needs to keep in mind that in the broadly based food marketing system it is inevitable that technology that improves economic efficiency becomes entwined with technology that improves such aspects as human health and the environment in the public's perception. A dilemma occurs when it is recognized that in most instances R & D efforts for the two types of technologies are competing for the same portion of a firm's limited research investment dollars. With additional regulatory pressures to develop and adopt environmentally beneficial technologies, research efforts devoted to improving economic efficiency that could contribute to a firm's profitability typically suffers. In an attempt to document the potential societal benefits from research investments, a few studies have estimated the benefit-cost ratios of research investments for food industry technology, with Jones (1973) projecting compounded annual internal rates of return ranging from 65 to 195 percent for utilization research. The point is that there are important societal benefits from both types of technological investment, and work is needed to develop guidelines to maintain a proper allocation of research investments. To focus on these critical issues in food marketing, it is helpful to briefly review the private sector and public efforts in this area.

PRIVATE SECTOR RESEARCH AND DEVELOPMENT

In discussing the private sector role in food marketing research, we are primarily concerned with the research activities of the estimated 30,000 food processing, manufacturing, wholesaling, and major marketing input supplying industries. A limited amount of applied marketing research is also conducted by the approximately 550 retail food chains. According to *Business Week,* the food manufacturing industry (the only segment for which research expenditure data are available) contribution to R & D is among the lowest of major industry segments (Table 7.3). Mueller et al. (1980) have found that only a small share of the innovations in food manufacturing could be attributed to food processors and food ingredient manufacturers. Bartels (1976) found that most of the research in the food manufacturing industry is in food processing and in the development of new "brand" food items. Other than the areas of product formulation, computer programming, and market research, R & D has generally been provided by the major input suppliers. Research by the suppliers does not show in the R & D of the food manufacturing industry. The available data also exclude some limited amounts of research in areas such as

TABLE 7.3. Comparison of industry R & D

Industry segment	R & D as percent of sales	R & D as percent of profit
Aerospace	3.7	93.0
Automotive	2.8	70.2
Electronics	2.6	56.1
Chemicals	2.5	41.3
Containers	0.9	27.7
Food and beverages	0.5	15.3
Fuel	0.4	8.6
Food and lodging[a]	0.1	2.0
All-industry composite 1979	1.9	34.4

Source: *Business Week,* July 2, 1979.
[a] *Business Week,* June 28, 1976.

physical distribution, which is not listed as company R & D. The retail food chains tend to emphasize site or market selection and limited merchandising studies and rely on equipment suppliers and others for research to improve their efficiency.

A problem in the effort to identify private sector food marketing research efforts is defining research. In the Current Research Information System (CRIS), basic research is distinguished as "research with the primary goal of gaining knowledge or understanding of a subject." Research that has as its primary goal the application of knowledge to meet a recognized need or to produce useful products would be considered as applied or developmental (USDA 1978b). Basic research is less predictable yet is necessary to supply the knowledge base for applied research.

Business Week reported estimated R & D expenditures for the food and beverage industries of approximately $30 million in 1979. While their survey does not include all firms conducting research in the food marketing sector, it does suggest a substantial investment by 40 firms as reported on their 10-K statements filed with the Securities and Exchange Commission. This estimate does not include the market research expenditures to evaluate product market potential and normal product improvement or quality control efforts. The survey also excludes food manufacturing firms that spend less than $1 million on R & D. Nevertheless, *Business Week* purports to include about 90 percent of all the privately funded R & D performed by all U.S. companies. One could question whether this is comparable for the food industry, but it does provide a benchmark.

The private sector tends to focus more heavily on short-run applied R & D projects versus the longer run more basic efforts. Shapley and Phillips (1978) indicate that while some high technology companies are fully committed to the importance and value to their companies of basic research, the more widespread view appears to be that such research is good and necessary but is not really the industry's job. While the current level of funding for basic research is small, the food manufacturing industry appears to be deepening its commitment to it (*Business Week* 1979). Concern over the effects of trace minerals, excess sodium, and fats in food is spurring food companies to divert dollars into basic research on nutrition.

While there could be some positive spin-offs for innovation in the future, this shift in emphasis in the food manufacturing industry might be considered defensive in response to existing (or fear of future) government regulation. Nason and Steger (1978) have shown that compound annual growth rates in R & D spending by the private sector for government regulation showed increases from 10 percent on product safety to 19.3 percent directed at research related to proposed legislation. When overall R & D spending is not growing this fast, other categories of research effort (basic research) that might have greater potential for technology development and innovation must suffer.

Meaningful statistics on food marketing R & D expenditures are difficult to find because of nondisclosure of proprietary information and differences in definition of research. The decline in productivity and the reports based on a few large firms suggest that industry R & D is focusing on short-term objectives and existing products. Industry R & D is a cost to be recouped out of income earned from future sales, not by prompt reimbursement from a federal customer (Shapley and Phillips 1978). This emphasis is increasingly being cited as a factor in the decline of productivity and innovation.

PUBLIC R & D

To describe government efforts to encourage technology development and adoption, it is useful to examine a study conducted for the U.S. Congress, Office of Technology Assessment (1978). The Center for Policy Alternatives prepared a classification system of government activities that identified 13 program areas:

1. Assessment of new and existing specific technologies.
2. Direct regulation of the production, marketing, and use of new or existing products.
3. Programs to encourage development and utilization of technology in and for the private goods and services sector.
4. Government support of technology for public services where consumers are the primary users.
5. Support for development of technology where the federal government is the primary user.
6. Support for the science base necessary for development of new technology.
7. Policies to affect industry structure that may affect development of technology.
8. Policies affecting supply and demand of manpower resources having an impact on technological change.
9. Economic policies with unintended or indirect effects on technological innovation.
10. Policies affecting international trade and investment.
11. Policies intended to create shifts in consumer demands.
12. Government policies responding to worker demand having impact on technological change.

13. Direct regulation of the research or development of new products and processes.

They concluded that three policy instruments emerged as the most effective government policies in influencing the rate and direction of technological change: regulation, federal R & D support, and government procurement of innovative technology-based products. This assessment did not examine agricultural research and extension in any detail (which other studies cite for unique contributions beyond those three), such as the conduct of a program of public sector agricultural research and the demonstration and education of technology benefits for farmers and agribusinesses.

The Agricultural Marketing Act of 1946 established a peculiar working relationship between industry, public agencies, and universities. The act declared it to be "the policy of Congress to promote through research, study, experimentation, and through cooperation among Federal and State agencies, farm organizations and private industry, a scientific approach to the problems of marketing, transportation and distribution of agricultural products similar to the scientific methods which have been utilized so successfully during the past 84 years in connection with the production of agricultural products. . . ."

Prior to 1946, little marketing research was conducted at the state level in state agricultural experiment stations (SAES). The Research and Marketing Act of 1946 stipulated that at least 20 percent of funds appropriated be used for marketing research.

The major thrust toward technology development and transfer in agriculture has been through federal and state agricultural research and extension. Hoover and Detwiler (1978) document a fascinating history of the development of agricultural utilization research that detailed the actions leading to establishment of some of the major federal research laboratories.

A comprehensive assessment of the public sector marketing research effort was undertaken by the USDA in 1978–79 under the term post-harvest technology (PHT). The assessment identified approximately $53 million expended by agricultural research (AR) and $50 million spent by other public agencies on PHT research in fiscal year 1976. AR represented the in-house USDA Science and Education Administration (1978a, 1979) research effort exclusive of the Forest Service and Economics, Statistics and Cooperatives Service. The other research included Cooperative State Research, SAES, and other colleges and institutions using available and comparable fiscal year 1976 Current Research Information System data (USDA 1978a,b). As with the private sector, there are some questions on the definition of research and what is actually marketing research, but the figures do serve as an estimate of the public sector commitment. To put them in further perspective, the marketing research funds represented approximately 15 percent of the agricultural research budget for the USDA, and approximately 37 percent of this was considered to be basic research (USDA 1978a,b).

It is significant that the food industry testified to Congress in the 1960s that much credit for its positive productivity performance in the 1950s could be given to USDA and land-grant university research and extension programs, and that curtailing these contributed to the declining productivity (Eavey 1964). The decline in marketing research effort has continued in the 1970s as documented by Metzger (1973) for the SAES and by the USDA Science and Education Administration for the federal and state programs.

PUBLIC VERSUS PRIVATE ROLE IN RESEARCH

Industry's unwillingness to conduct much of the R & D now done by the public sector was indicated by a USDA survey (1979). Food industry leaders concurred that there is a need for agricultural marketing research (performed by both private and public sectors) aimed at practical objectives fulfilling human needs, but the motivations differ between industry and the public agencies. National objectives, where the concern is to ultimately benefit the consuming public, differ from corporate objectives, which must include the expectation of profit to help each company in its own competitive market. Hence industrial research results would be kept proprietary in nature.

Industry leaders indicated that the risks of economic failure in the food industry are greater because food is subject to some unusual restrictions and conditions. The demand for food is subject to changes in dietary and social habits that can greatly alter the life cycle for processed products. In addition, technology improvements have led to development of more sophisticated measuring devices capable of detecting previously unidentified hazards in food products (which can result in new regulations) or to changed consumption patterns (which make food products less attractive investment risks). Food and feed products are subject to serious deterioration by insects, bacteria, molds, and other hazards. Publicly funded research is essential because the beneficial results are of general usefulness and because it is seldom possible for a single company to derive sufficient benefit to recover the research investment. The report concludes that many parts of the food distribution industry (such as fresh produce) lack the organization, capital investment, and capabilities of doing their own research.

At the same time the Cooperative State Research Service surveyed state agricultural experiment station directors (Babb 1976). They were unanimous in the opinion that private industry would not finance and carry out the needed PHT research to service societal needs. They agreed that the private sector was likely to devote research efforts to innovations for which they could realize the benefits and to leave the more basic research and national focus to the universities and the federal government.

TECHNOLOGY ASSESSMENT

Technology, by any definition, has impacts on people, society, the economy, and the environment. Technology assessment is essentially a systematic approach to the study of technological change. It is the identification, analysis, and evaluation of the potential impacts of technology on social, economic, environmental and political systems, institutions, and processes.

The primary, or first-order, impacts are those the technology is specifically intended to achieve. Assessment is less concerned with these than with second-order and greater impacts. These higher order consequences may be beneficial, detrimental, or neutral, but they are almost always unintended or unforeseen. The purpose of technology assessment is to reduce this area of surprise.

In most assessment programs, projects and products are usually evaluated through some form of benefit-cost analysis, but both costs and benefits are generally very narrowly constructed. Ordinarily the direct costs and the direct, immediate benefits to the sponsor or investor and to the user are considered. But if technology drives social change, this creates the opportunity for alternative lines of technological development and societal futures. Given these alternatives, society may be able to invest public and private resources selectively in lines of technological development that will maximize societal benefits and at the same time have the possibility of avoiding, minimizing, or controlling the undesirable action or decision. There are always trade-offs to be made. Thus anticipatory technology assessment offers a number of possibilities:

1. Desirable developments and applications can be encouraged and stimulated through a variety of incentives, including investment of funds in exploratory R & D.
2. By detecting unsuspected or unplanned potential benefits, new applications of existing technology can be encouraged.
3. When potential impacts are uncertain or impossible to evaluate fully, monitoring may be called for to provide early warning of detrimental effects.
4. Where detrimental impacts are inevitable but outweighed by potential benefits, controls can be instituted.
5. Projects or programs can be modified, relocated, or redesigned.
6. Unjustified, undesirable, or dangerous projects can be blocked before heavy resource investment has been made.

TECHNIQUES OF TECHNOLOGY ASSESSMENT

There is no magic formula and no universally accepted systematic methodology for technology assessment. The essence is to conceptualize a new technology or an innovative application of technology and to trace out the potential impacts, attempting to evaluate their probability, significance, direction, magnitude, and duration; to identify wherever possible technologies that indicate the need for either avoidance or enhancement; and to provide general policy options for use of the decision maker.

A number of techniques have been advanced for conducting technology assessments, e.g., by National Academy of Sciences and National Academy of Engineering reports and the MITRE methodology (U.S. Congress 1978). The selection of study methods is largely determined by the particular technology under assessment. The process can be reduced to seven steps; these should not, however, be regarded as a detailed prescription or methodology but rather as categories that should be considered in any policy-related assessment of technology. These categories are:

1. Description of subject technology and its parameters and development of the data base.
2. Description of alternative, supporting, and competitive technologies.
3. Development of state-of-society assumptions for present and future time periods.
4. Identification of potential impact areas.
5. Analysis of impacts in terms of affected parties.
6. Identification of possible policy options.
7. Assessment and comparison of alternative action options.

Technology assessment as an input to decision making is applicable and appropriate in the public and private sectors. Most of the activity and discussion of technology assessment in the United States has focused on government decision making.

Growing hostility to technological programs such as large public projects like highway and airport development intensified congressional suspicion of the process in executive agencies and its ability to provide adequate information about the impacts of government programs. In 1972 a bill was passed establishing an Office of Technology Assessment (OTA) to service the Congress.

The OTA's basic function is to help legislators anticipate and plan for the long-term consequences of technological applications and to examine the many ways, expected and unexpected, in which technology affects people's lives. The OTA has been active in the food marketing area. Since 1977 the office has completed studies in retail food grading, surface transportation policies affecting food distribution, emerging food marketing technologies, open shelf-life dating of food, and the role of the public and private sectors in PHT research.

A number of large corporations are investigating the use of technology assessment as part of their corporate planning, but there has been little action to date. Unlike government, business does not have the broad charter to promote the general welfare.

A major reason for business considering technology assessment is that the public is asking hard questions and may be prepared to make hard choices if environmental degradation, the energy crisis, and limitations on space and natural resources become more acute. It is in the best interest of business to anticipate these concerns and avoid heavy investment of capital and resources in technologies that are socially undesirable, obsolete, or so hazardous as to be rejected. It is also in the interest of business to anticipate and provide the know-how and technology for using scientific breakthroughs that offer new societal benefits or new approaches to old problems.

Business has most of the problems and constraints of government (and more) in seeking to perform technology assessment, i.e., lack of time, funds, and interdisciplinary expertise; justification of soft research when the potential pay-off may be years in the future; and lack of universally accepted methodology. In addition, corporations have the problem of making assessment results acceptable and understandable to top echelons of management and their stockholders.

It is misleading to think that technology assessment is a modern concept. Our growing awareness of the pervasiveness of the impacts of technology has led to inclusion in the assessment process of the indirect, social, political, and environmental effects that often accompany new technology. No longer is it sufficient to demonstrate that a new technique or product will result in greater speed, power, or efficiency, for recent history is replete with examples of technological advances that produce these effects while they degrade the environment, dehumanize people, or threaten the existence of humankind.

For these very reasons, technology assessment has an implied negative impact on technology adoption. This, however, overlooks the positive contributions whereby the assessment teams can identify nondesirable and tertiary effects and ameliorate them at an early stage. It also offers the opportunity to identify in the early stages technologies that have future benefit and to encourage their development and adoption.

SUMMARY AND CRITICAL ISSUES

Cordaro and Smith (1976) have summarized the current situation by indicating: "The state of R & D in the food industry, excluding agriculture, is bad." (The term "agriculture" was used by them to represent "farming.") They went on to suggest, "The result is wasteful practices in distribution, processing, and retailing at a time when this country cannot afford waste, particularly in energy. Consumers have to pay the steadily increasing cost of food, not because of the rising cost of raw products, but rather because the costs between the farm and the consumer have risen so much. A considerable amount of this cost might be eliminated by research aimed at technology development in distribution, processing, and retailing."

If anything, the marketing research situation has deteriorated since then. In addition, questions are being raised about the structure of the agricultural system. While much of this present effort is directed at the farm level, there are questions of greater economic importance about the marketing sector and interaction of the two.

Essentially, theorists are asking for the kind of food industry that would assure efficiency of production, assembly, processing, and distribution; provide appropriate services to consumers; and yet maintain acceptable competitive alternatives of procurement and sale in all segments of the food industry from producer to consumer.

There are still questions relative to innovation rates and levels of freedom in market competition. Economic research and theoretical developments are needed to clarify innovation's role and suggest promising directions. Most writings have dealt with an expanding growth industry (food processing and distribution); expanding populations and markets; population shifts; and relatively little government intervention into innovations, products, and processes. The importance of innovation in a relatively stable or maturing economy needs to be addressed.

When oligopolistic firms appear to have adopted a live-and-let-live policy, publicly funded research might serve to encourage new competition, enhance freedom of entry, and remove the lethargy toward innovation. Technology transfer can affect levels of competition and change normal operating margins; public demonstration of the computerized check-out system completely transformed the cash register industry by opening it to new competitors; the system is beginning to provide productivity and information benefits to distribution firms that have adopted it, as well as to consumers. In addition, technology can create new industries and alternative markets for agricultural producers. Electronic trading in livestock and meat marketing shows considerable promise for expanding the marketing opportunity for some agricultural commodities.

Furthermore, with increased demands on industry R & D to perform defensive research, economists will need to evaluate the societal costs of not doing R & D to improve economic efficiency in the marketing sector and (without market studies) of having less economic and technical information available in these areas.

Agricultural research pays off in benefits to people in all walks of life. As Kennedy and Gardner (1979) have indicated, "Innovation and research have done more in some areas to improve the public's health and to protect the consumer than any other factors. Innovation can actually simplify our tasks. We need to seek ways to foster innovation not only because it is good for the country, but also because it is good for consumer protection." Future productivity gains and the benefits stemming therefrom will depend on the level of research and education investment that private industry and the federal and state governments are willing to support. A depressed research base coupled with the time lags between innovation and adoption portends real challenges for the future. The greatest dangers for the future lie in political decision about scientific events without access to adequate data.

In conclusion, we would like to suggest the following high-priority research issues for study:

1. *Improved measures of marketing productivity.* Traditional measures of productivity have focused on labor productivity because it is the largest marketing input. Examination of other inputs such as capital, packaging, transportation, and energy could lead to the development of improved total marketing factor productivity analyses. This would provide more meaningful opportunities for comparison within and between industries and would help identify problem areas for focusing research efforts. It would also add to the creation and development of a better data base that would be used to reduce labor resistance to change and subsequent relocation of industries.

2. *Technology assessment model.* An interdisciplinary model is needed that

projects the costs and benefits of technology development and adoption. There have been several studies of the cost and benefits of agricultural research, and several have examined specific marketing segments. Typically, they have emphasized the more easily quantifiable economic benefit-cost analyses associated with the development of products or hard savings estimated from the adoption of new technology. Concerns about resource use and environmental impacts need to be incorporated in the models and weighed against the costs and potential benefits to society and industry.

3. *Assessing government impacts.* Patents, regulations, and financial incentives are three examples of government tools that influence innovation. The Department of Commerce reports that most of the current patent applications are coming from foreign sources. The National Science Foundation reports that private sector R & D funds have been diverted to meeting regulatory or defensive efforts. Government has not been able to use the exclusive rights inherent in patents to encourage commercial development of publicly funded research discoveries. The government has been reluctant to provide additional financial incentives for R & D. Multinational firms conduct some R & D in the United States and abroad and control its use and the location of its application (which may not be in the United States). Government roles need to be redefined and reassessed in terms of the application or impacts of these and similar tools to encourage R & D in the United States, or at least the tools need to be reexamined to determine if they can be made more effective.

4. *Market structure.* Changes in structure, market power, and electronic communication systems have altered the total agricultural production and marketing system to the point that producer access to markets has been limited. While attention has focused on the larger processing and distribution firms, much of the system is still characterized by relatively small assembly, processing, and distribution units. The relationship between industry composition, technology adoption, and innovation need to be examined, along with the effects of the changes on pricing, performance, and productivity at all levels from farm to consumer. This will be likely to require a comprehensive analysis of the changing marketing system and possibly new definitions or terms to describe new alignments and procedures. The role of innovation, especially publicly funded research, as a catalyst to encourage or maintain competition should also be examined.

5. *Public versus private sector marketing research.* Several studies have documented benefits from research investments, and a few have attempted to address the issue of who should conduct marketing research. Increasing marketing costs and declining productivity are such that we need to define questions of research responsibility and get on with the work. At present, researchers are reluctant to work on meaningful problems for fear of criticism that they are peforming industry's job. Studies need to define or delineate respective private and public sector responsibilities in marketing. This is extremely difficult because of variations in

firm size and research capabilities within and between specific agricultural marketing industries. An issue begging clarification is the extent to which the consumer benefits from publicly funded marketing research.

6. *Identifying potential innovations.* Questions have been raised about availability of adequate technology to solve many current marketing problems. Some systems have been developed that attempt to catalogue innovation efforts. There is need for common terminology to facilitate the exchange of information between social and physical sciences, especially for techniques that might have broad applicability. An information retrieval system with this data could help identify the potential applicability of innovative concepts early in development and speed up the adoption process. There is need for procedures to be used in evaluating potential innovations that could provide guidance to researchers or managers on where the biggest payoff is likely to be for allocation of research funds.

7. *Marketing system capabilities.* The government is quick to commend agriculture for its favorable contribution to our balance of trade but adds that the system could do more. While we have always exported agricultural commodities, the system was not really designed for present or anticipated volumes. An assessment should be made to identify the critical technology requirements needed to improve the export marketing of agricultural products and to assess the effects, if any, on the domestic marketing system. Given energy and potential equipment constraints, we need to appraise the capability to distribute products from distant concentrated production areas to population centers and through export ports. Transportation pressures and regional comparative advantages could shift for specific commodities in certain seasons.

8. *Resource limitations.* Resource shortages, particularly energy and water, are likely to have profound effects upon the organization, technology adoption, and economics of the food and fiber system. Systems analyses should be conducted to identify and quantify these effects and assign priorities and to recommend research needs and suggest policy alternatives. Changes in relative prices of resources will affect the demand for certain types of technology and the need for alternatives. Can economists help anticipate the changes in prices?

9. *Technology for reducing food losses.* There are estimates that between 20 and 30 percent of all food produced in the United States is lost or wasted each year. It is estimated that recovery of these losses could increase the available food supply 10–15 percent without bringing new land into production. Studies are needed to identify and quantify these losses as well as to assess the cost and benefits of alternative methods for recovering at least portions of them through improved harvesting, storage, packaging, and transportation technologies.

10. *Technology development in a maturing economy.* Projections for the future have typically assumed continued rapid growth. However, with lower population growth rates, declining availability of prime farmland for production, and declines in investment in R & D, concerns have been raised about the United States becoming a maturing economy. Although

still needed to maintain slower economic growth rates and to provide continually improving standards of living, the role of technology development and innovation could change. The impacts of a maturing economy and a presumed inability to feed the world population need to be assessed in terms of developing policy alternatives for R & D.

REFERENCES

Allvine, F. C. 1968. Diffusion of a Competitive Innovation. Proceedings, fall conference, American Marketing Association, pp. 341-51.

Anderson, Dale L. 1977. Changes needed in the organization of the nation's food industry to meet challenges ahead. *J. Food Distrib. Res.* (Feb.): 40-55.

Babb, Emerson M. 1976. Impacts of Federal Funding Requirements on Marketing Research at State Agricultural Experimental Stations. USDA, report to CSRS.

Bain, J. S. 1948. *Pricing Distribution and Employment: Economics of an Enterprise System.* New York: Holt.

Bartels, Robert. 1976. *The History of Marketing Thought,* 2nd ed. Columbus, Ohio: Grid.

Bloom, Gordon F. 1972. *Productivity in the Food Industry—Problems and Potential.* Cambridge, Mass.: MIT Press.

Business Week. Research and development spending patterns for 600 companies. July 2, 1979, pp. 58-77.

Cordaro, J. B., and Robert L. Smith. 1976. Productivity increasing technologies in the food industry: Their impact on society. *J. Food Distrib. Res.* 7:40-55.

Davis, Howard R. 1977. Innovation and change. In J. Salasin, ed., The Management of Federal Research and Development—An Analysis of Major Issues and Processes, MITRE Corp., Metrek Div.

Drucker, Peter F. 1957. *Landmarks of Tomorrow.* New York: Harper.

Eavey, Henry J. 1964. Hearing Record. House Subcommittee on Appropriations, 88th Congress, 2nd session.

Evenson, Robert E., Paul Waggoner, and Vernon W. Ruttan. 1979. Economic benefits from research: An example from agriculture. *Science* 205:1101-7.

French, Ben C. 1977. The analysis of productive efficiency in agricultural marketing: Models, methods, progress. In L. R. Martin, ed., *A Survey of Agricultural Economic Literature,* vol. 1, pp. 93-206. Minneapolis: University of Minnesota Press.

Jones, Harold B., Jr. 1973. Benefit-cost ratios and return on investment for agricultural utilization research. *South. J. Agric. Econ.* 5:89-97.

Hoover, Frank, and Samuel B. Detwiler, Jr. 1978. Agriculture utilization research. In R. Teranishi, ed., *Agriculture and Food Chemistry: Past, Present, Future,* pp. 25-34. Westport, Conn.: Avi.

Kennedy, Donald, and Sherman Gardner. 1979. Memo to the food and drug administration policy board. *Food Chemical News* 21(8):15-17.

Kivenson, Gilbert. 1977. *The Art and Science of Inventing.* New York: Van Nostrand Reinhold.

Lynn, Frank. 1966. An Investigation of the Rate of Development and Diffusion of Technology in Our Modern Industrial Society. Report of the National Commission on Technology, Automation, and Economic Progress, Washington, D.C.

Mansfield, Edwin. 1968. *The Economics of Technological Change.* New York: Norton.

Metzger, H. B. 1973. Marketing Research at State Agricultural Experiment Stations, Past, Present, Future. USDA, CSRS.

Mueller, Willard F., John Culbertson, and Brian Peckham. 1980. Market Structure and Technological Performance in the Food Manufacturing Industries. NC-117 publ., Madison, Wis.

Nason, Howard K., and Joseph A. Steger. 1978. Support of Basic Research by Industry. National Science Foundation, Washington, D.C.

Peterson, Willis, L., and Joseph C. Fitzharris. 1976. Organization and Productivity of the Federal-State Research System in the United States. In T. M. Arndt, D. G. Dalrymple, and

V. W. Ruttan, eds., *Resource Allocation and Productivity in National and International Agricultural Research,* pp. 60–85. Minneapolis: University of Minnesota Press.

Rogers, Everett M. 1962. *Diffusion of Innovation.* New York: Free Press.

Rosen, Stephen. 1976. Wherein future shock is disputed. *New York Times,* June 18, A23.

Rostow, W. W. 1962. *The Process of Economic Growth.* New York: Norton.

Scherer, F. M. 1967. Market structure and the employment of scientists and engineers. *Am. Econ. Rev.* 52:524–31.

_____. 1971. *Industrial Market Structure and Economic Performance.* Chicago: Rand McNally.

Schumpeter, Joseph A. 1954. *History of Economic Analysis.* New York: Oxford University Press.

Shapley, Willis H., and Don L. Phillips. 1978. Research and development. AAAS Rep. 3, American Association for the Advancement of Science, Washington, D.C.

Steiner, Gary A. 1965. *The Creative Organization.* Chicago: University of Chicago Press.

U.S. Congress House of Representatives. 1969. A Study of Technology Assessment. Report of the Committee on Public Engineering Policy, National Academy of Engineering to the Committee on Science and Astronautics, Washington, D.C.

_____. 1970. A Technology Assessment System for the Executive Branch. Report of National Academy of Public Administration to the Committee on Science and Astronautics, Washington, D.C.

U.S. Congress, Office of Technology Assessment. 1978a. Government Involvement in the Innovation Process. Contractor's report by the Center for Policy Alternatives, MIT, Cambridge, Mass.

U.S. Department of Agriculture. 1978a. An Assessment of Post-Harvest Technology. Science and Education Administration, Cooperative Research. Report to Office of Management and Budget, ARM-H-3.

_____. 1978b. Manual of Classification of Agricultural and Forestry Research. Science and Education Administration.

_____. 1979. Post-Harvest Technology Research Assessment. Science and Education Administration, Agricultural Research.

_____. 1980. Measurement of U.S. Agricultural Productivity. ESCS Tech. Bull. 1614.

U.S. Department of Commerce, National Bureau of Standards. 1976. Federal Funding of Civilian Research and Development, vol. 1, summary. NBS-GCR-ETIP 76–03.

U.S. Department of Labor. 1978. Productivity Indexes for Selected Industries. Bur. Labor Stat. Bull. 2002.

Webster, F. E., Jr. 1968. New product adoption in industrial markets: A framework for analysis. *J. Mark. Res.* 5:426–28.

WILLIAM G. TOMEK

8

ALTERNATIVE PRICING MECHANISMS IN AGRICULTURE

THE TERM "price discovery" has a specific definition in the literature of agricultural economics: it is the process of establishing prices (Thomsen and Foote 1952; Tomek and Robinson 1972). Price discovery is associated with transactions between buyers and sellers, but these occur in many ways. Thus pricing mechanisms involve numerous alternative physical and institutional arrangements for discovering prices and other terms of trade. Buyers and sellers may privately negotiate prices; prices may be established at auctions; an oligopolist may set a price that the buyer must accept if a transaction is made. The objective of this chapter is to provide guidelines for future research on pricing mechanisms in agriculture.

Price discovery in agriculture includes the pricing of farm inputs, farm outputs, food and fiber products in various stages of processing and marketing, and items at the retail level. Pricing may be either for forward or spot delivery. All these pricing phenomena are within the purview of this chapter, but little will be said about the pricing of farm inputs or establishing prices for manufactured products containing little farm product. In general, emphasis is on pricing at the farmer–first handler level, but with changing institutions the number of junctions formed by price in the marketing chain has declined. Thus the price received by a farmer may have been derived from the sale of a processed product by a cooperative at the wholesale level. Administrative coordination has replaced price coordination at some market levels for some commodities, and consequently this chapter cannot concentrate solely on pricing at the farm level.

WILLIAM G. TOMEK is Professor of Agricultural Economics, Cornell University. This chapter was written while he was a visiting economist with the National Economics Division, Economics, Statistics, and Cooperative Services, USDA. Walter Armbruster, Olan Forker, Marvin Hayenga, Richard Heifner, Allen Paul, and Randall Torgerson reviewed an initial draft of this chapter, and their assistance is gratefully acknowledged. The final version, of course, is the responsibility of the author.

IMPORTANCE OF RESEARCH

Background. Farm product prices depend on current and expected levels of the variables influencing them. Moreover, for many products, international as well as domestic factors influence prices. Thus the availability and quality of information plays an important role. The price discovery process can contribute to the quantity and quality of information available, to distribution of information to potential traders, and to evaluation of the information (as reflected in the decision to trade or not trade).

The development of pricing mechanisms in agriculture is characterized by diverse and sometimes contrasting trends. The number of participants in cash markets has declined, although the quantity marketed has increased. In general, prices are being established by fewer sellers and buyers. The volume of farm products sold through central marketplaces has decreased, especially the total quantity marketed.

The number of cattle marketed through terminal markets has declined both absolutely and relatively. In 1960, 34 million cattle were marketed in the United States, and the salable receipts in Omaha were 2 million head. In 1975, almost 57 million cattle were marketed, about 925 thousand in Omaha. According to data from the Packers and Stockyards Administration, meat packers purchased 46 percent of their cattle in central terminals, 16 percent at auctions, and 39 percent directly in 1960. By 1975 the respective percentages were 14, 20, and 66.

More products are being sold directly from farmers to processors and other outlets. Prices may be negotiated between individual farmers and buyers, but there is increasing use of formal contracts and formula pricing. For some commodities, farmers appear quite interested in organized bargaining for price and other terms of trade.

Although quantities marketed at central marketplaces have declined, the volume of contracts traded on futures markets has increased. More forward contracting is occurring. Another factor offsetting decreased use of terminal markets is increased use of the telephone, teletype, and computer. While certain methods of pricing have declined in relative importance, they have not always disappeared. For certain commodities in some regions, farmers may have more ways of pricing than 30 years ago. Fed cattle may be sold through a central market, auctions, and directly to meat packers; a buyer may come directly to the cattle feeder, or the feeder may call various potential buyers; the cattle may be priced on a live or carcass basis (Rhodes 1976). For many commodities, however, the number of marketing alternatives is extremely limited.

In addition, pricing mechanisms may be influenced by changes in retail marketing strategies for food and fiber products. For example, private label and retailer brands have tended to stress price competition, while brand-name manufacturers are interested in developing and controlling the distinctive characteristics of the product. This may require closer coordination between stages of the marketing system than for nonbrand-name products.

A variety of problems and concerns have arisen out of this milieu. Prices discovered at central (cash) markets may be less representative of current economic conditions as volume declines, but these are precisely the prices used as a base in the formulas and contracts of direct marketing. The decline in the number of buyers or the growth of farmer bargaining associations may cause prices to be biased relative to the competitive norm. With more private treaties and direct marketing, the amount of information available to the general public decreases. But increased trading on futures markets may contribute to the information base, although futures trading could attract ill-informed participants.

Two general hypotheses are implicit in the foregoing discussion and hence are implicit in the research on pricing mechanisms in agriculture. One is that a causal relationship exists between pricing mechanisms (or characteristics of the mechanisms) and performance as measured by certain criteria. A second hypothesis is that pricing mechanisms or their characteristics are themselves determined by market structure, government policy, and other variables and hence are subject to change. That is, price discovery mechanisms are endogenous variables in the marketing system; and if the performance of a pricing mechanism is somehow unsatisfactory, private or public decision makers may be able to influence development of appropriate mechanisms. Thus the general justification for research on the topic is to develop information as a basis for such decision making.

Using the foregoing hypotheses, research on pricing mechanisms may be classified as follows: describing characteristics or structure of pricing mechanisms, analyzing performance of pricing mechanisms, and identifying factors that change the mechanism. Presumably, the intent of research is ultimately to improve performance, and importance can be related to the magnitude of potential improvements in performance. In practice, however, it is often difficult to establish a relationship between a pricing mechanism and performance (holding other things constant). Farmers, for instance, have expressed concern about thin markets. But what are the consequences, if any, of a thin market? Indeed, what is a thin market (see Hayenga 1979; Tomek 1980)?

Improving Performance. Numerous performance critera exist, and participants in the pricing process give different (implicit) weights to them. Here the criteria are categorized into three groups: private and public costs, pricing and productive efficiency, and welfare considerations. While these categories are not mutually exclusive, they provide a convenient framework for discussing the importance of the topic and the scope of the problem area.

Perhaps the most obvious costs of price discovery are those associated directly with transactions, i.e., private transaction costs. These include the time of the participants in searching for the most advantageous trade, negotiation, and so on; physical costs of exchange, such as the marketplace and communication equipment as well as grading and assembly of the commodity; and costs associated with settlement of disputes or guaranteeing performance on contracts.

One (admittedly narrow) objective of research could be to find the pricing mechanism that minimizes private transaction costs. In comparing existing methods of marketing fed cattle with a teletype auction, R. Johnson (1972) has estimated that the change would save $44.5 million in buying and selling costs and $15 million in transportation costs under 1969 conditions. If similar potential savings exist for other products, this implies large returns to research on the topic. (Johnson's estimates seem rather large for 1969 conditions.)

From the viewpoint of individuals, costs of search could be reduced by using someone else's price. A formula might be negotiated that uses a base price established elsewhere. After the initial negotiation, these buyers and sellers would have essentially no search costs, but someone must bear the cost of establishing base prices. Thus research must consider costs for the entire pricing process, and it must also deal with the public good aspect of price information, the "free-rider" problem, and distribution of transaction costs.

Stated another way, traders can be classified as informed and uninformed (Grossman and Stiglitz 1976). The uninformed may be able to take advantage of the information contained in prices provided by the informed. Traders have an incentive to become informed only to the extent that the costs of errors in pricing exceed their costs of transactions. Beyond the information contained in existing prices, it may be very costly for an individual to become informed. An optimal amount of information might be defined in terms of the marginal cost of obtaining additional information relative to its marginal benefit (related to pricing errors, i.e., pricing efficiency). Research can be justified in terms of whether incentives to the individual result in a market optimum, or whether market failures occur and hence whether public subsidies may be justified.

In the real world, society has an interest in obtaining true prices beyond the traditional social costs implicit in pricing errors. Public programs require correct market prices; e.g., deficiency payments and tariffs may be based on market prices, and release of stocks held under price support programs depends on those prices. Thus decision makers not participating in the market also desire informed prices.

Efficiency criteria typically use the competitive model as a norm, hence efficient prices are those generated under perfect competition. Actual markets are not perfect; indeed, some are highly imperfect. Pricing inefficiencies arise when markets contain monopoly elements, governments intervene by introducing restrictions on trade, the cost of information is not zero, and so on. Hence, some analysts have questioned the usefulness of a norm based on perfect competition, but it also has been justified as an ideal that can be used as a standard of comparison. Marion and Handy (1973) discuss alternative measures of performance such as workable competition. Their report also provides a useful list of references on performance measures.

A measure of inefficiency is the bias, if any, in the level or differentials among prices relative to the warranted (competitive) level. Another measure of inefficiency in prices is the extent to which they are more variable than warranted by economic conditions. If imperfections increase price variability, then price risk is presumably increased and, given risk-averse behavior of producers, resource allocations are influenced. The conventional view is that a negative relationship exists between the level of own-price variability and product supply. Robinson (1975), among others, has argued that high prices attract additional investment and improved technology to farming, while low prices squeeze out the relatively poorer managers. The two hypotheses are seemingly inconsistent, though the Robinson view is somewhat longer run. In a noncompetitive market, prices may also be too inflexible relative to the competitive norm.

The idea of an efficient price includes not only the correct level but also the prompt reflection of all available information about the factors influencing prices. In a perfect market, prices would respond instantly. In practice, the speed and path of adjustment of prices are measures of relative efficiency. The speed of adjustment may differ for price increases and decreases. After a change in information, price may initially overadjust or underadjust followed by still further adjustments to the level warranted by the information.

Pricing systems could be designed to improve productive efficiency; i.e., the mechanism may be linked to the optimal scale of processing plants and to optimal assembly and distribution systems. For example, direct marketing to a processor based on a contract price might improve productive efficiency relative to the assembly and pricing of the product at a terminal market. Such a shift, however, may be inconsistent with competitive pricing. In each instance these are hypotheses subject to test.

A large number of feeder cattle are still priced through auctions, and a study by Williams and Farris (as cited by Raikes 1976) suggests that a fully integrated system from cow-calf operations to the feedlot would reduce costs $2-10 per head. Under current conditions, however, actual observation of feeder cattle by buyers at auctions may help establish accurate prices for specific lots of animals with varying quality relative to other pricing methods. That is, some buyers do not trust third-party descriptions. Moreover, auctions can achieve economical sorting of animals into even quality lots, and an integrated system would not necessarily avoid these costs. Thus further research on possible improvements in productive efficiency from a change in pricing methods seems justified.

Welfare notions are often implicit in criticisms of pricing methods. Sometimes a price is alleged to be too low relative to a concept of fairness rather than to the competitive norm. Indeed many farm programs are designed to raise prices above competitive levels. In contrast, economists have tended to look at welfare in terms of aggregate economic surplus under biased or unstable prices relative to the economic surplus attainable under more stable conditions. Numerous studies have explored the welfare effects of instability, and some have estimated the aggregated social costs associated with instability and hence the potential benefits from stabilization (e.g., Plato et al. 1978).

Hayami and Peterson (1972) have estimated the marginal social return from reducing the sampling error in estimating crop size. For example, in one model for wheat, if the sampling error is reduced from 3 to 2.5 percent, the marginal return is an estimated $36 million (1966–68 average data). Since the cost of reducing the sampling error is less than the estimated gain, the authors conclude that a net return to improved information exists. While their study pertains to returns to an improved statistical technique, it implies that improved pricing methods could also provide net benefits. The pricing mechanism distributes and processes information.

A public policy concern is the potential ratchet effect of unstable farm prices. Higher farm prices result in higher retail prices, but when farm prices subsequently decline, retail prices do not (it is alleged). In addition, increases in retail food prices may result in wage increases through escalator clauses in labor contracts. Thus temporary peaks in farm prices may be followed by irreversible increases in the general price level. This creates additional incentives for stabilizing farm prices.

Price policy also might be used to improve nutrition, increase economic opportunities for small farmers, or achieve other social goals. Clearly many different welfare criteria exist, and all could be used to justify the importance of the topic. Even in the narrower framework of cost and efficiency criteria, research on pricing mechanisms is important because efficient prices are desired at low cost. A perfect, fully informed price can be obtained only at an infinite cost. Thus, in practice, trade-offs are involved in evaluating performance. Perhaps the greatest general concern is that existing incentives may encourage pricing methods that tend to minimize private transaction costs but do not provide efficient prices.

CONCEPTUAL IDEAS IN PAST RESEARCH

Research on pricing institutions has not been based on a unified conceptual framework, since economic theory has had relatively little to say about the process of price formation (Morgenstern 1972). Since a comprehensive framework does not exist, this section is organized around the three research areas mentioned in the previous section, namely, the description and performance of pricing mechanisms and changes in them.

Describing Pricing Mechanisms. Economists have not agreed upon a unique classification scheme or nomenclature for pricing mechanisms. One approach has been to describe existing institutions without an explicit conceptual base. For example, Tomek and Robinson (1972) use five categories: private negotiation between individuals; trading at organized marketplaces, including auctions; bargaining between groups; formula pricing; and administered pricing.

A different approach is to relate the pricing method directly to what is commonly called market structure (Rogers 1970), which can be viewed along a continuum from pure competition to absolute monopoly. A simple categorization of pricing methods is competitive and noncompetitive (or supply-and-demand and administered pricing). Implicit in this approach is the use of market structure characteristics in making the classification of pricing mechanisms, although the explicit structural characteristics for classifying pricing schemes have not been spelled out.

Ward (1977) has made a valiant effort to integrate pricing mechanism literature with that of market structure. For classification purposes he asks, What interactions (contacts), if any, take place between buyers and sellers as price is established? Horizontal contacts are those among sellers (buyers). Vertical contacts are between buyers and sellers. If one seller has no contact with other sellers in setting price, then this is an "intrafirm" relation in Ward's terminology. If the seller likewise sets (administers) a price without negotiation with a buyer, this is an "intralevel" relationship. Thus, for example, an intrafirm-intralevel pricing arrangement would include a price set by a monopolistic seller or a formula established by a seller with no interaction with other sellers or with buyers.

Ward then asks, When will such pricing arrangements occur? The intrafirm-intralevel mechanism presumably requires monopoly power. Ward's criterion for classifying pricing methods, however, is whether interaction exists among participants either horizontally or vertically. In an industrial organization framework, one might say that the market structure determines the pricing mechanism (the conduct), but Ward appears to be adding another structural characteristic, the degree of interaction of participants. This approach has the virtue of providing a criterion for classifying pricing methods. The principal problem is the resulting complexity of the numerous classes. It would be difficult to relate price behavior, costs, and other performance criteria to the various pricing structures.

All the classification schemes have limitations as a conceptual base for predicting performance of the alternate categories of the mechanisms. For example, the seeming dichotomy of competitive and administered pricing really is not mutually exclusive. An elevator operator buying grain from farmers posts buying prices; in this sense prices are administered rather than negotiated, but such prices can change in response to competitive conditions. Administered prices in this instance behave like competitive prices. In addition, price discovery often takes place in two steps. Base or reference prices may be established by one method, such as at a central market, while many individual prices are established as differentials from the base. These differentials might be negotiated on an individual basis or fixed by formula. For example, the majority of beef prices at the wholesale level are established by a formula using the so-called *Yellow Sheet* price as a base, but a fairly large minority of prices are negotiated on an individual basis by buyers and sellers (USDA 1978).

Evolution of Pricing Mechanisms. There is no unified theory of why pricing mechanisms evolve and change. Some concepts have been borrowed from industrial organization and theory of the firm. Some research is commodity-specific and examines factors peculiar to that commodity.

One line of reasoning relates changes in the pricing mechanisms to changes in market structure, especially growing concentration of the buyers of farm products. Large buyers want to deal directly with large sellers; they may not want to buy small lots in a central market. Vertical integration may be a response to price and other risks.

A second but closely related concept is the changing nature of the decisions made internally by the firm and those made externally to the firm. As firms grow and become more complex, they take over functions previously performed by markets. Alchian and Demsetz (1972) suggest that "the firm takes on the characteristic of an efficient market in that information about the productive characteristics of a large set of specific inputs is now more cheaply available. Better recombinations or new uses of resources can be more efficiently ascertained than by the conventional search through the general market.'' That is, resources are allocated more efficiently in some instances via vertical and horizontal integration than via a pricing mechanism.

Economic theory contains some discussion of the boundary between firms and markets. Many years ago Coase (1937) argued that markets do not operate costlessly and that firms can perform some functions at lower costs. Thus, according to this theory, if market transactions are becoming less important, it is because they are becoming relatively more costly. Alchian and Demsetz (1972) emphasize that when team-oriented production can increase productivity, this provides a necessary condition for the definition of a firm. Productivity is increased by going to internal allocation decisions within a firm rather than external allocation decisions given by the marketplace. A broader question is that of a superior or optimal economic system (Hurwicz 1973). The question arises, for instance, whether socialism or capitalism is a superior resource allocation mechanism. But one can also be concerned about centralized versus decentralized ways of organizing firms or about market versus command systems of resource allocation. The current literature is abstract and mathematical, but those doing research on pricing mechanisms should be aware of it. Attempts are made to define specific mechanisms and to analyze (in a pure mathematical sense) their feasibility and optimality.

Both Gray (1964) and Williamson (1971) have made related arguments. Their emphasis is more on market failure and transaction costs as incentives for administrative coordination. Williamson, among other things, points to monopoly elements in markets as factors contributing to market failure. In addition, desires of firms for increased control over prices are motives for vertical integration. But Gray argues that, although the number of junctions in the marketing chain formed by prices has declined, the importance and efficacy of price in the remaining junctions may be enhanced.

Another factor influencing the price discovery mechanism is the character of the product being priced. Commodities vary in their bulkiness, perishability, homogeneity, and the form in which they are sold. A factor in the federal order pricing of fluid milk is presumably the perishability and seasonality of supply. Some commodities, like the grains, are sufficiently homogeneous that buyers are willing to make transactions on the basis of written specifications. In contrast, feeder calves are sufficiently variable in quality that many buyers either want to inspect the individual lot or use order buyers as a part of the pricing process.

Since farm products change relatively little, this might not seem like a factor that changes pricing mechanisms. But changes in technology alter relative costs of storage, processing, transportation, and communication. A commodity in a sense becomes less perishable, and the method of marketing and pricing changes to accommodate the change in the economics of storage and other factors. In addition, the nature of demand for the final product may change. The demand for butterfat in dairy products has declined; demand may shift from fresh to processed products (or vice versa); retailers may want potatoes or apples in bags rather than bulk. In each instance, the feasibility of alternative pricing mechanisms may be altered.

Still another factor influencing the evolution of pricing mechanisms is relative prices and the technology of the mechanisms themselves. When the price of labor and management has risen relative to the price of capital, it becomes more costly for price to be established by time-consuming haggling of individuals. Or, if fuel becomes more costly, physical assembly of a product at a central market is discouraged. Improvements in computer technology and communications systems makes electronic pricing systems economically more attractive. Ultimately, buyers may be able to inspect items via television monitors in conjunction with an electronic auction.

Performance Measures for Pricing Mechanisms. As the previous section suggests, numerous potential measures of performance exist. Studies of pricing mechanisms have used traditional cost concepts and applied them to the private transaction costs attributable to the mechanism (e.g., Kuehn 1971; R. Johnson 1972). The unmeasurable costs of pricing mechanisms, such as the loss of public information that may be associated with moving from one method to another, have received little attention.

A considerable literature on information theory has developed in general economics (e.g., Akerlof 1970; Rothschild 1973; Grossman and Stiglitz 1976; Figlewski 1978). These concepts, though not used extensively in research on pricing mechanisms, have the potential for integrating transaction cost and pricing efficiency concepts. In principle, the marginal cost of information can be equated with the marginal return from information, where returns are measured by the efficiency of prices. With this optimizing principle, one can ask about imperfections in existing pricing mechanisms, analyze the effects of subsidies or taxes on a pricing method, or consider whether a particular mechanism is inherently better than another. Both private and public costs and benefits could be included, but in practice the concepts are likely to be difficult to measure.

Cox's (1976) work on commodity futures markets is an example of an application of information theory to a pricing mechanism. He argues that, relative to dispersed trading and private negotiations, futures markets are an inexpensive method of searching for the best bid and offer. "It becomes worthwhile for more individuals to trade and thereby communicate information."

Many studies have estimated the pricing efficiency of a particular market, though usually using only one or two measures of efficiency. The general methodology is to develop a standard of comparison and then measure how actual prices conform to the standard. Such research examines a particular set of prices and (implicitly) one pricing mechanism. The mechanism is being compared to the standard and not to other mechanisms.

Perhaps the simplest measure of pricing efficiency used in past studies is the correlation coefficient. If two lots of a particular commodity are perfect substitutes (or perfect complements) in a perfectly competitive world, the correlation in price movements through time would be one. Thus, in this sense, actual correlations become a measure of the degree of perfection.

Another common approach is to develop the price differentials or behavior that should exist under competitive conditions and compare the actual differentials with the standard (e.g., Hassler 1953; Trierweiler and Hassler 1971; Bohall 1972). The standard is usually developed from cost relationships such as transfer costs among regions or from other prices such as closely associated grades.

The research on pricing fed cattle or hogs has estimated the errors in pricing live animals relative to the carcass value after slaughter (based on the grade and yield of the carcass). This approach assumes that the carcass value is the correctly determined competitive price (R. Johnson 1972; see also Hayenga 1971 and references therein). A related issue is the one of accurate grading and its cost. In a pioneering study, Farris (1958) has looked at the grading of wheat by country elevators relative to grading by an independent laboratory. He found wide differences for the individual samples. Elevators appeared to overvalue low-quality wheat and to undervalue high-quality wheat.

Studies of the behavior of short-time (say daily) price changes exemplify the use of a theoretical model as a standard of comparison. The efficient market hypothesis implies that futures prices should behave as martingales; thus deviations from the martingale model imply deviations from the perfect market (for definitions and review of concepts, see Fama 1970). Research using price series from futures markets typically finds only minor departures from the random walk hypothesis (e.g., Larson 1960; Smidt 1965). The deviations are sufficiently small that profitable speculation cannot be based on the systematic component (in the view of the researchers). In contrast, the decision-making aids used by many speculators and hedgers are based on procedures such as moving averages that implicitly assume nonrandomness.

Relatively few studies have attempted to estimate biases in price levels or heteroscedasticity of prices. This is related in part to the obvious difficulty of obtaining a standard of comparison. One procedure is to make price a function of causal variables including the pricing mechanism. Then the effect of the mechanism on price can be estimated. Garcia and Forker (1975) have followed this approach for egg prices at the producer level in New York State for a three-month period. While the results are not clear-cut, on balance the pricing arrangement appeared to have little effect on producer prices.

Schultz (1949) argued that prices established on organized markets fluctuate excessively (presumably relative to the variation justified by the competitive model). This is a viewpoint, however, rather than a conclusion drawn from formal research. Studies of the variability of prices have often employed a "before and with" methodology rather than using an absolute standard of comparison. Onion prices were analyzed using time periods before, during, and after the existence of a futures market for onions. Beef cattle, pork belly, and wheat prices have been studied before and with futures trading. This research suggests that the establishment of futures markets, if anything, tends to reduce the variability of cash prices; at least price variability is not increased as is commonly alleged (see Powers 1970; A. Johnson 1973). While futures markets may not increase price variability, commodities with organized futures markets may have larger price variability than commodities without futures. The existence of a futures market may be a response to price variability rather than a cause of it. Clearly, a comparison of price behavior across commodities and institutions can be dangerous.

The welfare measures of performance of alternative pricing mechanisms have involved several different concepts. At the level of the individual buyer and seller, a few studies have examined the distribution of pricing errors (Farris 1958; Hayenga 1971; R. Johnson 1972). In principle, the errors could be evenly balanced so that no net transfer of income occurs between buyers as a group and sellers as a group. In this case the problem is limited to distribution of income and reallocation of resources among individuals.

If, on average, prices are biased, sellers as a group may be overpaid or underpaid relative to the competitive norm. Both old and recent studies of the aggregate effects of government support programs are in this spirit. In the late 1950s and early 1960s, several studies looked at the income consequences for farmers of removing price support programs (e.g., Robinson 1960). More recently, the costs of price support programs and marketing orders to consumers and taxpayers have been analyzed (e.g., Heien 1977; Nelson and Robinson 1978).

Perhaps the principal conceptual tool used in aggregate welfare analysis is that of consumer and producer surplus. This has been applied to measure the effects of biased and unstable prices. A publication edited by Eaton and Steele (1976) contains some applications (see also Chayat et al. 1974; Plato et al. 1978). A paper by Just (1977) provides a useful discussion of important conceptual issues in measuring the welfare gains from stabilization.

ASSESSMENT OF PAST RESEARCH
This section reviews the accomplishments and failures of previous research on pricing mechanisms. Some research has been oriented toward marketing and pricing institutions in general and has been summarized in extension materials edited by Forker and Rhodes (1976). Such essays describe the general advantages and disadvantages of alternative mechanisms. Much of the existing research is oriented toward the price behavior of specific commodities or mechanisms, and while considerable attention is devoted to such studies in this chapter, the proportion appraised is small.

Pricing Mechanisms: Classification and Evolution. Past research has developed commonsense categories for pricing mechanisms, and these categories help economists communicate (e.g., Forker 1975). Two problems exist with the current classifications. First, no common, agreed-upon categorization exists; terminology and categories vary with the author. Second, categories should ideally provide a foundation for appraising performance and for analyzing evolution of the mechanisms. Thus the classification should be based on important differentiating characteristics of the mechanisms. (Some suggestions are made in the next section.)

An analogy exists with the structure-conduct-performance concept of the industrial organization literature. For pricing mechanisms, the received concept of "structure" or "mechanism" is not well defined. The pricing mechanism is perhaps analogous to conduct that flows from the market structure, though some esssays imply that pricing mechanisms have characteristics not determined by market structure. In addition, the "causation" may not run exclusively from structure to mechanism but rather may be simultaneous in character (see Campbell and Hayenga 1978). Structural characteristics help determine the nature of the pricing mechanism, but the nature of the mechanism may also influence structure. For example, farmer bargaining may be a response to the declining number of buyers, but the development of a bargaining group may in turn contribute to the structure of the buyers.

The difficulties and usefulness of defining precise categories can be illustrated by considering formula pricing, one category sometimes used for pricing mechanisms. From a descriptive viewpoint, the term conveys the idea of discovering price via a mathematical equation (a precise way to obtain "the" price). A transaction price equals, for example, a base price plus or minus differentials for location, quality, and other terms of trade. Usually the base price for the formula is available to traders at little or no cost, and the differentials are negotiated once with the intent of applying the formula to a series of subsequent transactions.

This conception of formula pricing suggests its potential benefits. Formula pricing can reduce the cost of search for price, and a formula, in a sense, reduces price risk. A known method is available for arriving at the transaction price prior to delivery of the product without the ambiguity implicit in haggling. The concept is useful in suggesting the incentives for shifting to formula pricing. (Problems exist but are not discussed here.)

Prices arrived at by formulas are not necessarily discovered in a uniform way. The frequency and method of changing the formula can vary greatly. Some formulas, like those in marketing orders for fluid milk, are changed infrequently via highly institutionalized rules. Others are changed frequently without formalized rules. In pricing beef carcasses, the frequency with which formulas change apparently varies with buyer-seller pairs. If formulas are negotiated frequently, the observed price behavior may differ little from the behavior with "pure" negotiated prices. Hence, and this is the important point, the concept "formula pricing" may not provide a basis for generalizing about price behavior under different pricing institutions.

The conceptual literature, outlined in the previous section, provides general hypotheses about the variables determining the evolution and change in pricing mechanisms. (Changes in the number of buyers and sellers is one such variable.) The literature contains relatively little on formal tests of such hypotheses. Numerous descriptions of trends in marketing and pricing systems exist. A report by Holder and Hepp (1978), for example, describes trends in the marketing and pricing of hogs. It also indicates the nature of existing alternatives open to pork producers. An objective of such research is to help farmers recognize the pros and cons of pricing alternatives.

Considerable research attention has been given to factors resulting in vertical integration and contracting for specific commodities (see Campbell and Hayenga 1978 for a summary and references). Past research has done an adequate job of analyzing the "whys" of historical events. It is not clear, however, whether sufficient conceptual and empirical knowledge exists to predict future changes in pricing mechanisms and their consequences. Existing research probably has failed to identify causal relationships with sufficient precision to be helpful to policymakers. For example, can the change in volume of sales on terminal markets (relative to total sales) be explained by measurable variables and are these variables controllable by public and private decision makers?

A major obstacle to research on changes in pricing methods probably is the lack of adequate data. For example, the pricing of beef and pork using the *Yellow Sheet* has generated much conjecture but little hard evidence, mainly because of lack of data. A USDA (1978) report analyzes individual transaction observations from invoices for beef carcasses in July 1977. The analysis is limited to a point in time and cannot consider the trend toward formula pricing. Also, the privacy of such data makes their collection and use difficult.

What was the speed of change to the current volume of pricing off the *Yellow Sheet?* Why did the change occur? What are the consequences of the change? How might this pricing method be influenced in the future? These seem like unsolvable questions unless better data become available. Thus past research has developed hypotheses about why pricing mechanisms change, but such research, it seems fair to say, has not been predictive or prescriptive.

Evaluating and Measuring Performance. Past research has considered cost, efficiency, and welfare measures. Usually these measures are not combined in a single appraisal of pricing methods. Costs are computed from the viewpoint of comparing alternative marketing systems for farmers or in comparing efficiency of size, for example, for auctions (as in Kuehn 1971; Haas et al. 1977). Such research provides information that may lower buyer and seller marketing costs. But the cost measures are divorced from price behavior.

Perhaps the most fundamental question is whether pricing efficiency is really related to the pricing mechanism. Over 30 years ago, Thomsen and Foote (1952) suggested that imperfections in the pricing mechanism are relatively unimportant in determining the level of prices. Nonetheless, concern has been expressed that pricing institutions influence price behavior (e.g., Rogers and Voss 1969; USDA 1978). Some farmers believe that futures markets influence price levels, and many potato growers in Maine apparently would like trading in potato futures made illegal (*Wall Street Journal* 1978).

It may be relatively easy to appraise whether spatial price differentials conform to the competitive norm, but the question of bias in prices is difficult to answer, partly because the warranted level is unknown. Moreover, with the passage of time, the warranted level changes. Current price may be correct given the then existing information but may be wrong in light of subsequent information. Thus two questions arise: How well does a price mechanism perform for a given level of information, and do pricing mechanisms differ in the amount of information provided at a given cost?

Much of the research on pricing mechanisms is commodity specific (e.g., Marion 1976) and tends toward descriptive essays. The empirical research that has been done usually finds rather small imperfections in pricing mechanisms, at least in pricing efficiency terms. Indeed, one of the paradoxes of research in this area is the gap between empirical results that find few problems of price performance and the numerous essays suggesting that important problems exist.

The price effects of trading in futures contracts is an example. Empirical research on futures markets suggests that they have beneficial effects on cash prices or that, at a minimum, they do not have negative effects. Yet allegations of adverse consequences of futures markets persist. On the one hand, critics of futures trading probably have not given sufficient credence to existing research results, but conversely, existing research has not always tackled the specific concerns of the critics. For example, the empirical research generally involves analyzing price behavior before and with futures. But critics are sometimes concerned with specific incidents of price behavior during the period of futures trading.

The pricing of eggs is another example. This problem area perhaps has received more research attention then any other single commodity. Egg prices are often established by formulas that use published quotations as base prices. The quotations are alleged to have been biased (downward), to have fluctuated too frequently relative to actual changes in supply and demand, and perhaps to have varied too widely. Unfortunately, past research has not established the magnitudes of these problems with precision (a number of studies are summarized by Rogers and Voss 1971).

Much of the work on egg pricing involved proposals of alternative mechanisms. This may have been influential in the development of Egg Clearinghouse Incorporated (ECI) and an associated egg market evaluation committee. ECI was created by producer interests to provide a market exchange function, and it provides price information. Whether the benefits of the information exceed the costs of the mechanisms have not been formally established. The costs seem small, however, and benefits would not have to be large to have a favorable benefit-cost ratio.

A number of studies of vegetables, grains, and the dairy sector have examined price differentials relative to the competitive norm. This research is useful in pointing out problems when they do occur as well as confirming that markets often work well.

Many papers have been prepared on the declining volume of trade at central markets or on thin markets (Hayenga 1979). The problem apparently exists for numerous commodities: cheese, butter, eggs, fruits, and vegetables and perhaps cattle and hogs. The various essays typically describe the changing volume of trade and call attention to the possibility of a problem, but as suggested above, most are not especially precise about the performance consequences. Presumably the potential for biased, imprecise, or manipulated prices is greater in a thin market.

Much of the empirical research on pricing cattle and hogs has dealt with the consequences of pricing live animals. Considerable scope for error exists relative to the true carcass value. Older research also pointed to the problem of price discounts for fat hogs not being a sufficient reflection of consumer demand for lean meat. This has been useful in defining the magnitude of the problem, estimating the benefits of alternative mechanisms, and encouraging alternative pricing methods.

Some empirical research has been oriented toward pricing mechanisms per se (rather than toward a commodity). Two noteworthy examples are Frahm and Schrader (1970) and Jesse and Johnson (1970). Frahm and Schrader have structured an experiment to compare price behavior for Dutch and English auctions. They investigated hypotheses such as "English auctions will result in greater transaction price variation." Though based on experimental rather than actual situations, price behavior is compared for two different mechanisms. The article also provides a useful reference list for those interested in the theoretical literature on auctions.

Jesse and Johnson have analyzed the price behavior of vegetable contracts between processors and growers in Wisconsin. They found that the net contract prices, after adjusting for contract provisions, tend to be equal. Thus the authors conclude that the pricing of contracts was competitive in the sense that similar prices were involved in all the contracts. This of course does not answer the question of whether the average price paid farmers is the competitive equilibrium price, i.e., whether processors have monopoly power.

A considerable literature exists on farmer bargaining and marketing orders, but again much of the research has been descriptive in nature. Chayat et al. (1974), however, estimate the benefits from increased bargaining power for egg producers. Jamison (1966) provides a comprehensive discussion of marketing orders with special attention to cling peaches, and an article by Geyer and Dahl (1978) summarizes the objectives and difficulties of collective bargaining by farmers. Much of the empirical analysis has involved estimation of demand functions to determine if the necessary elasticity conditions hold for price discrimination and to provide information for bargaining (e.g., Hoos and Seltzer 1952; Piggott 1976). A limited number of studies have provided critical appraisals of price discrimination schemes in agriculture (e.g., Smith 1961). There is little evidence on the price enhancement consequences of bargaining; however, Garoyan and Thor (1978) have found that "prices received by farmers for crops with negotiated prices were significantly higher in the period 1960–75 than for crops without price negotiation."

With respect to welfare criteria, a number of essays have described alternative stabilization mechanisms. In addition, much effort has been devoted to analytical models that describe the theoretical benefits under certain assumed conditions in terms of consumer and producer surplus. In my view, too much effort has been placed on theoretical models that are not sufficiently realistic. Each successive paper has involved a small variant of the prior (supply-demand) model. As Helmberger and Weaver (1977) point out, such models typically do not allow for private storage. One benefit of the theoretical models is that they help establish the limitations of empirical research. Measured benefits, for example, depend on the functional form of the equations used in the empirical analyses.

The social costs of price instability have been estimated, especially for the grains. The concept of economic surplus provides a single measure of the welfare and efficiency effects of stabilization. Such a measure of the "goodness" of a pricing mechanism may appeal to policymakers who are trying to make decisions about alternative mechanisms. A single number, however, cannot convey the details of the effects of a change in price behavior (see Just 1977 for other limitations). For instance, what are the consequences of more stable prices for the supply of a farm product? Will the supply function be shifted? Gray (1974) also argues that using a single (average) demand function to appraise the consequences of price stabilization is inappropriate. In his view, the demand function for small inventories differs structurally from the demand function for large inventories. He then hypothesizes that it is consumers, not farmers, who benefit from stabilization programs. Economic surplus measures probably are not practical methods of evaluation. In addition to conceptual difficulties, policy recommendations based on consumer and producer surplus are not likely to be acceptable to politicians (Cochrane 1980).

The competitive norm is implicit in most of the performance measures used in appraising pricing mechanisms, and the appropriateness of this criterion has received relatively little attention. One can take the position that monopolies and cartels have good features. They may stabilize prices; they may provide monopoly profits to disadvantaged groups that society deems deserving of such returns; they may be more dynamic, progressive, or innovative than purely competitive firms. Thus the real issue may not be whether mechanisms are efficient in the competitive sense but whether they provide equity for particular groups, stimulate innovation, and so on. Agricultural economists have a history of doing research oriented toward specific groups such as farmers. There is room and need for research using various criteria, and it can be oriented toward providing the information necessary for informed choices in a democracy (such as choosing to provide for collective bargaining for farmers).

AREAS FOR FUTURE RESEARCH

This section outlines promising areas for future research. The inclusion of the topic means that, unless explicitly stated to the contrary, research on the topic is feasible and important contributions are possible.

Methodology. Useful research on pricing mechanisms is limited in part by available methodology. One methodological constraint is the *ceteris paribus* comparison of alternative pricing mechanisms, except under experimental conditions. One approach is to compare existing behavior against a norm; the quality of the comparison depends on the relevance of the norm. A second approach is to study performance before and with the existence of a pricing mechanism. Clearly, the question arises, Is the change in performance, if any, really attributable to the change in the mechanism? If two or more pricing mechanisms exist at the same time for a given commodity (e.g., negotiated and formula pricing of carcass beef), they may be compared. But even assuming data are available for making comparisions, a statistical problem arises, i.e., holding other things constant in a nonexperimental setting. While improved comparison procedures would be helpful, they are likely to be difficult or costly to achieve.

A second difficulty of appraising pricing mechanisms is their lumpiness. If a new pricing system is proposed, such as an electronic exchange, the new system is not a marginal change from the existing methods. Economists have not been very successful in analyzing these big changes, but research in this area seems justified. For example, futures markets have been successful for some commodities but not for others, and similar differences in success exist for other pricing institutions. Why does a particular mechanism become relatively more important in the pricing of a product? Does the competitive model work for pricing mechanisms in the sense that the more efficient mechanisms tend to evolve with time? If not, can public policies influence the development of the best mechanisms?

Two preconditions that seem necessary for a successful futures market are technical feasibility and economic incentives. If the principal economic justification for futures trading is hedging use, the factors influencing hedging demand need to be studied. These factors include the quantity of physical product that might be hedged, the magnitude of the price risk that might be reduced by hedging and/or the potential size of profit enhancement from hedging, the availability of substitutes for hedging, the size and concentration of firms that might hedge, the cost of learning about and using futures, the degree of risk aversion among potential hedgers, and probably other variables (see Telser and Higinbotham 1977). In futures trading, technical feasibility centers on writing a contract that is useful both to buyers and sellers. Such preconditions for success can be studied by commodity, and analogous studies could be done for pricing mechanisms other than futures markets.

A third methodological problem is the need to classify pricing mechanisms into a few useful categories, which should be relatable to performance measures. In classifying mechanisms, the following structural characteristics are among those to be considered: (1) the market power of buyers and sellers including the number of each, their financial resources available for influencing price, and the alternatives open to each (influenced by such things as the perishability of the product and spatial monopoly power) and (2) the information content of transactions that may depend on frequency of transactions, ease of communicating terms of trade (grades and grading, communication systems, and degree of interaction among buyers and sellers), quality of information conveyed (proportion of informed traders), and timing of the pricing decision (before, at, or after delivery). For example, the important distinction between a formula and a negotiated price may not be the mechanics of arriving at a price so much as the difference in information conveyed by the respective transactions. A formula price probably conveys less information than a negotiated price. A formula tends to be passive with time, although the formula itself may have been negotiated. A negotiated price can be based on poor information, or one party to the transaction may have great market power. Thus a priori a negotiated price is not necessarily "better" than a formula price.

In the context of the suggested characteristics, four categories of pricing mechanisms exist: (1) buyers and sellers have no market power and the information content of their transactions is large and of good quality, (2) traders have no market power and poor information, (3) market power exists and information is good, and (4) market power exists and information content of transactions is poor. The first category is consistent with pure competition, but this need not imply that it is the preferred alternative from a performance viewpoint. In some instances, the performance implicit in the third category may be preferred. No one wants a transaction system that provides "poor" information, but the cost of improving information must also be considered.

A fourth area of methodological concern is measures of performance. In some instances, the need is to make existing criteria function and to develop the linkage between characteristics of the pricing mechanism and its performance. Relatively little attention, for example, has been given to measuring the public costs of changing pricing mechanisms. In addition, researchers need to consider the appropriateness of performance measures for their specific research problem. When is the competitive norm appropriate and when are other norms needed?

Studies of the costs of price discovery seem to implicitly assume a competitive market structure. Perhaps the costs of price discovery vary with the market structure. With many buyers and sellers, the cost of searching for price is potentially great, and the potential benefits of finding a low-cost mechanism are accordingly large. At the other extreme, with one buyer of farm products, the cost of searching for price by the farmer is small; there is no alternative. But with a small number of buyers the situation is less clear. Few alternatives exist, and the farmer probably can make a complete search of the alternatives. Price comparisons, however, may be difficult (e.g., the comparison of contract terms). In addition, the cost of defaults on the contract may be important, but the probability of default may be impossible to evaluate. Thus additional research on the private costs of price discovery appears justified. Moreover, public information probably is more limited in oligopolistic situations.

Thin Markets. With the declining volume of transactions on terminal markets and the increased dispersion of pricing, the concern is that price behavior is influenced adversely and price reports are based on inadequate information. The alleged imperfections in price behavior are especially important if other prices are established by formulas that use the terminal market prices. Research on this topic can be divided into four questions: What is a thin market, i.e., how might one be defined or recognized? What are the price effects, if any, of a small number of transactions? What is the mechanism linking market volume and price behavior? (This has implications for how one might study the effects of thinness.) And why have marketplaces become thin?

The link between the number of transactions and price behavior is partly a function of information. Each public transaction conveys information; and as volume declines on a public market, the quantity and quality of information is eroded. This implies less precision in price discovery. In addition, thin markets are probably more subect to manipulation than markets with large volumes; but given available data, the existence and effects of possible price manipulation probably are not researchable questions.

Tomek (1980) proposes defining a thin market by using Chebyschev's inequality; that is, how many transactions n out of the total volume N are required to estimate the true but unknown price (or price change) with a given level of precision? Or, turning the question around, given the existing marketplace volume, what is the implication for the precision of pricing? While empirical analyses may be difficult, considerable scope appears to exist for estimating whether particular markets are thin and whether various public and private price reports (such as the *Yellow Sheet*) are based on an adequate sample of prices to meet some specified level of pricing precision. In studying price behavior, the implicit standard of comparison is usually the competitive model; sometimes it is possible to compare price behavior in a market with a small volume with behavior in a market with a large volume, or one may be able to analyze prices as market volume declines.

If an existing market has price performance problems related to a lack of volume, the question of how to rectify the situation can be studied. The benefit of having a base price derived from a central market is not tied directly to the cost of participating in the pricing process. Of course, if the base price is erroneous, buyers and sellers would benefit from improved price discovery. It may be possible to estimate the public subsidy necessary to maintain a useful central market or to study alternate pricing arrangements for obtaining base prices.

New Developments. One proposal for solving the problems outlined above is installation of some sort of electronic pricing mechanism; a number of research topics can be related to new developments in the technology of pricing. The pricing or information system may be based on the teletype, telephone, or computer or some combination thereof. These systems, *ceteris paribus,* lower per unit costs of search and transfer of information. Considerable interest exists in electronic exchanges. A few exist, a number have been proposed, and the Agricultural Marketing Service has evaluated electronic exchanges. Little doubt exists about their technical feasibility, but the difficulties of building realism into a system probably have been underestimated. Moreover, the magnitude of the potential benefit of an electronic system over existing systems may be small. Low per unit costs depend in part on a large volume of use, and the incentives for using an electronic exchange require study. Research can estimate total costs of electronic pricing relative to other pricing mechanisms, and the discussion of methodology provides some suggestions for studying the volume of trading relative to pricing methods.

Resistance to change is likely to be a potential problem in adopting a new pricing (or information or grading) system, but this is a difficult problem to research. There is an analogy, however, with the literature on adoption of new techniques by farmers. Is the resistance to change based on a correct assessment of the benefits and costs of the new mechanism? Do public benefits (not reflected in the private incentives) justify subsidies for a new pricing mechanism (or maintenance of an existing one)? Or will adoption occur after an initial learning phase? (The growth pattern, or lack thereof, for new futures markets may be a fertile area for study.)

Interest in electronic pricing often arises because price behavior with existing mechanisms is thought to be unsatisfactory. Thus another research question is the magnitude of possible improvements in price behavior with a shift in pricing mechanisms.

A different kind of new development is in describing products. New measurement techniques may permit improved descriptions of a product's economic value; additional information may improve consumers' decision making; new uses for old commodities require changing methods of valuation. Thus, like research on pricing mechanisms, questions about the benefits and costs of new grading (information) systems can be answered. A better system can reduce transaction costs and improve price signals. Even if few consumers at the retail level use the information, its availability may improve the competitive behavior of sellers, and the researcher must appraise the full range of benefits and costs.

Market Structure and Bargaining. Two long-standing problem areas are the effect on farm prices, if any, of buyer market power and the welfare consequences of farmer bargaining, marketing orders, and the like. Although the topics are old, there is little empirical evidence that buyer concentration has reduced farm prices or that farmer bargaining has raised prices. Additional evidence would still be useful.

Mechanisms for improving farmers' welfare include marketing orders, bargaining legislation, and cooperative action (for bargaining and/or processing). The objectives of such mechanisms are usually to increase farmer returns and/or stabilize them. This may be accomplished through supply control or price discrimination or by increasing the marketing services provided by farmers (or the cooperative).

In this context, an obvious research question is the potential of a mechanism for raising or stabilizing farm incomes. Notwithstanding the numerous empirical studies available, relatively few are applicable directly to proposed bargaining legislation or marketing orders, especially those at state or regional levels. Another question is, If farmer bargaining is successful in increasing farm prices, does the increase come from processor profits (because the monopoly power of the seller offsets the power of the buyer) or from higher retail prices or both? If processors are part of a relatively competitive industry, will a regional bargaining unit impair the economic viability of the processors within the region?

In my view, some research done from the viewpoint of farmers can be justified. In a democracy, legislation is often enacted to benefit particular groups. It is important, however, for research to estimate the welfare implications of various proposals for all sectors of the economy.

If a farmer cooperative is the only buyer (say of fruit for processing) in a particular region, prices from private firms are not available as a basis for transfer prices from the co-op to its members, and farm prices may be computed as a residual after deducting costs from a wholesale price. In this context, a number of interesting research questions arise. What is the economic feasibility or justification for establishing a cooperative processor where a proprietary firm has failed? In general, what are the implications for returns received by farmers (the mean, the variance, the time lags in payments) relative to the alternatives? The question is academic if the farmers have no choice in sales outlets; but if alternative payment arrangements exist, there could be a trade-off between the time lag implicit in paying the net farm price after wholesale sales have been made and the stabilizing effect of receiving a pooled return based on the average of wholesale prices.

Similar questions may be analyzed with respect to profit-sharing contracts between farmers and private firms. For instance, are grower returns increased or stabilized? What are the incentives for such an arrangement to a buyer? What commodities are suited to such an arrangement? Pricing through contingency contracts is another fruitful subject for research. In this case, contract terms are negotiated subject to certain contingencies (e.g., acts of God), and research can consider the benefits and limitations of such an arrangement.

Delayed and Forward Pricing. It is becoming more common for prices not to be established at the time of the physical delivery of the commodity (whether or not a cooperative is involved). Farm prices may be established before, at, or after delivery. Farmers may set (approximately) their selling price through hedging in futures or via forward cash contracts. Buyers often hedge forward contracts in futures. Hence futures markets are important in establishing prices under these arrangements.

Such pricing arrangements provide the farmer with greater flexibility in timing of the pricing decision; price need not be established just at the time of physical delivery. At the firm level, research effort could be devoted to helping farmers make pricing decisions. Models might be developed to predict forthcoming prices, but it is not clear whether government and university economists can provide accurate information with sufficient timeliness to be useful to decision makers. Nonetheless, research can help establish optimal levels of hedging, given existing levels of yield and price risk, and research can aid farmers in comparing hedging in futures with forward cash contracts.

At a more aggregate level, if prices are being established at times other than delivery, questions might be raised about the changing information content of prices and the possible effects on flow and storage of commodities. Perhaps prices reported for a particular day (or other time period) will have little relation to quantities delivered in that time.

Empirical analyses of basis (cash-futures) relationships are needed not only for formal hedging strategies but also for the price discovery role of futures. Relatively little research has been done to explain and predict basis relationships. Economists have perhaps underestimated the magnitude and importance of basis variability. The ability of hedges to reduce the variability of returns and/or increase their level depends on basis behavior. Moreover, since futures prices are established for a particular quality at a specific location, knowledge of basis relationships is important in discovering spot prices in other locations and for other qualities.

Additional research on the price effects of futures trading is justified by the increased use of futures in establishing spot prices for farmers and by the general concern about futures markets. Does the behavior of cash prices differ with and without futures? For instance, are the random component, the seasonal component, and the like reduced (or increased) by futures trading? Or does the existence of a futures market result in speculative runs in prices that otherwise would not have occurred? An important subject from the viewpoint of market regulation is the ability to identify effects of price manipulation. This is likely to be a difficult research question, but perhaps a model of normal price behavior can be developed as a standard of comparison.

The pricing of services can be a difficult problem, and futures markets may have an increasing role to play in this area (Paul 1979). The difference between the current price of a contract for future delivery and the current cash price has long been recognized as a price of storage. In an analogous way, prices of feeding services for cattle (or other livestock products) and processing margins for soybeans can be defined by using appropriate futures prices. While some attention has been given to analyzing prices of services, more seems justified. Indeed, even the price of storage has received little empirical analysis under 1970 conditions.

Transmission of Prices. Farm-retail margins are not commonly associated with the pricing mechanism. The allegation of a margin being too large is usually associated with monopoly power. But one issue related to price discovery is the efficiency of transmission of prices through the marketing system. If the farm price increases, how rapidly is this reflected in retail prices? Is the response at the retail level more or less than that justified by the farm-level increase? Are the responses to increases and decreases different? Can empirical models of price transmission be developed that will be useful in monitoring market performance? Given the pricing policies of food processors and retailers, are prices providing adequate guides for producer and consumer behavior?

Price controls have been applied occasionally to food and fiber products, and with high rates of inflation they may be applied in the future. Agricultural economists ought to be better prepared to analyze the economic consequences of price controls, although the empirical base for study is limited. Nonetheless, analytical models can be sharpened, and experiences such as those in the early 1970s can be studied. For example, if retail or wholesale prices are controlled, what are the implications for marketing margins, for farm prices, for future supplies?

Several methodological problems make the study of margin behavior difficult. To study the transmission of prices, the observations probably should pertain to weeks, but sometimes only quarterly data have been available. It may be possible to examine the problem with the monthly data that are available (e.g., Heien 1980). It is also difficult to obtain short-time observations for the prices of marketing services to hold other things constant in the study of the transmission of farm price changes. In addition, since the nature of the transformation from the farm to the retail product varies so widely from product to product, quite different models should be built for each commodity. It seems unlikely that similar models can be applied to such diverse products as eggs (little processing) and bread (highly processed with a small farm component). Moreover, the market structure and pricing mechanisms vary greatly from product to product. Presumably a large number of different commodities would have to be analyzed.

Summary. Researchable areas include the analysis of new pricing mechanisms such as computerized exchanges, the longer standing issues of declining number of buyers and farmer bargaining, arrangements that permit timing of price discovery to differ from timing of delivery (e.g., futures markets), effects of thin markets, and transmission of prices through the marketing system. Specific research questions in each area often center on price behavior and costs of price discovery. Another important question is the welfare effect of alternative pricing institutions. In the area of methodology, we need improved understanding of the variables influencing development and evolution of pricing mechanisms, an improved classification scheme for pricing mechanisms, and better ways of measuring performance.

The list of research problems is not exhaustive, and with time the relative importance of topics will change. No author can identify all the important research needs of the future. It is hoped the reader has been stimulated not only to undertake research on some of the topics outlined in this chapter but also to add to the list of topics. In addition, the chapters on industrial organization, alternative control mechanisms, and information and data systems discuss research issues that are especially complementary to this chapter.

REFERENCES

Akerlof, George A. 1970. The market for "lemons": Quality, uncertainty and the market mechanism. *Q. J. Econ.* 84:488–500.
Alchian, Armen A., and Harold Demsetz. 1972. Production, information costs, and economic organization. *Am. Econ. Rev.* 62:777–95.
Bohall, R. W. 1972. Pricing Performance in Marketing Fresh Winter Lettuce. USDA, Mark. Res. Rep. 956.
Campbell, Gerald R., and Marvin L. Hayenga, eds. 1978. Vertical Organization and Coordination in Selected Commodity Subsectors. NC-117 Work. Pap. 20, University of Wisconsin, Madison.
Chayat, M., O. D. Forker, and D. I. Padberg. 1974. An econometric determination of the welfare impact of giving bargaining power to farmers: A case study of the egg industry. *Search* 4:1–43.

Coase, R. H. 1937. The nature of the firm. *Economica* 4:386–405.

Cochrane, Willard W. 1980. Some nonconformist thoughts on welfare economics and commodity stabilization policy. *Am. J. Agric. Econ.* 62:508–11.

Cox, Charles C. 1976. Futures trading and market information. *J. Polit. Econ.* 84:1215–37.

Eaton, David J., and W. Scott Steele, eds. 1976. Analysis of Grain Reserves, a Proceedings. USDA, ERS-634.

Fama, Eugene F. 1970. Efficient capital markets: A review of theory and empirical work. *J. Finance* 25:383–417.

Farris, Paul L. 1958. The pricing structure for wheat at the country elevator level. *J. Farm Econ.* 40:607–24.

Figlewski, Stephen. 1978. Market "efficiency" in a market with heterogeneous information. *J. Polit. Econ.* 86:581–97.

Forker, Olan D. 1975. Price Determination Processes: Issues and Evaluation. USDA, FCS Inf. 102.

Forker, Olan D., and V. James Rhodes, eds. 1976. Marketing Alternatives for Agriculture: Is There a Better Way? National Public Policy Education Committee Publ. 7 (13 leaflets produced by the New York State College of Agriculture and Life Sciences, Cornell University, Ithaca).

Frahm, Donald G., and Lee F. Schrader. 1970. An experimental comparison of pricing in two auction systems. *Am. J. Agric. Econ.* 52:528–34.

Garcia, Philip, and Olan D. Forker. 1975. Egg Pricing and Marketing Arrangements at the Producer Level in New York State. Cornell University AE Res. 75-3, Ithaca.

Garoyan, Leon, and Eric Thor. 1978. Observations on the impact of agricultural bargaining cooperatives. In Agricultural Cooperatives and the Public Interest, pp. 135–48. NC-117 Monogr. 4, University of Wisconsin, Madison.

Geyer, L. Leon, and Dale C. Dahl. 1978. Collective bargaining for farmers. *Minn. Agric. Econ.* 599:1–8.

Gray, Roger. 1964. Some thoughts on the changing role of price. *J. Farm Econ.* 46:117–27.

⸺. 1974. Grain Reserves Issues. Paper given at USDA Agricultural Outlook Conference, Washington, D.C., Dec. 9–12.

Grossman, Sanford J., and Joseph E. Stiglitz. 1976. Information and competitive price systems. *Am. Econ. Rev.* 66:246–53.

Haas, John T., Paul C. Wilkins, and James B. Roof. 1977. Marketing Slaughter Cows and Calves in the Northeast. USDA, FCS Res. Rep. 36.

Hassler, James B. 1953. Pricing efficiency in the manufactured dairy products industry. *Hilgardia* 22:235–334.

Hayami, Yujiro, and Willis Peterson. 1972. Social returns to public information services: Statistical reporting of U.S. farm commodities. *Am. Econ. Rev.* 62:119–30.

Hayenga, Marvin L. 1971. Hog pricing and evaluation methods—Their accuracy and equity. *Am. J. Agric. Econ.* 53:507–9.

⸺, ed. 1979. Pricing Problems in the Food Industry (with Emphasis on Thin Markets), NC-117 Monogr. 7, University of Wisconsin, Madison.

Heien, Dale. 1977. The cost of the U.S. dairy price support program: 1949–74. *Rev. Econ. Stat.* 59:1–8.

⸺. 1980. Markup pricing in a dynamic model of the food industry. *Am. J. Agric. Econ.* 62:9–18.

Helmberger, Peter, and Rob Weaver. 1977. Welfare implications of commodity storage under uncertainty. *Am. J. Agric. Econ.* 59:639–51.

Holder, David L., and Ralph E. Hepp. 1978. Cooperative Strategies for the Pork Industry. USDA, Mark. Res. Rep. 1097.

Hoos, S., and R. E. Seltzer. 1952. Lemons and Lemon Products: Changing Economic Relationships, 1951–52. University of California Agric. Exp. Sta. Bull. 729, Davis.

Hurwicz, Leonid. 1973. The design of mechanisms for resource allocation. *Am. Econ. Rev.* 63:1–30.

Jamison, J. A. 1966. Marketing orders, cartels, and cling peaches: A long-run view. *Food Res. Inst. Stud.* 6:117–42.

Jesse, E. V., and A. C. Johnson, Jr. 1970. An analysis of vegetable contracts. *Am. J. Agric. Econ.* 52:545-54.

Johnson, A. C., Jr. 1973. Effects of Futures Trading on Price Performance in the Cash Onion Market, 1930-68. USDA, Tech. Bull. 1470.

Johnson, Ralph D. 1972. An Economic Evaluation of Alternative Marketing Methods for Fed Cattle. University of Nebraska Agric. Exp. Sta. SB 520, Lincoln.

Just, Richard E. 1977. Theoretical and empirical possibilities for determining the distribution of welfare gains from stabilization. *Am. J. Agric. Econ.* 59:912-17.

Kuehn, J. P. 1971. Costs and Efficiencies of Model Livestock Auctions in West Virginia. West Virginia University Agric. Exp. Sta. Bull. 606, Morgantown.

Larson, Arnold B. 1960. Measurement of a random process in futures prices. *Food Res. Inst. Stud.* 1:313-24.

Marion, Bruce W., ed. 1976. Coordination and Exchange in Agricultural Subsectors. NC-117 Monogr. 2, University of Wisconsin, Madison.

Marion, Bruce W., and Charles R. Handy. 1973. Market Performance: Concepts and Measures. USDA, Agric. Econ. Rep. 244.

Morgenstern, Oskar. 1972. Thirteen critical points in contemporary economic theory: An interpretation. *J. Econ. Lit.* 10:1163-89.

Nelson, Glenn, and Tom H. Robinson. 1978. Retail and wholesale demand and marketing order policy for fresh navel oranges. *Am. J. Agric. Econ.* 60:502-9.

Paul, Allen B. 1979. Some basic problems of research into competition in agricultural markets. *Am. J. Agric. Econ.* 61:170-77.

Piggott, R. R. 1976. Potential gains from controlling distribution of the United States apple crop. *Search* 6:1-21.

Plato, Gerald, Richard G. Heifner, and John W. Murray. 1978. The Social Costs of Instability in the U.S. Wheat Industry. Paper presented at the American Agricultural Economics Association meetings, Blacksburg, Va., August 7-9.

Powers, Mark J. 1970. Does futures trading reduce price fluctuations in the cash markets? *Am. Econ. Rev.* 50:460-64.

Raikes, Ronald. 1976. Feeder-cattle marketing channels and exchange arrangements. In B. W. Marion, ed., Coordination and Exchange in Agricultural Subsectors, pp. 102-12. NC-117 Monogr. 2, University of Wisconsin, Madison.

Rhodes, V. James. 1976. Exchange arrangement: Fed beef cattle. In B. W. Marion, ed. Coordination and Exchange in Agricultural Subsectors, 218:113-22. NC-117 Monogr. 2, University of Wisconsin, Madison.

Robinson, K. L. 1960. Possible effects of eliminating direct price support and acreage control programs. *Farm Econ.* October:4813-20.

_____. 1975. Unstable farm prices: Economic consequences and policy options. *Am. J. Agric. Econ.* 57:769-77.

Rogers, George B. 1970. Pricing systems and agricultural marketing research. *Agric. Econ. Res.* 22:1-11.

Rogers, George B., and Leonard A. Voss. 1969. Pricing Systems for Eggs. USDA, Mark. Res. Rep. 850.

_____, eds. 1971. Readings on Egg Pricing. University of Missouri, College of Agriculture, Bull. MP 240, Columbia.

Rothschild, Michael. 1973. Models of market organization with imperfect information: A survey. *J. Polit. Econ.* 81:1283-1308.

Schultz, T. W. 1949. *Production and Welfare of Agriculture.* New York: Macmillan.

Smidt, Seymour. 1965. A test of the serial independence of price changes in soybean futures. *Food Res. Inst. Stud.* 5:117-36.

Smith, Roy J. 1961. The lemon prorate in the long run. *J. Polit. Econ.* 69:573-86.

Thomsen, F. L., and R. J. Foote. 1952. *Agricultural Prices,* 2nd ed. New York: McGraw-Hill.

Telser, Lester G., and Harlow N. Higinbotham. 1977. Organized futures markets: Costs and benefits. *J. Polit. Econ.* 85:969-1000.

Tomek, William G. 1980. Price behavior on a declining terminal market. *Am. J. Agric. Econ.* 62:434-44.

Tomek, William G., and Kenneth L. Robinson. 1972. *Agricultural Product Prices*. Ithaca: Cornell University Press.

Trierweiler, John E., and James B. Hassler. 1971. Measuring efficiency in the beef-pork sector by price analysis. *Agric. Econ. Res.* 23:11–17.

U.S. Department of Agriculture. 1961, 1976. *Resume*. Statistical Issues, Packers and Stockyards Administration.

_____. 1978. Beef Pricing Report. Agric. Mark. Serv.

Wall Street Journal, Aug. 30, 1978, p. 28.

Ward, Clement E. 1977. Analytical Framework for Price Discovery Research. Unpubl. manuscript.

Williamson, Oliver E. 1971. The vertical integration of production: Market failure considerations. *Am. Econ. Rev.* 61:112–34.

WALTER J. ARMBRUSTER
JOHN W. HELMUTH
WILLIAM T. MANLEY

DATA SYSTEMS IN THE FOOD AND FIBER SECTOR

THE DATA SYSTEM is an important element of the U.S. food and fiber sector, providing a basis for efficient market operation. While data are used by public officials (macro) in measuring the size of the agricultural sector and monitoring its performance and developing policies affecting it, one of the most important functions of data is to serve private (micro) decision makers. Data lie at the heart of the crucial agricultural price discovery process.

Changes that have occurred in marketing methods, structure, and pricing practices have affected data accuracy, reliability, accessibility, timeliness, and comparability. The nature and extent of these fundamental changes call for reexamination of the food and fiber data system.

The preceding chapter has dealt with developments in pricing mechanisms that have important implications for the agricultural data system. The purpose of this chapter is to examine the functions served by data, concepts behind past research, results of that research, and promising areas for future economic research on data systems for the food and fiber sector.

SCOPE OF DATA SYSTEMS RESEARCH

Data are the numbers that measure or describe economic phenomena and

WALTER J. ARMBRUSTER is Associate Managing Director, Farm Foundation; **JOHN W. HELMUTH** is Agricultural Economist, Committee on Small Business, U.S. House of Representatives; and **WILLIAM T. MANLEY** is Deputy Administrator, Marketing Program Operations, Agricultural Marketing Service, USDA. The authors are indebted to the other authors of this volume, particularly Bill Boehm, Steve Heimstra, and Bill Tomek for helpful suggestions on earlier drafts of this chapter. Judy Brown, graduate student at Michigan State University, also provided valuable comments. Harold Breimyer offered review comments that were very useful in revising the chapter.

form the basis for economic decisions. Data are analyzed to convert them into information for decision makers, whether they are firm managers or public policy officials. Bonnen (1977) provided a useful definition differentiating data from information. "Data are not information. An information system includes not only the production of data but also analysis and interpretation of these data in some purposeful policy decision or problem solution context. The demand for data is generated by the need to make decisions on problems, but decision makers rarely use raw data."

Our focus is primarily on data series and economic research related to data appropriate to describing economic phenomena. The term "agricultural data system" is used for convenience to mean "data systems related to the food and fiber sector." The focus on data series and improvements leaves an extremely broad topic to be covered. Economic research questions relate to data collection, data distribution, costs and benefits of data systems, and evaluating existing data series in the face of changing agricultural and marketing structures. No matter how the scope of this chapter is delineated, other areas of concern could be added or substituted for some that are treated.

This chapter emphasizes the role of economists in researching questions dealing with the importance of a publicly funded data system, concepts underlying the agricultural data system, dynamics of the data system and user needs associated with the changing market structure, trade-offs between precision of measurement and costs involved, and economic outlook and analysis of data.

IMPORTANCE OF RESEARCH ON THE AGRICULTURAL DATA SYSTEM

The need for research related to various aspects of the data system arises from continuous calls for improvement. Commodity interests and researchers seek refinement in existing data series and gathering of additional series. Policymakers need data specifically related to the programs they are charged with administering. Usually a problem is identified, then a call for relevant data is heard. Agricultural data are seldom collected as an end in themselves.

Agricultural data gathering has increased rapidly along with that for other segments of the economy. While the volume of data gathered on agriculture has long been substantial, additions or expansions are costly. Therefore, it is important that additional data be collected where it will provide the most return for the dollar invested, and that appropriate deletions or revisions be made regularly. Research by economists should help determine the trade-off between precision of measurement and costs involved and adjustments needed to keep abreast of the changing economy. The ultimate goal of research and changes in data is to improve resulting decisions derived therefrom. Care should be exercised that discontinuations or revisions do not make long-term comparisons impossible..

The agricultural data system is largely provided from public funds and this is necessarily the case in a relatively unconcentrated industry such as production agriculture. Decision makers who need the data may be in private firms or in the public domain. And the public, including potential participants in the marketing system, has a need for data to facilitate their understanding of this important segment of the economy.

While there may be doubt about how much data is necessary, the amount actually collected is constrained by economic trade-offs; i.e., how much additional benefit is gained from providing an additional set of observations? There are also constraints on data collection provided by considerations such as concern over the invasion of privacy and respondent burden. But the role of data as a public good and the necessity of a publicly funded mechanism for providing it are sometimes overlooked.

As the market becomes more concentrated or less public, it becomes increasingly less clear where the line is to be drawn on the amounts and types of data collected. It is still relatively clear that data needs to be collected by public bodies, given its public good characteristics. Large organizations are capable of generating their own data series to use for their decision processes. However, since the data thus collected are private goods, they may give distinct advantages to large corporations or bodies capable of supporting a data collection function relative to the remainder of the society who would not have access to such data (Hirshleifer 1973; Arrow 1974; Reimenschneider 1977). Thus increasing levels of concentration may work to the detriment of the public data system.

THE PRESENT DATA SYSTEM

The current agricultural data system consists of publicly supported data gathering organizations, increasingly augmented by private sources and publications. The broad range of the USDA statistical series is described in an eleven-volume publication focusing on agricultural prices and parity; agricultural production and efficiency; gross and net farm income; agricultural marketing costs and charges; consumption and utilization of agricultural products; land values and farm finance; farm population and employment; crop and livestock estimates; farmer cooperatives; market news; and foreign production, consumption, and trade of agricultural products (USDA 1969–72). Current market prices and quantities are reported by the USDA's Agricultural Marketing Service through the Federal-State Market News Service. Reports are issued orally via radio, television, and telephone answering devices or in written form for bulk mailing or release to newspapers or other printed sources. These current data cover most commodities on a timely basis, tracking market activities within recent days or weeks. More detail on the coverage of Federal-State Market News Service reports can be found in the eleven USDA (1969–72) publications.

Monthly, quarterly, or annual data series on prices and quantities for most agricultural commodities are published through the USDA's Economics and Statistics Service (ESS). These data series are not reports of actual transactions, as in the case of the *Market News,* but rather represent aggregate quantities and average prices for various locations or regions. Thus there is a conceptual difference in the way *Market News* and ESS data series are generated and consequently how they should be interpreted. For further detail on the content and concepts behind the ESS data series, see USDA (1969–72).

Some data, along with substantial interpretive information, are provided by situation reports from the ESS. These periodically released reports for most major commodities explain the current situation and the outlook for the coming 3–12 months.

A number of data series not related to specific commodities are also maintained by the ESS. These statistics track farm income, marketing margins, the marketing bill, the cost of a market basket of food, and similar aggregate measures important for monitoring activity of the food and fiber sector in total and for analyzing policy concerns. More detailed descriptions of the ESS data series are also contained in the eleven-volume USDA report.

The regular USDA data series, provided on a continuous basis at considerable cost, are widely known. In addition, the USDA reports foreign trade data and publishes one-time or nonrecurring reports on special topics. Such statistical reports also provide a wealth of data for use in the agricultural decision process.

Other government agencies also publish data series useful in agricultural decisions. Of particular note is the periodic Census of Agriculture as well as reports from the Federal Reserve Board, the U.S. Department of Commerce, and the Farm Credit Administration. Government-industry organizations (e.g., marketing order administrative committees) also generate and publish substantial amounts of data for some commodities.

Private market news sources such as *Cattle Fax,* the *Yellow Sheet,* the *Pink Sheet,* and the *Egg Clearing House* are well known. Trade magazines provide a major or sole source of information on a number of agricultural marketing activities; e.g., information on packs and stocks of processed fruits and vegetables are available only from trade magazine sources. Trade associations provide an important function in gathering data for certain types of marketing activities with which their members are concerned.

Most such private data sources serve a complementary role to the publicly generated series. In fact, a number of private services publish data mostly gathered by the USDA and supplemented with their own. Without public funding for the very expensive data gathering function, many private organizations would not be able to sell data and information services at the modest prices now charged. In fact, most would probably not exist.

Much useful data for agricultural decision making are generated by the commodity futures markets. While these institutions are private organizations, the public nature of their activities and the data generated make them important sources.

A number of private sector efforts are concerned with forecasting agricultural sector activity. Credit institutions often provide short-term data analyses through periodic newsletters. Firms using econometric models have received much attention. Subscription to the modeling results are in some cases maintained by private firms primarily for the convenience of the data provided in conjunction with these modeling efforts. Sometimes overlooked is the fact that the USDA makes extensive use of models in data projection work. The success of such data forecasting efforts for the agricultural sector, whether by private organizations or the USDA, is largely unanalyzed. For a general assessment of econometric modeling efforts for individual commodities and sectors of the agricultural system, see King (1975), Johnson (1977), and Just (1977).

Data System Concepts. The concepts underlying the agricultural data system may be thought of in micro- and macroeconomic terms. At the micro level, the individual farm firm and individual marketing decisons in the perfectly competitive setting are the concepts upon which the data base was established. At the macro level, data series aid decisions affecting major segments of the agricultural system and provide input for policy analysis, assuming aggregation of the competitively established individual decisions.

The farm firm concept led to establishing data series based on individual firm decisions. Agricultural production occurred on numerous farm operations. Each responded independently to data reported on production, marketing volume, and prices. Having made a production decision, the individual farmers then sold harvested quantities through market channels such as terminal markets, local auctions, buying stations, local elevators, or whatever was appropriate to the particular commodity. The buyers involved were generally making individual decisions based on their knowledge of nearby markets and data reported from similar but more distant markets.

The aggregate impact of these numerous individual producer and marketing firm decisions was reflected in national and some state or regional data series. Market prices and quantities moving through central terminal markets could be observed. Production estimates could be obtained by sampling individual farms and quantities marketed through the relatively limited outlets; farm income could then be meaningfully estimated from these estimates of the sale price and quantity produced.

In the setting of numerous individual producers and buyers, data facilitate market operation by making better informed pricing decisions possible. Pricing efficiency in the marketing system requires ready access to data (Houck 1977). The efficient discovery of price at a given location or for a given time would be greatly impaired without knowledge of prices discovered in other locations or at previous times. Accurate, unbiased information is a necessary input for numerous individual decisions, which together comprise the market. Accurate market decisions are important to each industry within the U.S. agricultural economy and to world trade. Pricing efficiency leads to efficient resource allocation, a desirable social goal.

As marketing methods and the market structure change, they affect the need for and collection of data. The shift away from central markets makes the sources of data much more difficult to identify, and the data may become less comparable or additive. While in a central market location it is possible to contact the same pricemaker several times throughout a day; transactions occurring elsewhere may be infrequent, widely scattered, and subject to different influences. This increases the cost of data gathering and probably increases variability of the data reported. Even in a central market setting, data reported are at best representative of specific transaction prices, since it is not possible to observe and record all transactions. In different locations involving separate observations by those recording information, several factors are likely to introduce "noise" into the reported data. Prices established at different locations represent a different bundle of characteristics such as produce form, location, and quality. Ideally these will be accounted for in the process of reporting, but it is doubtful that locationally dispersed price reporting is as accurate as multiple observations at one central market location. As market exchange mechanisms become more individualized through contracts negotiated for each sale or as prices become more rigid, it becomes increasingly difficult to report comparable data that are useful for other contract negotiations. The quality of the commodity involved and the location may be adequately described, but the other contract terms may vary enough to make reported data less useful as measures of central tendency. Yet availability of data as indicators of the effects of market forces remains extremely important (Hayek 1945).

An optimal data system may be defined as one that is dynamic in adjusting to changing agricultural production and marketing conditions. It should provide unbiased estimates or measures of economic activity (timely enough for the decision-making process) in enough detail to facilitate the microlevel decisions by numerous entities yet in the aggregate reflect the macrolevel relationships between the agricultural economy and major segments of the general economy. More detailed discussion of the concept of an optimal data system is presented later in this chapter.

Assessment of Data System Research. Extensive reviews of data series, their underlying concepts, structural changes in the food and fiber system that have impact on the relevance of data, and a detailed evaluation of research related to data have been comprehensively treated in the literature cited below. We use this previous work as a point of departure for recommendations concerning future research.

Upchurch's (1977) review of agricultural economic information systems considered current commodity statistics, farm data, production resources and costs, marketing-related data, data on farm-related business, food consumption and nutrition, and international data. He provides an exhaustive list of references to analyses of data availability and needs, methodological and conceptual problems, and theoretical and applied analyses of the data system. Some of the references deal with broad measures, while others are commodity specific. Trelogan et al. (1977) have reviewed technical developments in agricultural estimates technology and include references to research upon which changes in the USDA data gathering methodology were based.

The American Agricultural Economics Association (AAEA) through its Committee on Economic Statistics has generated interest in data systems and sponsored work related thereto. Bonnen (1975), Hildreth (1975), and Brandow (1976) have evaluated questions related to the validity of various types of agricultural data. Implicit in their analyses of the data system were evaluations of the related research accomplishments and failures. One consistent theme of these reports is the increasingly obsolete conceptual base for many of the currently used statistical series. They developed a number of recommendations, some of which were general and dealt with conceptual problems while others were detailed suggestions for improving or replacing current series.

Focusing on agricultural market data collection, dissemination, and its use in decision making, Kroupa et al. (1976) provide an annotated bibliography of available publications from 1955 through 1975. The reports listed have a heavy emphasis on application and operation rather than on theoretical concerns.

The survey articles and task force reports cited above provide extensive evaluation of strengths and weaknesses in the data systems for the food and fiber sector. Drawing upon them, it is possible to identify a number of research areas that hold promise for helping to improve the data system.

Though general recommendations have been developed for improving the data system to keep abreast of current needs, actual changes in data series have not been rapid. This may be the result of the difficulty of designing satisfactory series to reflect changes in the food and fiber sector. To the degree the sector has evolved away from competitive markets, the call for new or improved data will not be likely to emanate from within the sector, reflecting a desire to keep information secret and suggesting some degree of monopoly power. New data series relating to new marketing methods may be resisted by the established operators of the sector, who may lose an advantage if data are provided on new elements of the system. This environment highlights a sense of urgency for additional research to identify needed changes and to translate that need into practical approaches for data gathering.

CHANGING MARKET STRUCTURE AND ITS
IMPACT ON THE DATA SYSTEM

Some of the agricultural data base is more accurate today than in the past because certain biological or physical variables are slow to change, but sampling methods used to measure them are subject to continual refinement. For example, Trelogan et al. (1977) point out that implementation of area-frame sampling was completed by 1965 and provided a sound basis for sampling throughout the United States. In combination with objective yield models based on actual plant measurements and small sample plots, the accuracy of crop forecasts has greatly improved. The increased capabilities in measurement and data processing combined with conceptual stability has increased the quality of such data. But that is not generally the case for the nonbiological segment of the agricultural data system.

Changes occurring in the food and fiber sector in the past several decades have made the agricultural data system increasingly obsolete (AAEA 1972; Bonnen 1975, 1977). The definition of the farm firm is the basis for most of the agricultural statistics, but the concept has become less useful relative to the food and fiber structure that has emerged. Sundquist (1970) concludes that while research pronouncements have long recognized the fact, our statistics have failed to explicitly recognize "that the equivalences assumed between a place, a farmer, or a marketing firm and a decison making unit or an operating firm in the food and fiber industry no longer are valid."

Increasingly Obsolete Data Systems. Obsolescence of the data system may arise for two reasons: rapid changes in organization and structure of the food and fiber industry may outpace changes in the data system, and shifts in the agenda for both public and private food and fiber policy raise new questions for which the data system is expected to provide input (Bonnen 1977).

Elsewhere, Bonnen (1975) argues that where extreme changes occur in the organization and nature of the food and fiber sector, traditional concepts underlying production and marketing statistics become most obsolete. In industries where contracting and vertical integration have gone the farthest (e.g., broilers, eggs, processing vegetables), the problems are particularly acute. For beef, price discovery has been more difficult and data more ambiguous as negotiated transactions have been replaced by formula trading and the structure of the industry has become increasingly concentrated (U.S. Congress 1980).

Conceptual obsolescence brought by changes in the agenda of food and fiber policy is illustrated by the fact that the environment, energy, and world food supplies have all become important parts of the agricultural policy agenda (Paarlberg 1978). Bonnen (1975) argues that we have mostly redefined or "made do" rather than conceptually redesigning the older data system to deal with the new questions. Existing data systems tend to take on a life of their own. Once established, it is difficult to discontinue a data series to free resources for a new, possibly badly needed, series.

The AAEA Committee on Economic Statistics (1972) states that our data have become obsolete because the system is not built on concepts matching the institutions that have developed since the data series were originally established. "There has been substantial shift toward functional specialization in the food and fiber industry, but our statistics have been built around the firms which performed the functions when the industry was conceptualized. . . . The need exists to distinguished between functions being peformed and the units of observation and to insure that the statistical system provides for effective monitoring of the functions."

Reimenschneider (1977) points out that "changes in economic structure toward greater concentration in and of itself need not cause additional problems in defining and measuring price as long as the same methods for arriving at the price between buyer and seller are used. However, this is often not the case." In fact, changes in the price discovery mechanisms are at the root of many concerns relating to our current data system. Some of the major changes in price discovery mechanisms, associated with changes in the marketing system, include increased individual negotiation, often involving contracts in some commodities; deferred and delayed price contracts for grain priced by formula (basis) off the futures markets; formula pricing, particularly for meats and eggs; group bargaining for price and other terms of trade; and privately administered prices, such as transfer pricing in integrated operations.

For example, changes in the egg industry pricing structure have stimulated much research on alternative pricing mechanisms and related questions concerning information and price data. Rogers (1971) points out, "Many problems with and questions about the prevailing egg pricing system were due to inadequate market information. And, the need to improve our total market information program exists regardless of the exact pricing methods which emerge in the near future."

As market structure changes and the marketing system involves more internal transfer within vertically or horizontally integrated organizations, data on the relevant market segments necessary to monitor performance become increasingly difficult to obtain. For example, as contracting replaced open market mechanisms, data gathering became more difficult (Paul et al. 1976; Helmuth 1977). And, "as firms become fewer, larger, and more highly integrated, data . . . assume the status of trade secrets. At some point in the process of integration there emerges a powerful incentive to withhold" (Heifner et al. 1977). Many publicly released company reports contain microlevel data that are so aggregated that identification of relevant information is impossible. Proposed requirements for product line reporting meet with stiff opposition from those involved. Given variation in company practices, it may be nearly impossible to obtain identical bundles of activities reported, since accounting systems are designed for internal company use in making microlevel decisions.

Changes in price discovery mechanisms lead to problems with nonprice data. Data on volumes, qualities, income, costs, and other economic parameters related to marketing decisions and policies also become obfuscated. Houck (1977) identifies a direct tie between prices and these other economic parameters, arguing that prices are indicators of performance and measures of change in performance: "For this purpose, prices are viewed as per-unit monetary values by which to aggregate the quantity and worth of goods and services in the economy. Then calculations can be made and conclusions can be drawn about the size and distribution of incomes, costs, expenditures, gross ouput, inventories, taxes, etc."

As these changes in the marketing system have occurred, alternative pricing mechanisms have developed. The use of forward pricing developed with the growth of contracting. Delayed pricing is another increasingly popular mechanism used in conjunction with contracts or open market sales involving informal contracts (Paul et al. 1976). Transfer pricing has become a necessity as cooperatives and other firms vertically integrate through a number of functions.

The agricultural and food system continues to be faced with a changing marketing structure and shifting distribution of market power compared to the predominantly central, open auction markets of earlier times. Objective data are important for monitoring the changing marketing system. Potential entrants, competitors, suppliers, and purchasers are interested in determining levels of profits, costs, volumes traded, and other nonprice characteristics for any particular agricultural industry. Analysts wishing to determine the distribution of benefits and costs throughout the economy need accurate data, particularly for the impacts of extramarket regulations, whether economic or relating to health, safety, and the environment. Monitoring market performance for the effects of organizational structure and control requires detailed data.

Accurate data are needed for macrolevel policy analysis. In many cases, the same data serving microlevel functions also are the basis for analyzing policy alternatives. One example, related to data gathering and policy analysis for government decision making, involves the international grain trade. Difficulties have been encountered because knowledge of total export commitments are lacking, poor data on foreign production are used in analyzing supply-demand conditions, and marketing channels have shifted faster than price reporting practices have been adjusted (Heifner et al. 1977). Efforts to improve grain export commitment data are plagued by an *ex ante–ex post* problem that originates because of the basic nature of the international grain trade. Timely *ex ante* commitment data for policy decisions inherently reflects over- or undercommitments by grain firms in response to perceived worldwide supply-demand conditions and thus do not provide an accurate picture of actual export flows. On the other hand, accurate *ex post* data on actual shipments often are not timely enough for policy decisions and certainly not for microlevel selling decisions. Regardless of how well export commitment data are refined, they can only serve as an estimate of actual *ex post* flows (USDA 1979b).

In addition, the export commitment data that are reported by the USDA only reflect sales by firms operating within the United States. Since the inception of the USDA reporting system in 1973, international grain firms have shifted export sales activity to their overseas affiliates to avoid the export reporting system (USDA 1979a). Thus the grain markets represent a situation where voluminous data are available about U.S. grain supplies, voluntarily supplied by U.S. farmers in cooperation with USDA, while only sketchy and unreliable data are available on the supplies from other grain producing countries, and worldwide demand data are almost totally shrouded in secrecy.

The basic macrolevel economic accounts associated with the food and fiber sector are also significantly influenced by the changing market structure (Carlin and Handy 1974):

> The farming subsector is currently viewed as a single national family farm for accounting purposes. The national family farm encompasses both household and business activities. This concept provides the basic format for the farm income series and the balance sheet for the farming sector—all of which were developed in the early 1940's. Processing and distribution activities associated with domestically produced food are monitored with the marketing bill . . . a series designed to explain the difference between consumer expenditures for domestically produced food and its farm value. These three series cover the current aggregate economic activity of what we call the food and fiber sector. Although these series have been and are being improved upon, they still maintain the basic generic properties given them at conception.

While there are problems with the national family farm and the marketing bill concepts, they remain useful and provide the basis for data series that answer many questions. But Carlin and Handy feel that they do not accurately portray the size, structure, and performance of the food and fiber sector and its role in the national economy. Thus they do not adequately deal with the questions that increasingly are being asked.

Economic accounts for the food and fiber system serve two primary purposes: to show the size of the system and its contribution to national product and income and to provide insight to the structure and activities of the food and fiber sector, including welfare measures that bear on farm policy.

Weeks et al. (1974) argue that the basic economic accounts for the food and fiber system should provide a comprehensive description of the nation's food and fiber economy. They propose a system of accounts for the entire economy as the most efficient method of providing aggregate value statistics. The food and fiber system economic accounts should be consistent with the general national accounts. Quantity, price, and population data should be provided to supplement the data on economic stocks and flows.

The interdependencies within the data system between the food and fiber system and the rest of the private economy, which produce new demands for data, are also one of several sources of generic stress on the current data system identified by Weeks et al. Such new demands, in conjunction with data needs for public policy decisions, create significant total stress on the food and fiber data system. A second generic stress that has been identified is that of having access to basic data. This problem becomes more difficult as data increasingly come from larger firms who have a vested interest in secrecy and as the data requests focus more on financial conduct and performance. Finally, even after an effective optimal data system is identified, generic stress arises from the need for committing adequate resources to provide a data base.

Weeks et al. also identify some existing, specific stresses on the current data system. These include lack of public data in a form readily useful by our current data system, changing time frames for which data are desired, and fitting the agricultural data system components into the broader and expanding federal data systems. There is a need to retain flexibility through changes in the data system, while still preserving the basic conceptual properties of aggregate measures.

Particular care is called for when using the national economic accounts to measure the well-being of farm producers. For example, Guebert (1978) points out that the date base underlying the farm income estimates can only support aggregate income measures. These farm sector data are best used in the spirit of the national accounts. Particular care must be exercised in making any per farm representation, since there are continuous changes in the resource mix of family labor, capital, and management.

An Optimum Data System. As the changing market structure of agriculture and changes in operations of the food and fiber sector place increasing stresses on the current data system, a new optimum data system should be developed. Such an optimum system must be defined in terms of the purpose to be served, concentrating on data generation and distribution. Clearly, a "perfect" information system assumed in the perfectly competitive model of economic analysis, where all participants are fully informed, would be infinitely costly. The institutional complexities of powerful firms dealing with many commodities probably make detailed descriptive statistics unattainable, at least on a voluntary basis. An optimum data system must be specified in terms of objectives to be reached subject to various constraints. An optimum does not imply one comprehensive system but rather a number of compatible data series that make up meaningful segments of the whole.

An optimum data system faces tremendous demands. It must provide data that are unbiased, timely enough to be useful in the micro decision-making process, and in sufficient detail to provide the basis for decisions by numerous entities and in its aggregate must reflect the relationships between the agricultural economy and major segments of the general economy. Further, it needs flexibility and a mechanism to assure continuous adjustments to keep current with the ever-changing economy. However, the virtues of continuity in data must also be weighed in determining when changes are warranted. For each separate purpose, different data may be needed. At the least, the end purpose may affect the cooperation necessary to voluntarily obtain data.

Thus an optimum data system could focus on such private market objectives as maximizing a firm's profits or the utility of consumers and the necessary data for their decisions. But from the perspective of public decisions and the good of society, it is important to consider broader objectives and constraints. Less detailed data may be needed in that case than for individual decisions. The question of who needs the data for what purposes must be weighed relative to the costs of obtaining and processing it into usable information. Questions of the public right to know versus private interest in protecting secrecy become involved.

Information theory deals with certain aspects of the problem of an optimal data system. For example, Ozga (1960) deals with the process of the equilibrium level of knowledge, Stigler (1961) deals with the trade-off between search costs and gains from improved decision making, and Demsetz (1969) examines the efficient allocation of resources to information production; but none deals directly with the issue of defining an optimal data system for a major sector of the economy.

Eisgruber (1973), discussing computer supported managerial information and decision systems, identifies as desirable a system that would attempt to deliver information to decision makers so that they would select alternatives that lead to outcomes having higher net payoff than would otherwise occur. Eisgruber further argues that:

> The development of a comprehensive *concept* of an information system will not necessarily result in the development of *one operational* system. Instead, it is more likely that the comprehensive concept will lead to the conclusion that, due to differing circumstances, different *operational systems* are appropriate and that the pursuit of one definite operational system is not as productive as the concentration on individual improvements. However, the comprehensive concept will aid in deciding which of the many possible individual improvements should receive priority.

Realistically, the development of an optimal data system will not come about by scrapping the existing system and starting over. What is needed are some major changes in the underlying concepts upon which many of our data series are based, better matching of the data obtained with the theoretical concepts they purport to represent, elimination of some data series to free scarce resources, and development of new series to cover emerging areas of concern.

Upchurch (1974) suggests that present farm, marketing, and census data series must provide the basis for any new data system. But an internally consistent data system is needed to identify logical relationships within the agricultural sector and between it and the national economy. Data in the system should have common units of observation; be logically consistent; and be usable for projections of future levels, structures, and macroeconomic parameters. He also endorses the challenge of Trelogan et al. (1977) to agricultural economists to develop their specifications for more accurate data. While Upchurch (1974) focuses on economic data for the food and fiber industry, the new policy agenda for agriculture identified by Paarlberg (1978), which includes consideration of nutrition, health and safety, and other consumer issues, requires a broader approach. Upchurch recognized this need when he pointed out that "the relationships between the food and fiber industry and the rest of the economy are more intricate, more intimate, and certainly more critical for the industry itself, as well as for suppliers and consumers, than ever before."

What then should the objective function of an optimal agricultural data system for the food and fiber industry include? We propose that an optimal data system should:

1. Be dynamic in adjusting to the changing economic environment.
2. Be unbiased.
3. Be "designed to give empirical content to an image of the functional relationships that define the total system itself" (Dunn 1971).
4. Be designed from an overall systems approach to be internally consistent and minimize overlap or duplication.
5. Be based on industry subsectors defined by type of product or service, type of institution, function, geographical location, type of resources employed, or type of ownership.
6. Have units of observation and aggregates defined to provide linkages and consistency with accounts for other sectors.
7. Provide "the raw material for identification and diagnosis of problems . . . and permit evaluation of alternative courses of action" (Upchurch 1974).
8. Be timely enough to aid decision makers, both public and private, in making better micro- and macroeconomic decisions, thus facilitating market operation and resource allocation.
9. Provide detailed measures of value added; returns to resources; and commodity, service, and resource flows.
10. Be designed to collect the minimum amount of data necessary, while obtaining maximum accuracy.
11. Be based upon and designed to accommodate the uses of the data.

12. Permit monitoring of food system performance.

The development of such an optimal data system is constrained by:

1. Fixed resources in the short run.
2. A need for timeliness of data.
3. A need to insure confidentiality, balanced against the public's right of access to data.
4. The need to rely on voluntary cooperation in most areas.
5. A lack of ability to obtain data in many important areas (e.g., lack of product-line data in integrated firms).
6. Concern over "freedom of information" rules and fear of disclosure of data to competing firms.
7. Objections by interest groups to having data related to their activities made publicly available (e.g., certain food system performance accounts are now published in aggregate form but without enough detail to analyze intermediate market operations).

These objectives and constraints define general characteristics of an optimal data system. What is now needed is basic research to chart the course from the existing system toward the optimal.

PROMISING AREAS FOR FUTURE RESEARCH

Ideally, any research on the agricultural data system should be part of a unified, coordinated effort directed at an optimal system. Such an effort would require massive resources, and probably the best that can be expected are individual research projects directed at specific problems. Identifying characteristics of an optimal system may increase awareness of the issues and their relative importance. In the absence of a centrally directed, coordinated effort, certain specific research areas are identified in which research by economists is needed.

Research areas related to data systems are discussed under general topics touched upon in the preceding sections. A number of researchable areas related to pricing and collecting and disseminating data to facilitate the pricing and marketing process are identified. Some areas of research related to economic accounting for farm and marketing activities are also treated. Finally, research needs related to data involving current and emerging policy concerns are summarized.

A measure of the value of information is repeatedly called for by those evaluating budgets for data gathering and may represent the most pressing area in need of research. It is a difficult concept to measure and attempts to do so have met with limited success. Data collection is becoming increasingly costly. In specific instances, some estimates of the minimum value of information have been made (Moulton et al. 1974), but that approach is not definitive. Research efforts to identify ways to calculate the value of information are needed, so that data collection efforts may be concentrated in the most beneficial areas.

Whether by government officials or private firm managers, policy decisions are often made in an atmosphere of crisis or in quick-response situations. While thoughtful long-range planning is an important part of the decision-making environment, unforeseen exogenous events inevitably raise questions that demand immediate response. These questions often address the impact of an unforeseen development on some part or all of the food and fiber sector. For the data system to effectively serve this need, it must provide, on a continuing basis, a detailed description of the current economic environment that can serve as the basis for impact analysis. One essential element of such analysis is cost data on relevant factors from production through retailing. Unfortunately, while at one time cost studies received important emphasis in agricultural economics, they have fallen out of favor. This trend has been reflected back to the data system and has resulted in increasing scarcity of current cost data. Any movement to improve the data system should recognize this scarcity. Research is needed on cost components throughout the food and fiber sector, and on the best data for meaningfully measuring the costs of various activities in the sector and between it and other sectors of the economy.

Tomek treats the next three topics in Chapter 8, focusing on the implications for price discovery as well as pricing mechanisms and processes. However, their impact on the data system make them priority research topics deserving some additional attention here.

There is increasing concern over thin markets and their impacts on market efficiency, distributional equity, and the relationship between structure and performance. There continues to be growing evidence that thinness is a problem in the markets for more and more commodities. Hayenga (1978) edited papers from a symposium in which economists and food industry representatives considered some of the issues surrounding thin markets and research needed on the topic. Questions regarding data validity, availability, needs, and interactions between data and the pricing mechanisms require substantial research.

Electronic marketing is often suggested as a possible cure for thin markets. Electronic exchanges are designed to provide access to a market for a number of buyers and sellers at scattered locations without having to physically assemble the commodity, thus resulting in transportation efficiencies. Anticipated results of electronic marketing may be hypothesized from economic theory (Henderson et al. 1976). Given proper safeguards for participant identities, the transaction data available from an electronic exchange can be complete and almost instantaneous. The ability of such electronic exchanges to operate depends on the existence of good product description. The implications for needed data reporting versus the integral provision of data needed to operate the system require research attention.

Forward pricing and delayed pricing are two methods becoming more widely adopted for use in contracting for future delivery of production. Identifying what data are needed to assess and monitor this marketing mechanism, whether it can be obtained in the conventional manner, and how should it be reported are all areas requiring further research.

Supply and price data often receive disproportionate attention from agricultural economists compared to demand data. For example, a significant demand-side research opportunity exists to study the present international grain trading system from a pricing efficiency and resource allocation viewpoint by comparing the present data system with one that would provide accurate, timely, and complete demand data for U.S. export sales. U.S. producers sell in a market where supply data is public knowledge and export demand data is secret. Given the resultant price uncertainties from such a market, the total costs of incorrect resource allocation decisions need to be measured. Also, measurements are needed on the impact of mandating an accurate and timely export demand data system, and on aggregate revenue gains to producers from improved marketing decisions resulting from more complete demand data.

Mandatory reporting is sometimes suggested to improve thin markets. Presumably, obtaining data from all transactions, whether they be in readily observable markets or not, would stimulate efficiency in the marketing system. Data on quantities traded and inventories may be as important as price data, especially in thinly traded markets. The conflict of the public need to know versus the private individual's desire for secrecy is at the heart of the matter. Mandatory reporting becomes increasingly relevant to consider as markets become more closed or move further from competitive structures. The public right to know versus the need for protecting appropriate confidentiality must be carefully weighed. But mandatory reporting issues require research before sound decisons can be reached.

Improvement of early warning capability generally (inventories, international trade changes, demand shifts, government policy changes), domestically and for foreign countries, is an area requiring further research. Efforts to improve yield and crop production prediction through satellite technology are reported by the ESS. Advance warning of impending shortages in one area could stimulate additional production in other areas of the world. If such improvements in yield data are to be of maximum benefit, they also require open data exchange between countries on a more cooperative basis. Economic research on the impacts of such improved estimation on the level and stability of prices for major internationally traded commodities is needed. We have already mentioned the need for research into the costs and benefits from improved export sales reporting.

The proper role of futures markets as a data source needs further investigation. While these markets have become increasingly important, there are no published data series on their use by producers.

In addition, the usefulness of these markets to producers has not been adequately determined. However, a few research projects in the mid-1970s looked at the use of futures markets by producers and other elements of the marketing system (Leuthold 1975; Paul et al. 1976; Helmuth 1977). Studies have also attempted to identify random or nonrandom patterns of futures market prices, relate prices in the futures markets to the cash markets, and otherwise explain operations of the futures market (Gray and Rutledge 1971; Mann and Heifner 1976). There appear to be large gaps left untouched in analyses of the usefulness of futures markets and the proper role of futures market pricing in the marketing mechanism. Do futures market prices relate in a consistent manner to spot market prices for both storable and nonstorable commodities? Or should they relate in such a manner? Do futures markets provide a useful source of data for producers or only for processors and handlers? Do futures markets provide large firms a convenient vehicle to profit from inside information?

Retail price reporting is frequently cited as needed by those concerned with consumer prices. Results are reported by Devine (1978) for some Canadian experiments in reporting retail prices, but a number of questions deserve additional research attention. Are retail prices made sufficiently available through public newspaper advertising for the average consumer's needs? How detailed and specific should publicly reported retail price data be to be useful to the public? Are retail price series needed for public policy decisions and for outlook work to complete the data system from production to consumption? Should retail price reporting replace reporting on some of the intermediate levels of the marketing system? Will retail price reports change the long-run marketing strategies of retail food chains? If resources are a problem for data gathering, could retail price reporting be done accurately, given a large variety of sizes and brands?

The limited data available on retail food prices, which are used in calculating price spreads for individual food products, have been reduced by changes in the Bureau of Labor Statistics (BLS) Consumer Price Index (CPI). The changes improved CPI coverage of population and varieties of products available, but the BLS stopped publishing actual retail food prices for the United States and 23 cities. The loss of retail food price data directly affects the calculation of the ESS market basket statistics and price spreads for individual foods, which are derived from retail prices and farm values for individual products (Badger and Johnson 1978). New procedures will be used for calculating total market basket statistics, but future data series on individual product price spreads will depend on the BLS again publishing food prices. Are there other satisfactory sources of retail price data to use in the calculations, or is the price spread data important enough to warrant extra resources for obtaining it?

In addition to the price spread series, the other economic accounts discussed earlier also need research directed at improving their usefulness in the food and fiber sector. It has been argued that the farm concept should be replaced with an establishment concept and the marketing bill concept with a product concept. The establishment and product concepts emphasize the performance of the industry and its role in the national economy (Carlin and Handy 1974). They propose that a new set of economic accounts could be developed for agriculture based on a value-added system much like our current U.S. national income and product accounts. But implementing such a new and consistent accounting system will require detailed research on individual elements. Can satisfactory, measurable transactions or functions be identified to provide a meaningful set of data that improves our knowledge of the functioning of the food and agricultural system? Defining food system performance accounts and specifying meaningful content, while considering the political pressures they may need to withstand, is a research area that is difficult but necessary. Research related to the amount of product line data necessary for effective marketing system decisions, whether from voluntary or mandatory data collection, is needed.

There is a scarcity of data series measuring components of the marketing system between the farm level and the consumer level. Basic marketing bill data for intermediate levels is lacking or weak. This allows much room for speculation on the relationship of farm level prices to retail level prices but provides little basis for analytic answers to the questions raised.

Research is needed to evaluate the effectiveness of federally authorized marketing programs (e.g., research and promotion programs funded through check-offs of producer receipts, long-run impacts of marketing orders). Research and promotion programs provide information on the importance of a commodity and its nutritional value. However, little if any data is available for evaluating their true impact. Research may need to focus first on defining data to be collected as a basis for measuring program effectiveness.

More attention may be needed on the problems of "freedom of information" rules and the fear of primary survey respondents that competitors will obtain data. Can economists provide useful input into designing data collection systems in which the "freedom of information" concern can be mostly alleviated by the nature of the data collected?

The relationship of data to price behavior in the marketing system needs further research. What data is supplied and in how much detail could have some impact on price behavior. The trade-offs between accurate data and the costs of obtaining it and its possible interaction with the pricing system have received little research attention.

Questions about outlook data—and whether they influence prices, are counterproductive, or are neutral and necessary for functioning of the marketing and pricing system—also require research. Improving the specificity and accuracy of economic outlook projections deserves further analysis related to possible benefits expected from additional costs and efforts. The importance of integrating foreign with domestic outlook data to provide a comprehensive picture needs further research. Finally, the role of government versus that of private agencies using government-collected data in providing specific, short-term forecasts or outlooks could be researched to guide policy decisions.

In addition to the individual areas identified above and the research needed on a wide range of questions related to the economic accounts for the food and fiber sector, a number of new policy agenda items indicate the research needed to define appropriate data series for policy analysis. These are areas for which data are not currently available or at least do not readily match the questions being raised. New data series may be needed to provide a consistent input for policy analysis. Research will be needed to identify information needs and the basic data series that should be compiled for use in decision making, both private and public.

Some current policy items seem likely to be long-lived enough to justify such data gathering efforts. The impacts of regulation are hotly debated, often without much supporting data on costs and benefits associated therewith. Questions on food quality and safety are generally not well addressed with data currently available on quality of products by USDA grade categories. Nutrition policy questions may require new data series if physical scientists can develop analyses connecting nutrition with measurable physical data. Energy use in agriculture and its production by agriculture may be areas for which better data series are needed, possibly to monitor production of crops or use of residues to supplement current energy sources. New data series related to cultural and genetic treatments in integrated pest management policies may be important to supplement current data on chemicals used for pest control.

In addition to the above examples, economists must focus research attention on data needs generated by other changes in the policy agenda. Research can guide decisions on which new data series are needed and which old ones can be dropped, thus providing a dynamic agricultural data system in keeping with user needs.

ORGANIZATIONAL CONSIDERATIONS

Problems related to respondent burdens and one-time surveys are associated with data gathering. It may become increasingly important for researchers to tie their data requests for specific projects to data series gathered on a regular basis. If data series are to be revised, thought should be given to possible research needs and the potential of gathering an additional marginal amount of data to provide useful research input. This is especially important for evaluating the value of information to protect the long-run viability of the data gathering system upon which the agricultural and food system depends.

Rewards for professional activity on more fundamental research such as generating needed cost data upon which quick-response policy analyses depend are needed to maintain an updated data system for the food and fiber sector. To reward only abstract, mathematical research while ignoring fundamental data needs can only be harmful to all research as well as the usefulness of the data for decision making. High-quality data may be more important in determining the validity of research results than the econometric methods used in analysis. Finally, the magnitude of the research problems involved and the widespread implications of possible findings make cooperative efforts between research institutions, data gathering organizations, and data users more important than ever.

REFERENCES

American Agricultural Editors' Association (AAEA), Committee on Economic Statistics. 1972. Our obsolete data systems: New directions and opportunities. *Am. J. Agric. Econ.* 54:867–75.
Arrow, R. J. 1974. Limited knowledge and economic analysis. *Am. Econ. Rev.* 64:1–10.

Badger, J., and A. Johnson. 1978. The new CPI: Its impact on USDA's food price analyses and forecasts. In *National Food Review*, pp. 6-8. USDA, ESCS, NFR-4.

Bonnen, James T. 1975. Improving information on agriculture and rural life. *Am. J. Agric. Econ.* 57:753-63.

_____. 1977. Assessment of the current agricultural data base: An information system approach. In G. G. Judge et al., eds., *A Survey of Agricultural Economics Literature*, vol. 2, pp. 386-407. Minneapolis: University of Minnesota Press.

Brandow, G. E. 1976. Review and Evaluation of Price Spread Data for Foods. Report of task force sponsored by Committee on Economic Statistics, American Agricultural Economics Association, and USDA, ERS.

Carlin, T. A., and C. R. Handy. 1974. Concepts of the agricultural economy and economic accounting. *Am. J. Agric. Econ.* 56:964-75.

Demsetz, Harold. 1969. Information and efficiency: Another viewpoint. *J. Law Econ.* 12:1-22.

Devine, D. G. 1978. A review of the experimental effects of increased price information on the performance of Canadian retail food stores in the 1970s. *Can. J. Agric. Econ.*, 26(3):24-30.

Dunn, Edgar S., Jr. 1971. The national economic accounts: A case study of the evolution toward integrated statistical information systems. *Surv. Curr. Bus.* 51:45-64.

Eisgruber, Ludwig M. 1973. Managerial information and decision systems in the U.S.A.: Historical developments, current status, and major issues. *Am. J. Agric. Econ.* 55:930-37.

Gray, Roger W., and David J. S. Rutledge. 1971. The economics of commodity futures markets: A survey. *Rev. Market. Agric. Econ.* 39:57-108.

Guebert, Steven R. 1978. Farm income: Data and concepts in review. In Proceedings of Workshop on Farm Sector Financial Accounts, April 14-15, 1977. USDA, ESCS, AER-412.

Hayek, F. A. 1945. The use of knowledge in society. *Am. Econ. Rev.* 35:519-30.

Hayenga, M. L. ed. 1978. Pricing Problems in the Food Industry (with Emphasis on Thin Markets). NC-117 Monogr. 7, University of Wisconsin, Madison.

Heifner, Richard G., James L. Driscoll, John W. Helmuth, Mack N. Leath, Floyd F. Niernberger, and Bruce H. Wright. 1977. The U.S. Cash Grain Trade in 1974: Participants, Transactions, and Information Sources. USDA, ERS, AER-386.

Helmuth, John W. 1977. Grain Pricing. Commodity Futures Trading Commission Econ. Bull. 1, Washington, D.C.

Henderson, D. R., L. F. Schrader, and M. S. Turner. 1976. Electronic commodity markets. In O. D. Forker and V. J. Rhodes, eds., *Marketing Alternatives for Agriculture: Is There a Better Way?* National Public Policy Education Committee, Publ. 7, New York State College of Agriculture and Life Sciences, Cornell University, Ithaca.

Hildreth, R. J. 1975. Report of Task Force on Farm Income Estimates. Sponsored by the Economic Statistics Committee of the American Agricultural Economics Association and USDA, ERS.

Hirshleifer, J. 1973. Where are we in the theory of information? *Am. Econ. Rev.* 63:31-39.

Houck, James P. 1977. Concepts of price: Implications for agricultural data collection. In Proceedings of Workshop on Agricultural and Rural Data. Ser. A, pp. 1-19. AAEA and USDA.

Johnson, S. R. 1977. Agricultural sector models: Discussion. *Am. J. Agric. Econ.* 59:133-36.

Just, R. E. 1977. Agricultural sector models and their interface with the general economy: Discussion. *Am. J. Agric. Econ.* 59:137-40.

King, G. A. 1975. Econometric models of the agricultural sector. *Am. J. Agric. Econ.* 57:164-71.

Kroupa, E. A., C. Burnett, and J. K. Johnson. 1976. Agricultural Market Information: Collection, Dissemination, and Use in Decision Making (an Annotated Bibliography). Department of Journalism, University of Wisconsin, Madison.

Lee, John. 1972. Discussion—Our obsolete data systems: New directions and opportunities. *Am. J. Agric. Econ.* 54:857-77.

Leuthold, Raymond M. 1975. Actual and Potential Use of the Livestock Futures Market by Illinois Producers. University of Illinois Agric. Exp. Sta. Bull. 141, Champaign.

Mann, Jitendar S., and Richard G. Heifner. 1976. The Distribution of Short-run Commodity Price Movements. USDA, ERS Tech. Bull. 1536.

Moulton, K., Alfred Levison, and Peter Thomas. 1974. The Feasibility of Measuring Benefits of the California Federal-State Market News Service. University of California, Spec. Publ. 8003, Davis.

Ozga, S. A. 1960. Imperfect markets through lack of knowledge. *Q. J. Econ.* 74:29–52.

Paarlberg, Don. 1978. Agriculture loses its uniqueness. *Am. J. Agric. Econ.* 60:769–76.

Paul, Allen B., Richard G. Heifner, and John W. Helmuth. 1976. Famers' Use of Forward Contracts and Futures Markets. USDA, ERS, AER-320.

Raup, Phillip M. 1959. Structural changes in agriculture and research data needs. *J. Farm Econ.* 41:1481–91.

Reimenschneider, Charles H. 1977. Economic structure, price discovery mechanisms, and the information content and nature of USDA prices. In Proceedings of Workshop on Agricultural and Rural Data, Ser. A, pp. 1–19. USDA, Stat. Rep. Ser.

Rogers, G. B. 1971. Better information—An essential for better egg pricing. In G. B. Rogers and L. A. Voss, eds., *Readings on Egg Pricing,* pp. 101–6. College of Agriculture, University of Missouri, Columbia.

Stigler, George J. 1961. The economics of information. *J. Polit. Econ.* 69:213–25.

Sundquist, W. B. 1970. Changing structure of agriculture and resulting statistical needs. *Am. J. Agric. Econ.* 52:315–20.

Trelogan, Harry C., C. E. Caudill, Harold F. Huddleston, William E. Kibler, and Emerson Brooks. 1977. Technical developments in agricultural estimates methodology. In Lee R. Martin, gen. ed.; G. G. Judge, Richard H. Day, S. R. Johnson, Gordon C. Rausser, and Lee R. Martin, vol. eds. *A Survey of Agricultural Economics Literature,* vol. 2, pp. 373–85. Minneapolis: University of Minnesota Press.

U.S. Congress, House of Representatives, Committee on Small Business. 1980. Relationship between Structure and Performance in the Steer and Heifer Slaughtering Industry. 96th Congr. 2nd Sess., Committee Print.

U.S. Department of Agriculture, Statistical Review Board. 1969–72. Major Statistical Series of the U.S. Department of Agriculture: How They Are Constructed and Used, vols. 1–11. Agric. Handb. 365.

_____. 1978. Economics, Statistics, Cooperatives: Program Results and Plans. ESCS-42.

_____. 1979a. Report of the advisory committee on export sales reporting to Secretary of Agriculture Bob Bergland.

_____. 1979b. U.S. export sales reporting system: Pertinent issues. In Report of U.S. Department of Agriculture Interagency Task Force to Congress.

Upchurch, M. L. 1974. Toward a better system of data for the food and fiber industry. *Am. J. Agric. Econ.* 56:635–37.

_____. 1977. Developments in agricultural economic data. In Lee R. Martin et al., eds., *A Survey of Agricultural Economics Literature,* vol. 2, pp. 305–72. Minneapolis: University of Minnesota Press.

Weeks, E. E., G. E. Schluter, and L. W. Southard. 1974. Monitoring the agricultural economy: Strains on the data system. *Am. J. Agric. Econ.* 56:976–83.

RANDALL E. TORGERSON

10

ALTERNATIVE OWNERSHIP AND CONTROL MECHANISMS WITHIN THE FOOD SYSTEM

AMERICAN AGRICULTURE has been historically differentiated from other major industries in the economy. This difference, changes in the economic organization of agriculture notwithstanding, has been largely associated with the biological nature of the production process and the atomistic structure of the farm production sector consisting of large numbers of relatively small, geographically separated farm firms and households. The disparity in size and marketing power faced by these two groups of market channel participants has led to numerous demands and a body of law and institutions designed to help them acquire, gain, and maintain a degree of marketing power and control in the food system. Similarly, corporate firms and other associations and labor unions connected with the food industry have sought to achieve and expand their power in the marketplace by various mechanisms and management strategies.

The purpose of this chapter is to identify ownership and control mechanisms in the food system as a fertile field for marketing research. Focus is upon the genesis of market power in our mixed economic system and how firms extend themselves through various organizational and ownership arrangements, both vertically and horizontally, and through control mechanisms both internal and external to the firm.

The concepts of ownership and control include horizontal and vertical dimensions through mechanisms such as various ownership forms (single pro-

RANDALL E. TORGERSON is Administrator of Agricultural Cooperative Service, U.S. Department of Agriculture. Appreciation is expressed to my colleagues Jack Armstrong, John Dunn, James Haskell, and Dave Holder for their helpful comments during preparation of this chapter. Also, reviewers Harold Breimyer, Dale Anderson, Bruce Marion, and Jim Shaffer made many helpful suggestions. Responsibility for errors or interpretations are the author's.

prietorships, general and limited partnerships, closed and public corporations, cooperatives, multicooperative organizations, joint ventures), contracts, influence over market rules, forward selling arrangements, and others. The basis for power is found in property rights. Control, the authority to exercise power in making decisions and the exercising of property rights, manifests itself in a variety of forms internally in organizations and externally through industry norms and public institutions. This chapter first gives some insights into the evolution of conceptual ideas on ownership and control. It reviews the literature with special emphasis on vertical integration and horizontal professional associations used by farmers. It then turns to performance dimensions of vertical integration, results of subsector research and inquiries into the economic organization of agriculture. Promising research areas are then explored.

IMPORTANCE OF RESEARCH ON OWNERSHIP AND CONTROL

Agriculture is a dynamic industry. Rapid technological change has affected the economic organization of farm production units as well as the industrial organization of farm input production, food processing, and distribution. The impact of technological change has been to cause a wider disparity in relative size and marketing power between agricultural production and the farm input, processing, and distribution sectors. In the process, the character of marketing and the structure and functions of market channel participants have adapted to each other in accommodating marketing power and technological change. The number of firms has dropped dramatically in the production, processing, and distribution sectors. A rash of firm mergers and consolidations in the 1960s and 1970s has led to the preponderance of more integrated and diversified corporate marketing entities. Food price increases in the mid-1970s also brought about some changes in the household consumption sector as well as new legislation destined to impact future marketings (i.e., passage in 1978 of a new law establishing a National Consumer Cooperative Bank that parallels the Cooperative Farm Credit System in making loans available to nonfarm cooperatives).

The upshot of these developments is a changing configuration in the cast of claimants in the marketplace both as to their internal organization and decision-making procedure and as to the nature of market competition and the external regulatory environment that characterizes the world in which they live. In the midst of this change in market structure the mechanisms for and methods of distributing power are changing. The impact is immense upon the individual in choosing how we organize our social and economic system. Whereas the American dream sought fulfillment in making the individual the singular focus of our collective concern, the shifts from decentralized to centralized organizational systems make it more difficult to retain that focus and to influence the basic underpinnings of society itself as well as the nature of markets.

As described by Boulding (1968), two methods of distributing power exist. One gives wide distribution to property and, generally, the means of making a living. It is the dispersed self-regulating economic system that classical and neoclassical thinkers dwelt upon. J. D. Shaffer contends that Boulding's first classification perpetuates the neoclassical myth of an unregulated market. All markets are structured by property rights, which represents power distribution (notes to author). The second method yields to a concentrated power and syndicalism (Breimyer 1977) in which we impose legal and constitutional limitations upon those who possess it. Breimyer describes syndicalism as the industry-by-industry codification of rules. (The author is indebted to Breimyer for citing Boulding's dichotomy.) Recent trends have pushed the character of our mixed economic system more from the dispersed to the concentrated, from the decentralized to the centralized.

An important aspect of this trend has been increased specialization and vertical coordination at the food production and processing end of the market channel and increased conglomeration at the processing and distribution end. Motivation for increased integration appears to have been a variety of efficiency and technological gains, including risk reduction, brought about through closer coordination, through ownership of production resources, or by contracting. Vertical integration, the ownership of successive stages in the marketing process, is one vehicle of vertical coordination that is frequently used to obtain ownership and decision control. In some cases, growth through vertical integration may enhance the integrator's market power, i.e., the ability to obtain and maintain control over one or several factors that influence prices and incomes. As one participant's power is increased, another's is often reduced. As these new organizational forms take the place of traditional firms and firm functions, a new set of policy issues and challenges is brought to farmers, consumers, and researchers. Questions of access to markets for producers, need for meaningful raw product transfer prices, availability of market intelligence, demands for new bargaining legislation or other price discovery rules, and questions of returns adequate to cover production costs or to achieve parity prices are increasingly heard from the atomistic production sector. Each is symptomatic of the stresses of structural change occurring in this sector and of associated changed market power relationships.

Research is necessary in the area of marketing organization and firm behavior to better understand the (1) motivation and strategies for firm growth and power accumulation; (2) dynamics and competitive influence of vertical integration; (3) growth and competitive impact of conglomeration; (4) influence of collective action upon income distribution, efficiency, progressiveness, and competitiveness of markets, and (5) influence of various government regulatory, antitrust, and other programs on the performance of markets and general economic health of the business community. Most importantly, marketing research should assess the organizational learning process as it pushes on new frontiers of structural design and encounters problems associated with the internal and external environment. If, for example, conglomerates cross-subsidize new areas of entry with the intention of driving competitors out and establishing primacy for themselves, what influence does this have upon firms or farmer cooperatives that are organized essentially along commodity lines? Questions of this type are at the heart of the alternative ownership and control mechanisms in the food industry. Since they have received only limited treatment from the agricultural economics profession, they represent fertile ground for future marketing research.

EVOLUTION OF CONCEPTUAL IDEAS ON OWNERSHIP AND CONTROL

In its broadest sense, Alderson (1957) has provided a conceptual framework of how firms acquire, gain, and maintain a degree of marketing power and control. He was preceded by Commons (n.d.; 1950; 1957a,b) and followed by Galbraith (1957, 1967) in this analytical treatment. Each will be highlighted for his contributions to this evolving area of inquiry.

Alderson places the roots to his approach in recognizing that marketing is a phenomenon of group behavior, a thread held in common with Commons and Galbraith. Marketing exchange taking place in the trade channel involves interactive groups in which individuals are bound together functionally. Whereas interactive groups are readily identified in local trading centers, the national market is viewed as a more loosely bound functioning entity.

The distinctive feature of Alderson's treatment is use of functionalism to identify the system of action and why it works. A common thread is found not only with other social sciences but the biological sciences as well. In this system of thought, marketing functions are identified to show how they apply in one situation after another. The pragmatism embodied allows for their use in interpretation of marketing phenomena and in solution of actual marketing problems. Furthermore, by starting with the study of organized behavior systems, marketing systems are seen as capable of adapting themselves to sudden changes in the environment.

Each marketing participant thus constitutes a group whose interrelated activities constitute an organized behavior system. Power and communication are fundamental and interdependent within the system. More importantly, power explains the tendency of these behavior systems to survive and expand. Understanding subsystems (power and communication) is essential to evaluating three growth stages in an organization's fight for survival—establishment, expansion, and consolidation.

The key to Alderson's concept is the power principle with which organizations function to promote their power to act (i.e., the exercise of authority over decisions). Maintenance and enhancement of power is viewed as an inherent goal for any organized behavior system. Corollaries of the power principle are action to expand control over additional sources of energy, to increase the system's capacity for various aspects of its operation, and to increase the efficiency of use of resources already under its control. A crucial element within the power subsystem is status, i.e., the set of rules or standards that govern its operation and the sanctions of coercion or persuasion used to control it. The expanding behavior system tends to require an increasing specialization of its members to maintain the effective functioning of the system as a whole, a notion that was to be later developed and elaborated by Galbraith. Critical to long-term survival, the behavior system will tend to survive as long as the footing it occupies endures, because of the collective action growing out of the status expectations of its components.

The interaction of these power centers in the marketplace is therefore one of each trying to grow or maintain its position and thereby adjusting itself relative to others through negotiation, the process of reconciliation among power centers. By extending itself through agreements, the behavior system (firm) can expand its sphere of influence. It seeks "closure" of the system by increasing its position to control the channel and routinize the system from the point of view of its own management objectives and profit centers. In fact, the facility for effective negotiation can greatly enhance a firm's growing power.

The Alderson framework is thus an important contribution to assessment of power and control in the marketplace because it suggests why, among other things, farmers desire to gain vertical control over functions on which they are dependent. By using a functional approach and developing the concepts of power and communication as primary subsystems, Alderson has produced a model that goes beyond descriptive analysis of the power centers by permitting market problem solving as well as analysis of firm growth and expansion through use of these subsystems. This approach can provide the conceptual foundation for further work and elaboration in the food system.

Another major contributor to the study of alternative ownership and control mechanisms was Commons (1950, 1957b), who studied the structure of social action and analyzed the economic system as a system of human relations. Commons, widely acclaimed as the intellectual leader of the institutional school of economic thought, developed many ideas (later drawn upon by Alderson and Galbraith) in his conceptualization of the market economy. To Commons, economic institutions (collective action in control of individual action) were the basic elements that made up the framework of social action. It was this collective action taken as a whole that yielded an economy of pressure groups such as corporations, labor unions, trade associations, professional associations, and cooperatives, each pursuing its collective will. Functioning of the economy, according to this school, could be best understood if one determined the motivations for power and control of pressure groups and the working rules that governed their interaction.

Commons observed that the market mechanism did not of itself bring fair results to different groups in the economy since a disparity in bargaining power among groups existed. To balance power, pressure groups had to push through necessary reform. These groups were viewed as an indispensable means for achievement of a just and rational economic order and were thus seen as the determinants of the outcome of most economic phenomena in the political economy (i.e., the marketplace). They constituted an occupational parliament (USDA 1976; Carter and Johnston 1978) more representative than legislators representing districts.

More than any other scholar, Commons (1957a,b) identified property rights as the legal foundation of capitalism. This included not only the ownership of one's personal possessions, land, and the product of one's labor but also labor itself.

One method used by Commons for following the evolving interpretation of property rights was study of court decisions in which conflicts between parties were resolved. Many researchers since have followed this practice of investigation, not only for inquiry into the issues involved but also for the intelligence on firm operations and structure that otherwise might not be made available to the public.

The rules governing job rights, negotiation over the value of one's labor, and collective bargaining became embedded in codes (working rules), many of which were determined politically through legislation or administrative rule making. These rules prescribed the parameters in which individual pressure groups operated as well as the rules that governed the relationships between them. In short, they described how the "rights, duties, liberties and exposures" are created and work in the relations of the individual to society. Rules governing transactions, classified (Commons 1957a) into three types—managerial, bargaining, and rationing—were the basis for Commons negotiational theory of the political economy.

In a more recent treatise, Shaffer and Torgerson (1976) describe market rules as the set of rights and obligations established by law, custom, and covenant that define relations among members of a community in respect to exchange of goods and services. These include not only the laws of property and contract but also rules of entry and exit, licenses, grades and standards, collective bargaining rights, patents, franchises, dealerships, restrictive covenants, tariffs, price regulations, and the like.

The fundamental problem to which Commons directed much of his analysis was the resolution of conflict among individuals (people and other legal individuals), pressure groups, and governments. The source of conflict was often found in the power relationships between contesting groups. In *Legal Foundations of Capitalism* (1957b) he distinguished between the change in the concept of property "from physical things to the exchange-value of things, from a concept of holding things for one's own use to withholding things from others' use, protected in either case by the physical power of the sovereign."

Two sources of power and wealth thus became apparent: holding something useful for one's own use and exchange and withholding from others what they needed but did not own. The latter, regarded as a neglected economic concept, was seen as an evolutionary development of the concept of property from "slavery, feudalism, colonialism and a sparse population, to marketing, business and the pressure of population on limited sources." Bargaining power thus became the willful restriction of supply in proportion to demand in order to maintain or enlarge the value of business assets. This was contrasted to producing power that is the willing increase of supply to enlarge the wealth of nations. The resolution of conflicts between market channel participants through rules governing negotiations thus became a major contribution to studies of ownership and control.

Like Alderson and Commons, Galbraith has contributed two somewhat diverse conceptual approaches to evaluating alternative ownership and control mechanisms. Following Commons, he expresses little faith in the market regulator in the classical sense, since the concept of willing seller and willing buyer is upset by the existence of private power. In Galbraith's work (1957) on countervailing power, he suggests that competition, which was formerly viewed as the regulator of economic activity, has been curbed and superseded. Instead, the self-generation of competitive activity has been eliminated by barriers to entry in imperfectly competitive markets and by the convention against price competition. The competitive model, superseded by the tendency of power to be organized in response to a given power position, thus becomes replaced by a self-generating phenomenon in which newly self-generated power acts to curb a previously existing power. Many applications of this concept can be found in farmer movements and development of cooperative marketing activity in agriculture even as they are now being experienced in the consumer sector through consumer lobby groups, buying clubs, and cooperatives. The relevancy of this approach is found in the generation of collective action among market channel participants in response to power relationships and in the motivations for horizontal or vertical growth by organized behavior systems.

Galbraith's second contribution (1967), perhaps the most significant one, concerns the makeup of the "new industrial state." In this treatise Galbraith states that "technological imperatives" have brought great changes to modern economic life with the ultimate consequence of "diminished effectiveness of the market." As a result, "the market is replaced by planning."

The technological imperatives to which he refers are the requirements of large-scale production, invention, and innovation that lead to very large industrial complexes and high market concentration. These large enterprises are perfect mechanisms for planning production, product development, and marketing strategy. Market power is not only an end result, but a prerequisite to success of the system.

Planning within the modern industrial enterprise is carried on by the "technostructure," which involves specialized technicians and professionals required for effective group decision making. Increasingly, the government is seen as a participant in the planning process, since many jobs are too big for even the largest firms. The government stabilizes aggregate demand, underwrites expensive technology, restrains wages and prices limiting inflation, provides technical and educational manpower, and buys up to a fifth of U.S. economic output. In short, Galbraith sees the government as playing a central role in economic planning with less reliance on the market.

Galbraith (1973) specifically argues, as Boulding, that we have two systems, one in which power is contained by the market and the second in which power of firms extends to the market and thereby to resource use. Neoclassical and neo-Keynesian economics assume away the existence of power in the marketplace and thereby ignore the power or planning system that accounts for over one-half of economic output. In so doing, the commitment of corporate bureaucracy to its expansion, to aggrandizement of its increasing presence, and to profits only sufficient to assure autonomy of management (technostructure) leads to its becoming a political instrument that rivals the state. Therefore, Galbraith argues that to deny the political nature of the modern corporation is to disguise reality. Excessive resource use in one leads to inequality in income between the two systems and to disparities in the market system, since the firms can protect themselves from adverse action by planning. The state according to this thesis is the prime objective of economic power.

The most insightful aspect of Galbraith's treatise is the role of professional management and technicians in the planning process and operations of corporate enterprise. These insights correspond closely to views of some behaviorists (Cyert and March 1963; Thompson 1967) who observe that technological pressure toward reduction of uncertainty generates pressure for organizations to grow and to expand their control. This growth is guided by the nature of technology (thus by the role and power of the technocrats) and by the task environment. This is of particular consequence for analyzing firm goals and growth strategies approached from the perspective of whose goals should be attained (management versus owners) and how these competing goals in enterprise, where ownership is separated from management, are compromised. Control aspects of these organized behavior systems, given the role and influence of technocrats, are helpful in analyzing the causes of conglomerate growth and are of particular importance in assessing management versus owner-user goals in voluntary organizations such as farmer cooperatives and professional farm associations.

The conceptual ideas on ownership and control presented by these three scholars each focus on firms as behavioral systems that seek to negotiate within their environment on a basis of market position and relative market power. Claimants to market benefits vary, based on their skill in negotiating within the system; the professional expertise they have at the helm; and the marketing power they can mass, maintain, and use in pursuing various decision strategies. Invariably, the nature of firms that emerge in the changing market economy creates separate systems and a different character to the market itself.

AN OVERVIEW OF PAST RESEARCH

The literature regarding organization and control is massive and diverse because of the nature of the subject and the general lack of a cohesive research thrust by agricultural economists in the area. Theoretical and empirical contributions come from several disciplines, including the business schools, sociology, and economics. This review is basically limited to agricultural economics inquiry and is organized around (1) the economics of collective action in agriculture, which includes organizational structure and functional analysis, the economic theory of cooperation, and the economics of collective bargaining; (2) vertical integration and coordination; (3) horizontal integration; and (4) the economic organization of agriculture. While pertinent aspects of law, market structure research, and intrafirm decision making are important aspects of this subject matter, they are treated elsewhere in this book and are therefore not addressed here.

Economics of Collective Action

ORGANIZATION STRUCTURAL AND FUNCTIONAL ANALYSIS. As a starting point for assessment of past research on ownership and control in the food system, we begin by examining structural and functional research, a form of inquiry advocated by Alderson (1957), relating to farmers' countless efforts to organize and control their industry. History is replete with the precipitation of farmer movements that have sought to protest the conditions faced by farm operators and have often resulted in formalized organizational efforts to exert influence in the marketplace. An early observer of these developments, Hibbard (1921) saw a distinction in two different kinds of organizational efforts used by farmers. The first is the establishment of general or all-inclusive organizations. Examples of these were the agricultural societies, clubs, and fairs. These were followed by other organizations that were a little more restricted in the matter of their membership but still not confined to farmers whose interests were in any narrow sense identical. Examples of these were the Grange, the Alliance, the American Society of Equity, the Farmers Educational and Cooperative Union, and the Nonpartisan League. These later types were considered more professional in character. The second type of organization was designed for a specific purpose such as the cooperatives, alone or federated. According to Hibbard, the first of these could be termed farmer movements and the second could be called agricultural cooperation.

Robert Liefmann (1932), the general industrial organization economist, also identified categories of economic and trade federations that could be distinguished in the economy. The three categories pointed out by him were trade federations (technical or professional associations), associations, and corporations or companies.

The professional associations had the predominant characteristic of being rather loose organizations aimed at defending the economic interests of their members against other economic interests or the government. The economic operations of members were assumed varied; therefore, the circle of persons who entered into such unions was also somewhat varied. The most important organizations of this type were those formed by members of the same profession or trade such as the Federation of Agriculturists and the German Agricultural Society.

The second classification included the common interests of the professional associations (as associations of entrepreneurs) and a mutual obligation to carry out or refrain from particular acts that were conducive or nonconducive to a common aim. In such associations, the members would agree to surrender some liberty of action or submit to control through joint agreements.

The third type of organizational effort was the business corporation or company, which was considered a combination of individuals who undertook some economic activity. Cooperative societies were regarded as a part of this category because individuals of the same type endeavored to carry out a joint undertaking. Liefmann argued that since cooperatives were somewhat different from unions or associations and from the ordinary industrial company, it was better to treat them as a special form of enterprise parallel to the industrial or trading companies.

The observations of these early students were supported by yet a third. In his dissertation, the Russian-trained economist Emelianoff (1942) directed his analysis toward examination of a cooperative economic type of association around which all actually existing varieties of cooperative associations are centered. His objective was to concentrate attention on analysis of the economic structural and functional essentials of cooperative organization. In this effort to develop an economic theory of cooperation, Emelianoff argued that researchers had to examine the elementary cooperative unit, for in this form all basic economic characteristics of the cooperative organization were concealed.

Emelianoff distinguished organizational forms according to whether they were aggregates of economic fractions or of economic units. The aggregate of economic fractions shared in common the aggregate structure of the organization and fractional economic aims of its members. Accordingly, profit seeking could be considered an aim of a group of entrepreneurial fractions. Within this framework, the general farm organizations or professional associations were considered as entrepreneurial fractions. All such groups were considered organizations of farmers, not of farms. In contrast to these professional associations, cooperatives were aggregates of economic units coordinating their acquisitive or spending functions, but in which each participating economic unit retained its economic integrity. In this manner, Emelianoff built upon the works of Liefmann in an attempt to decipher the structural attributes of organization forms in the farm sector.

Together, the observations of Hibbard, Liefmann, and Emelianoff provide the long-standing premise that basic structural types of organizations can be identified in representing the economic interests of farmers and their farm firms. Assuming this structural identification, it is then left to researchers to identify the economic needs and the structural and functional relationships that permeate it.

Following these early writers, Nourse (1945), the cooperative theorist and philospher, developed a theory that cooperatives provided a procompetitive influence on the economy as a whole. According to Nourse, the mission of cooperatives was not to be the commercial Napoleon but rather the economic architect of the competitive economy. Cooperative presence pushed market performance closer to the competitive norm and thereby contributed to the public interest by providing a yardstick to measure performance of other market channel participants.

Olson (1965) has also elaborated on certain aspects of group action by farmers by observing that professional associations have often resorted to economic activity to provide a beachhead that undergirds their other activities. Examples are use of insurance subsidiaries or farm supply affiliates that generate revenues used to sustain lobby and related representational endeavors. Torgerson (1971a) took a broader approach that builds on Hibbard, Liefmann, and Emelianoff by showing that a structural and functional relationship exists between basic organizational types. He suggests that a "cooperative systems" approach that analyzes the interrelated bundle of organizational activities used by farmers should provide the basis for researching ways to maximize farmers' marketing power.

Within the broad framework of the economics of collective action in agriculture, the two basic organizational forms, cooperatives and professional associations, with exceptions noted, have generally received separate analytical treatment. Development of the economic theory of cooperation has gone farthest in application of the economic theory of the firm to the cooperative business organization as a special case. Treatment of the professional associations has generally been lacking except in analysis of negotiated pricing. Separate treatment on each follows.

ECONOMIC THEORY OF COOPERATION. Going beyond his structural and functional distinction between organizational types, Emelianoff was among the first to attempt a comprehensive economic theory of cooperation. He began by identifying that nonstock, nonprofit, and capital stock cooperatives were deprived of the fundamental test of enterprise (entrepreneurial income) and therefore are not acquisitive economic units (enterprises). In this respect Emelianoff drew upon an earlier student of cooperation, Liefmann, who claimed that cooperatives did not undertake profit maximization for themselves and therefore could not be considered an enterprise.

Having determined that cooperative associations were not an enterprise, Emelianoff went on to conclude "that the only comprehensible and indisputable test of the cooperative character of an organization is their aggregate structure." As such the cooperative was conceived to be an aggregate of economic units in which each retained its economic individuality and independence. A vertical relationship was identified, with the cooperative viewed as an off-farm extension of the farm firm. The aggregate form was an "agency" of the associated units (through which the economic units conducted their business activities) that functioned in the capacity of a branch or part of the associated economic units. As seen in this context, the cooperative was a special department or branch of the individual economic units.

In this manner, Emelianoff closely approximated the conceptualization of a cooperative maintained by Nourse (1922). It was Nourse's thesis that membership in a cooperative rested on a personal basis in which people voluntarily "coalesced." The basis for this combination was their community of interest and their ability and willingness to assume the significant financial and other obligations implicit in mutual undertakings. Cooperatives were a means by which participants could more efficiently perform certain economic functions that were integrally related to the economic activities of the participants. Both a horizontal and a vertical relationship of the cooperative to participating members was implied.

Following the initial works of Emelianoff and Nourse, two channels of thought developed in the continuing analysis of cooperative organizations and how they affected control and served as tools of farm operators. The point of divergence between them has been the difficulty of conceptualizing exactly how the cooperative organization should be defined and handled theoretically. One group has insisted on following the Liefmann and Emelianoff approach of viewing the cooperative as an aggregate of economic units that did not constitute an enterprise. Phillips (1955) and Aresvik (1955) have followed this line by using the farm firm as the basic unit of analysis. The second group chose to recognize the cooperative as an enterprise and used it rather than the farm firm as the basic unit of analysis. Robotka (1947) and Helmberger and Hoos (1972) have essentially followed this line with a progressively sophisticated application of the theory of the firm (Barnard 1938; Helmberger 1964).

The divergence in these two channels of evolving cooperative theory has focused essentially on the relationship of the cooperative member and the farm firm to the cooperative as an organized behavior system. This interrelationship is of importance to the questions of organization and control mechanisms in agriculture, since questions of policy treatment hinge on the nature of the cooperative itself in determining the future of a dispersed agriculture. In the more recent contributions to the evolving economic theory of cooperation, Helmberger and Hoos have lifted themselves above the argument of whether an enterprise exists by assuming, from organization theory, the existence of a "peak coordinator" defined as a person or group of persons that specify the ends of the firm and engage in action to secure its attainment. The concept of a "peak coordinator" is drawn from earlier work (Barnard 1938). This assumption of interaction at a top level between a cooperative board and its management allowed application of the theory of the firm to the cooperative as a special case. The essential difference recognized in this application is that the cooperative, unlike other corporate enterprise, emphasizes returns to its member patrons, whereas the corporation emphasizes satisfactory profit levels and in the process seeks to minimize payments for its inputs. Cyert and March's (1963) concept of large-scale organizations behaving as political organizations with multiple objectives and claimants presents another dimension to cooperative theory that has received only limited exploration in the cooperative systems context (Torgerson 1971a).

COOPERATIVES AS INSTRUMENTS FOR MARKET POWER. Several researchers have identified structural weaknesses of cooperatives as instruments for achieving market power. Emelianoff (1942) noted that by their nature cooperatives are saturated with internally disruptive influences. Galbraith (1957) identified basic structural weaknesses: the cooperative is a loose association of individuals; rarely includes all the producers of a product, cannot control the production of members, and has less than absolute control over farmers' decisions to sell. Mueller (1960, 1979) has also observed that a cooperative, acting alone, cannot achieve market power, since it is not an adequate instrument for achieving control over supply. Market power, according to this analysis, depends on a high degree of horizontal concentration or product differentiation in some level of distribution or production.

Certain cooperative operating practices, in contrast, have come under close review. Youde (1978) found that, consistent with hypothetical relationships derived from cooperative theory, a few centralized cooperatives restrict membership to maximize returns to members. Similarly, Masson and Eisenstat (1978) have argued that by controlling market outlets cooperatives may exclude marketing opportunities for those "free rider" farmers who wish to remain outside the cooperative. In their view, the cooperative is properly an instrument for developing only countervailing power, not original market power. Without classified pricing, they find that vertical integration by way of cooperatives leads to the economic ideal through countervailing monopsony power and output expansion that cancel out the effects of final-product monopoly. Appropriate policy conclusions, then, are to facilitate vertical integration by cooperatives as in the dairy industry. This conclusion is similar to the "competitive yardstick" theory of cooperation advocated by Nourse (1945).

ECONOMICS OF COLLECTIVE BARGAINING. Organized activity by professional organizations (single or cross commodity) are rudimentary forms of bargaining cooperatives that generally operate on the horizontal plane in the price bargaining or political lobbying arenas. Basic work on bargaining cooperatives has been done by Hoos and Helmberger (1965) and on professional associations by Torgerson (1971b). The approach in each is identification of the process for attainment of full value for raw farm products (consistent with supply and demand conditions) through negotiation with buyers. Particular emphasis on this form of horizontal integration has been in association with marketing mechanisms such as marketing agreements and orders. However, attention in recent years has been augmented by increased use of contracting for a variety of crops and livestock. While initially focusing on perishables such as milk, fruits, and vegetables, use of piece-wage contracts in the poultry industry and other forms of contractual coordination have given added impetus. These developments have been chronicled in the report of the National Commission on Food Marketing (1966) and proceedings of the National Bargaining Conferences published by the USDA annually since 1951.

A flurry of analyses accompanied a farmer movement in the 1950s that resulted in the National Farmers Organization's collective bargaining program and use of withholding actions. Efforts to define bargaining and the scope of farm bargaining activity resulted in books by Ladd (1964) and Roy (1970). The North Central Policy Extension Committee headed by Breimyer and Barr (1971) also identified institutional arrangements and conditions necessary for effective bargaining. More recently, Bunje (1980) has given a practitioner's assessment of bargaining association management and negotiating strategy.

Madsen and Heady (1971) have estimated effects of bargaining programs on production, net farm incomes, and food costs for specified price levels. Clodius (1957, 1959) found the exercise of bargaining power of farmers to be dependent on the physical presence of cooperatives of adequate power and that growth through merger was a determining factor in survival.

Recent research on bargaining has been limited to a few important studies. Lang (1978, 1980) has identified nonbargaining services provided to members by bargaining associations and benefits from bargaining as seen by management and members. Reduction of marketing costs, securing of markets, quality control, coordination of planting time, and stabilized prices all ranked high as tangible benefits. This suggests major benefits through coordination that go beyond the simple "high price" objective. Garoyan and Thor (1978), however, found that crop growers with bargaining associations fared better than those without them, even where operating cooperatives are an important dimension of market structure. This indicates limits to the ability of operating cooperatives, previously noted, in raising industry price levels. Knutson (1971) found five prerequisites to successful bargaining, including product control, recognition, organizational discipline, efficiency, and production control. The mechanisms and features of exclusive agency bargaining have been described by Shaffer and Torgerson (1976). Skinner (1981) and Blum have assessed the impacts of Michigan's exclusive agency bargaining law and have found no negative impacts on market channel participants. A state circuit opinion has upheld the constitutionality of the law (Brown 1981). Finally, recommendations for design of a system for effective bargaining in the processed-potato industry on an interregional basis were made by the Farmer Cooperative Service (Phillips et al. 1977).

Vertical Integration and/or Coordination. The primary focus of economists studying vertical coordination in agriculture has been on the interaction of adjacent levels at the exchange or transaction points in commodity flows from farm to consumer. Much of the work has centered on exchange between the producer and first handler or processor levels. Only recently have efforts been made to view interstage exchange mechanisms in the context of the complete farm-to-consumer chain.

Mighell and Jones (1963) took a broader definitional approach by defining vertical coordination as "the general term that includes all of the ways of harmonizing the vertical stages of production and marketing. The market price system, vertical integration, contracting and cooperation singly or in combination are some of the alternative means of coordination." Most authors have concentrated on why one coordinating mechanism might be chosen over another. The shortcoming of these efforts has been a general lack of theoretical focus.

Bain's (1959) industrial organization theory has served as a base in explaining why firms facing buyers and suppliers behave differently under alternative market structures. This approach, while informative, does not fully explain all the reasons why firms or producers might choose nonmarket means of interstage coordination, nor does it trigger public policy concerns until very high concentration levels are achieved.

One of the most compelling arguments for vertical integration is derived from the business schools and is oriented toward developing the concept of product acquisition and delivery systems. Alderson's (1957) treatment, noted earlier, draws on the notion of firms exercising their power through growth and control over the marketing channel and seeking to routinize the system to their own advantage. There is a convergence of ideas in the literature from the business schools on this issue in the micro sense, with the macro concepts of market power described by Commons (1950) and Galbraith (1973). In this respect, Farris (1972) notes that many firms with extensive marketing distribution systems to reach consumers are in an advantageous position to control the channel and routinize the system from the point of view of their own management objectives and profit centers. Firms seek "closure" of the system and use the latest in technology, information, and marketing power to achieve their objectives. By commanding a vast delivery system extending to consumers, these firms are alert to new procurement sources and distribution changes that permit them to adjust their routines in a fashion that is advantageous to them. Similarly, they can adopt technological innovations that extend their control over sources of instability and market uncertainty.

In this connection, Goldberg's (1968, 1972) treatment of agribusiness coordination through use of joint ventures provides an interesting dimension. A firm with a widespread merchandising program can take advantage of its high returns from this activity and reduce market uncertainty by controlling its source of supply through the joint venture. However, this usually leaves the lower margin production and processing end of the channel to farmers and cooperatives. By controlling final product markets and majority control of the joint venture, the distributing firm has effectively routinized the system to its own advantage.

Prices act as the primary coordinating device in the theoretical situation of numerous single-product firms completing costless transactions on the basis of perfect information and knowledge. Real-world deviations from the assumptions of this theoretical model create incentives for exchange to be coordinated with mechanisms other than market price. Several authors have suggested a number of incentives for seeking alternative vertical arrangements (Trifon 1959; Padberg 1966; Paul 1974; Mueller 1978). These authors and others differ as to the relative importance of the various incentives, but most recognize that no single incentive is of greater importance in all situations. The incentives may be grouped into four interrelated but identifiable categories: technological change or physical complementarity, imperfect markets in adjacent stages, reduction of risk and uncertainty, and reduction of transaction costs. These will be treated separately, but it must be kept in mind that decisions to adopt some form of nonmarket price coordination are usually made as a reaction to several incentives.

INCENTIVES FOR VERTICAL COORDINATION

Technological changes or complementarity. A firm may desire to coordinate activities of adjacent levels more closely to achieve cost savings that arise from the physical complementarity between levels or to decrease input costs by adopting new technology. The classic example of steelmaking is often used to demonstrate the savings that may be obtained through interlevel coordination. By placing adjacent processes at a single location, energy may be saved through elimination of the need to reheat materials.

In agriculture, the technological complementarity between levels may best be seen in minimization of storage costs when output quantity at a lower level is closely coordinated with input quantity requirements at the next level. When input supplies are highly variable, a firm (particularly one with high fixed costs) is hard-pressed to maintain operation at a minimum-cost level. Mueller and Collins (1957), in an examination of the grower-processor relationship in the fruit and vegetable industry, conclude that maintenance of a steady supply is a primary reason for processor integration into growing. Henry and Rauniker (1960) point to the great effect on net returns of quantity changes for levels in the broiler industry with high fixed costs.

By integrating adjacent levels, decision making becomes centralized, enabling management to maximize joint profits by exploiting the complementarity between processes in terms of input quantity and quality. Blaich (1960) views the integration of vertical levels to be a long-run planning activity, which may result in lower overall costs through intrafirm coordination of activities.

Firms may choose to integrate backward into production to introduce technological innovations that may result in lower input costs. Some authors (Henry and Rauniker 1960; Padberg 1966) indicate that firms integrate to adopt new techniques because of the inability of producers to finance these capital-intensive investments. Farris (1972) and Trifon (1959) believe that coordination through the marketplace does not provide the incentive for producers to adopt technological innovations, leaving this role to integrated processing firms by default.

The continuing trend toward specialization has increased the need for vertical coordination. Paul (1974) points out that specialization through technological improvement has made the specialized operation more vulnerable to changes in the market, thus providing the incentive to coordinate supplies. Specialization tends to narrow input specifications for processing firms. Trifon (1959) indicates that the open market system will not be likely to provide precisely specified inputs, thus the firm is induced to coordinate input production. Specialization through technological innovation also leads to increased market price and production risks for firms whose resources are not easily converted to other enterprises, as is the case in broiler production (Marion 1976).

Stigler (1951) argues that the vertical structure in an industry depends on the size of the final consumer markets. He believes that in "new" industries integration may be necessary to ensure supplies at each level. As the market grows, economies of scale may be achieved at some levels, leading toward specialization and disintegration (i.e., a life cycle concept of the market). Blaich (1960) appears to support this view in his prediction of the separation of corn production from hog operations in the traditional hog-corn farm as a result of improved technology. In summary, use of vertical integration, joint ventures, and contracts result from changes in market size that dictate changes in the optimum size of firm or enterprise combinations and in technology (Marion 1976). They are seen as mechanisms for shifting functions from one stage to another and spreading risk where increased specialization evolves and as instruments to spread cost-reducing technology.

Imperfect markets in adjacent stages. The structure and behavior of firms in a market that deviates from the perfectly competitive norm may act as an incentive for firms to undertake vertical integration. A firm faced with monopolistic suppliers of inputs may attempt to avoid monopoly prices by providing its own (Trifon 1959). A firm faced with monopsonistic buyers may wish to integrate forward to provide alternative market outlets. Padberg (1966) attributes the increase in farmer cooperatives to the desire to provide a less monopolistic market outlet for their products.

Trifon (1959) suggests that vertical integration results in as well as flows from monopolization at one level. A firm may integrate backward to establish control over supplies upon which its competitors depend. Walsh (1968) similarly found that vertical integration, while seldom conferring original market power, can serve to strengthen horizontal power. This can lead to foreclosure of markets or supplies for competitors, price squeezing, and possibly restructuring of an industry. It is also argued that advertising by manufacturers is preselling the firm's products, a form of vertical coordination that moves the merchandising function from the retailer to the manufacturer (Marion 1976). Paul (1974) asserts, however, that regardless of organizational arrangement, the long-run forces of supply and demand will be the final arbitrator.

Mueller (1978) has examined the ability of multiindustry commodity firms to subvert the market through manipulation of consumer demand. Cross-industry subsidization by conglomerate firms alters the concept of market competition based on efficiency. Interlevel coordination of activities within an industry becomes increasingly important for single-industry firms faced with competition from multiindustry organizations.

Reduction of risk and uncertainty. Many authors (Mueller and Collins 1957) have credited the reduction of risk and uncertainty with playing a major role in encouraging a firm to undertake some form of vertical coordination. Logan (1969) has developed a multiple-objective utility function postulating that firms seek to maximize profits subject to risk constraint. By measuring risk as the variability of returns for a process, he has shown that a firm may wish to vertically integrate if by doing so it could reduce its risk, even though its total costs and profits may remain unchanged.

Risk is generally recognized in two forms, price variability and supply variability. The risk of supply variability for processing firms is that they may not be able to obtain adequate quantities to operate at an efficient level. This is a particular problem for firms dealing with highly seasonal and perishable commodities. Mueller and Collins (1957) have shown how vegetable processors, by growing part of their needs and overcontracting with independent producers for the rest, could effectively shift the risk burden of oversupply to the producers, while assuring adequate volumes during low harvest periods. The National Commission on Food Marketing (1966) similarly argued that the reason broiler production was not integrated through ownership was the availability of underemployed farmers who were willing to accept low returns to labor because of limited alternatives.

Contracting may serve to lower price risks for producers and firms by establishing a fixed price or supply in advance. The reduction of risk in a fixed-price contract is short term in nature, and there is considerable doubt that any long-run reduction of risk results (Paul 1974). If, through contracting, the producer or firm is forced to make capital investment, its resulting position of dependency may raise the overall risk level.

Vertical integration may be used to reduce risk of input cost variability if the cost of producing inputs is less variable than their open market price or if profits at successive stages tend to be inversely related over time.

Reduction of transaction costs. A frequently mentioned incentive to vertically coordinate through nonmarket means is that transactions through the markets can often be costly and information transmitted through the markets is incomplete. When a firm is able to lower its transaction costs and obtain better control over the quality of its inputs by avoiding the traditional markets, it will be induced to contract or integrate. Transactions were the unit of inquiry used by Commons (1957a) and were classified as bargaining (transactions in the marketplace), managerial (transactions in the working group), and rationing (transactions between the individual and the state).

The costs of obtaining complete information, given the ever-increasing array of prices and commodities, makes an exhaustive information search prohibitive (Paul 1974). Williamson (1971) and Alchain and Demsetz (1972) indicate that nonmarket coordination improves the flow of information. They go a step further when they state that integration is preferable to contracting in an information flow sense. With all transactions internalized, communications are between acquaintances with less likelihood of misunderstanding. Disputes may be settled by fiat rather than through expensive haggling.

Handy and Padberg (1971) imply that the existence of meaningful grades and quality standards tends to limit nonmarket coordination by transmitting more accurate information to the markets. When recognized grades and standards do exist, more efficient market price systems may be introduced, as exemplified by the electronic exchange program for cotton.

Market prices may not relay the same information to the parties in exchange. Purcell (1973), in a study of Oklahoma feeder-packer relationships, shows how market prices are subject to different interpretations by feeders and packers, resulting in a failure to coordinate supplies between the two levels.

SUBSECTOR SYSTEMS RESEARCH. There has been much interest in using a subsector approach in studying various methods of vertical coordination in agriculture. This approach involves evaluation of structure and behavior vertically and horizontally over the entire commodity flow from farm to consumer. The rationale behind the subsector approach is that vertical coordination at one level is not independent of other levels in the production-distribution system. Objectives of firms at one level will affect their behavior in dealing with firms at adjacent levels. Shaffer (1968) has made a case for subsector research to discover barriers to improved performance and problems faced by participants identifying means to removing the barriers and solving the problems. The concept is one of researching a total interrelated system.

Much of the effort of the NC-117 project has taken up this challenge and been directed toward subsector analysis for various commodity groups. Research has been done for the broiler, egg, dairy, cattle-beef, sweet corn, snap bean, citrus, and tart cherry subsectors (Campbell and Hayenga 1978). Each subsector was evaluated in terms of its structure and how that affected vertical coordination within the subsector. By comparing subsectors, it is believed that hypotheses concerning structure-coordination-performance relationships may be formulated.

Certain common subsector traits were found. Increased farm/firm size occurred in all sectors relating primarily to economies of scale. Concentration has occurred primarily at the distribution level. The number of available market outlets for producers has generally declined.

In all subsectors there has been an increasing tendency toward product differentiation at the consumer level. Brand name differentiation has generally resulted in an increase in coordinating activities by manufacturers attempting to improve quality control.

Most research at the subsector level is involved in ascertaining and describing structure and relationships among the levels. A sound analytic approach is lacking, as is the case in research concerning adjacent level exchange. Incentives for vertical coordination are the same whether using a micro or a macro subsector approach. The real benefit to a subsector analysis appears to be the clearer picture it provides of the dynamics of commodity flows. The incentives for a firm to contract or integrate forward or backward can be better viewed in the context of interlevel relationships through the entire subsector rather than a myopic concentration on a single exchange point.

A major attribute of subsector research, not yet fully exploited, is study of the total interrelated system, with particular emphasis on various marketing mechanisms, rules, and decision-making processes and their impacts on efficiency, production, prices, distribution of income, and other performance criteria of the system as a whole. These areas are briefly listed to provide insights to future productive areas of inquiry.

OWNERSHIP FORMS. Ownership, the legal right to possession of resources, can be held in five different ways: by an individual or family, through a partnership, by a group of persons who form a cooperative, by a group of persons who invest in a corporation, or by government (Henn 1970).

Each method of ownership has certain characteristics. In addition, certain legal modifications or arrangements that deviate from these basic forms can be used to meet various ownership and/or control objectives. Examples of modification include subsidiaries, franchises, dealerships, joint ventures, and limited partnerships, among others. Matthews and Rhodes (1975) have identified basic tax avoidance and other reasons that have led to use of limited partnerships in agriculture by nonfarm investors. Meisner and Rhodes (1974) have conducted an intensive study of conditions that led to nonfarm investment in beef feeding schemes and the impact this had on structure of feedlot ownership and location of the industry.

Goldberg (1972) has focused on use of cooperative-corporation joint ventures. Subsequent litigation has raised questions about the workability of such arrangements, although this use on an intercooperative basis requires further analysis.

The presence of public stock corporation membership in cooperatives has been called by some writers a "manifest contradiction" of cooperative purpose and principles that is laden with policy implications (Breimyer and Torgerson 1971). A study of this phenomenon, however, indicates that its use is presently limited in larger cooperatives to fruit and vegetable and poultry associations and in smaller cooperatives to farm supply and grain cooperatives (Schneider et al. 1978). The economic performance of cooperatives with such membership and voting control aspects remains to be studied before further consideration of this phenomenon can be assessed.

Studies of the makeup of ownership patterns in public stock corporations and use of limited partnerships, Webb-Pomerene Associations, Subchapter S corporations, land trusts, and closed family corporations in the food industry are necessary to better understand the ownership trends of the production, processing, and distribution components of various subsectors (USDA 1981). Krebs (1978b) has made initial efforts at tracing such ownership phenomena.

OWNERSHIP VERSUS CONTROL. The separation of management from ownership raises a series of "benefit for whom" questions of increasing importance for analyzing ownership and control in the food industry. Hoffman (1970, 1980), while adhering to the concept of "economic determinism," nevertheless views with considerable alarm the "avalanche" of conglomerate mergers and the trend to "monopoly capitalism" envisioned by Marx. Baumol's (1959) suggestion that managers are motivated by total revenue maximization subject to a minimal profit constraint (a level of competitively acceptable earnings to stockholders while leaving the rest for expansion investment) is of particular significance here. Krebs (1978b) has initiated the painstaking study of interlocking directorates as a control mechanism in public firms. More research is essential for better comprehension of the nature of control and objectives of publicly held companies.

Control issues are likewise of crucial importance in voluntary organizations. Separation of management from ownership in cooperatives, a traditional practice, raises many interesting questions as cooperatives grow internally and through merger, including diversified growth. This is of particular significance when management is brought in from noncooperative firms without properly understanding the basics of cooperation. A recent study (Ward et al. 1979) indicates that the vast majority of cooperatives use a one-member, one-vote system. A continuing challenge that remains is reconciling democracy with efficiency in large-scale voluntary organizations.

A significant dimension of control that has emerged in Europe is the participation of labor union representatives as voting board members in public stock corporations and some cooperatives.

Comparison of benefits and performance under alternative control mechanisms remains a promising area for further research.

TAXATION AND OTHER FINANCIAL MATTERS. Among the many rules governing the conduct of firms, perhaps none has had as great an impact on the structure of food and other industries as taxation (USDA 1981). Matthews and Rhodes (1975) have identified tax implications of various ownership forms upon production agriculture. It can be also argued that among the top experts in the technocratic structure of conglomerates is a cadre of tax attorneys who define ways to manipulate the acquisition and disposal of companies to the advantage of the parent firm. Raup (1970) argues that preferential tax treatment of capital gains and related income tax rules favor corporations over cooperatives even though the latter are based on a single tax principle. Raup also makes a case, as does Brown (1979), that inflation may place nonfarm corporations in a better relative position in capitalizing their business than cooperatives financed by retained earnings. Both these arguments require substantiation.

The ability of firms to mobilize capital effectively is a major functional test of business success. It is also, as Paul (1964) suggests, fundamental in structuring organizations since "markets become restructured only as new decisions are made on how capital is to be used." The market power implications of the ability of firms to acquire and use capital is therefore a major determinant of their staying power and growth as noted by Cook (1978).

COORDINATION BY MARKETING MECHANISMS. Institutional devices such as marketing agreements and orders and marketing boards provide public sanction for farm-related marketing activity and regulation of handlers in the marketing process. Numerous studies of marketing orders have assessed their impact and focused on specific provisions. Jacobson (1978) discounts marketing orders as a source of cooperative market power since they do not provide price enhancing opportunities through supply limiting provisions. He posits an alternate hypothesis that cooperatives would develop additional power in the absence of market orders. However, Masson et al. (1978) argue that marketing order regulations codify market power schemes. In contrast, Boynton (1977) found that the dairy subsector is well coordinated under marketing orders and that full supply contracts make a positive contribution to efficiency as seen by both producers and handlers. Black (1978) regards marketing agreements, which are formal written commitments between a member and a cooperative, as leading more to greater cost efficiency than to monopoly pricing power.

The role of marketing boards based on experiences in Canada and other countries has been receiving more discussion. Shaffer (1968) states that this approach, a publicly sanctioned and regulated cartel, should receive high research priority in the future. Others have noted that marketing boards are not consistent with our pressure group economy.

Whatever the marketing mechanism, these publicly sanctioned entities require systems research to further assess their merits (from the producer, handler, and consumer perspective) in improving market performance.

COORDINATION BY CONTRACT. It is possible to coordinate a vertical system completely by contracts. Ward (1977) has described such an alternative for the beef subsector. More often, however, contracts have been less pervasive and have been initiated by the channel participant possessing greatest market power in an input acquisition or supply coordinating role. Mighell and Hoofnagle (1972) have reviewed and classified contracts according to provisions that used 1960 and 1970 data. Rhodes (1969) has done similar analysis for the turkey industry, Rogers (1976) for the poultry industry, Campbell (1976) for the vegetable canning industry, and Nicols and Sporleder (1976) for the cotton industry.

Despite numerous studies, contract types need to be researched concerning the transfer of entrepreneurial prerogatives. Exchange relationships under contractual arrangements need to be assessed along with the performance of different types of these arrangements. Of particular significance is the possibility of new systems for exchanging contracts, as recommended by Holder and Sporleder (1976), and new means of reporting contract terms. The possibility for organizational failure also needs to be assumed in systems so completely integrated that pricing efficiency can break down because of lack of a base on which to calculate transfer prices.

POLITICAL INVOLVEMENT. If, as Galbraith (1967) postulates, the obsession of growth-oriented firms is the state, then assessment of ownership and control of the food system needs to include analysis of political activity by market channel participants. Indeed, a case can be made that the political nature of control has been on the increase. This includes activity affecting property rules, regulations, and legal obligations that affect how firms behave and conduct themselves.

Relative assessment of political power presents itself on legislative issues where lines are clearly drawn. The activity leading to passage of the Agricultural Fair Practices Act of 1967 provided one such opportunity (Torgerson 1970). The same opportunity will be present concerning debate over good faith bargaining legislation first introduced in 1971. Hearings on the National Farm Bargaining Act, H.R. 3535, were initially held on July 10–11, 1979, and provide another opportunity for this assessment. Other exercises of power in association with administrative rule making and politically administrated decisions are more difficult to ascertain but nevertheless require research. A start can be made by determining resources spent for political action through public affairs divisions of firms, advertising aimed at public attitude rather than product differentiation, corporate political contributions, use of political action funds, and the like. More macro analysis along industry lines also would be useful (Hadwiger and Talbot 1968).

Further development of each of these areas can greatly augment subsector analysis in the food industry and lead to a deeper understanding of the roots of power and control.

Horizontal Integration. Analysis of horizontal integration by the agricultural economics profession, aside from that concerning group action by farmers and market structure analysis focusing on firm concentration at various levels, has received scant attention in the literature. The work covering group action by farmers has been focused at the producer–first handler exchange level where most farmer cooperative activity is found (French et al. 1980). Horizontal merger activity by dairy cooperatives in the late 1960s stimulated research on pricing activities and other forms of conduct by the profession as well as by outsiders in the regulatory agencies (Knutson 1971; Eisenstat et al. 1974; Kwaka 1975; Boynton 1977; Babb et al. 1979; Babb 1980). Results of analyses are conflicting, thereby suggesting the need for further analysis.

Conglomerate enterprise, a relatively new form of horizontal integration and market power, first received attention from Edwards (1955) and later Mueller (1977) and Parker and Connor (1979). Mueller (1978) has identified advertising-achieved product differentiation and application of conglomerate power to restructure industries through subsidized advertising and plant expansion as examples of the exercise of vast market power by conglomerates. Marion et al. (1977) have studied performance of the food retailing industry and found substantial monopoly overcharges. Connor (1980) reported on the impact of structural change on economic performance of the food manufacturing industry.

Analysis of use of franchising, dealerships, and subsidiaries in food distribution has primarily been the domain of business school research. Yet this form of horizontal market extension has yielded an increasingly significant away-from-home market for food consumption that requires attention by agricultural economists from a systems perspective.

Merger activity as one means of achieving corporate control may be the most efficient. A market for market control exists. According to Manne (1965) its advantages are "a lessening of wasteful bankruptcy proceedings, more efficient management of corporations, the protection afforded noncontrolling corporate investors, increased mobility of capital, and generally a more efficient allocation of resources." A conflict is seen with horizontal integration limits in assessing the boundaries of this market. Heflebower (1963) maintains that it is difficult to distinguish whether a firm's merger activity is conglomerate or horizontal and that definitive economic criteria for both types remain to be developed.

Economic Organization of Agriculture. Research on the economic organization of agriculture is difficult, and research dollars devoted to this area have been relatively small. Nevertheless, it is important because public policy decisions affecting the well-being of farmers and consumers are going to be made either by legislative action or by allowing trends to continue unabated.

Breimyer (1965) was among the first to point out the gravity of policy and other issues associated with the changing economic organization of agriculture, particularly as it impacts the farm firm. His treatment not only views vertical integration processes with considerable alarm but also anticipates many of the organizational issues necessary for farmers to survive in a world that is moving away from open market organization. He suggests that there is no aspect of individual freedom more prominent today than freedom through group action. Relative to farm operators, he says that "the minority who remain aloof from a group action that benefits all may not be expressing their own freedom so much as they are violating that of others who worked in their joint behalf."

The issues that Breimyer anticipated have since been developed on a variety of fronts. The Missouri school of which he is a part has done much to focus on the structure of agriculture (Breimyer 1964, 1965; Rhodes 1969; Torgerson 1970; Breimyer and Torgerson 1971; Meisner and Rhodes 1974; Matthews and Rhodes 1975). Rhodes points out four paths that the structure of agriculture may follow, including a family farm–open-market agriculture, a family farm–collective bargaining agriculture, a corporate-integratee agriculture, and a corporate-farmhand agriculture. (These comments were made by Rhodes at a talk presented at the annual meeting of the National Council of Farmer Cooperatives on January 14, 1969, which has been widely used by the author and other economists to identify changing structural paths in agriculture.) Whereas the first two alternatives visualize a dispersed ownership by farmers with varying amounts of group action, the last two envision corporate domination through contractual or vertical ownership of the production process and use of unionized farm labor. Added to these alternatives might be a cooperative production agriculture and a government-owned collective agriculture, neither of which is very developed in this country. Municipal land trusts are found in some communities, but these too are limited.

Farris (1970) observes that the common bonds tying farmers together horizontally are disappearing and becoming weakened as vertical systems claim increasing attention. To cope, he suggests that family farmers can attempt one of three approaches. The first is to counter trends by increasing coordination through fine-tuning marketing mechanisms that offset the need for coordination through management control. Similar options of fine-tuning open-market mechanisms, more use of negotiated pricing, and further vertical coordination and integration have been addressed by French et al. (1980). A second alternative is for farmers to take control through managed coordination of their own system. More aggressive efforts by cooperatives include innovation of new buying and selling arrangements and initiation of new efforts at developing new products, market development, selling and promotional activities. Finally, farmers can acquiesce in being shifted into the hands of nonfarm firms and become piece-wage operators or employees.

Shifts in farm ownership, size, and modes of coordination have been identified by Emerson (1978). A major conclusion from three alternatives analyzed is that conflict between farmers and consumers over price levels would not be a primary consideration in decisions concerning the future structure of agriculture. Rather, the impact would be on total receipts to farming for various size groups, the level and distribution of net farm income, and the viability of rural communities.

In a speech to the National Farmers Union on March 12, 1979, Secretary of Agriculture Bob Bergland called for a national dialogue concerning the structure of production agriculture, leading to the 1981 farm bill. His report based on structure hearings and other analysis was published in 1981 (USDA). Following this talk, considerable analysis was forthcoming from the USDA concerning farm structure and the benefactors of farm programs. The U.S. Senate committee on Agriculture, Nutrition, and Forestry (1979) grouped farms into three size categories: large farms (over $100,000 in sales), moderate sized farms ($20,000 to $100,000 in sales), and small farms (under $20,000 in sales). While small farms represented 70 percent of the total number, they accounted for less than 11 percent of total off-farm sales in 1974. The moderate sized group represented 25 percent of the farms and 37 percent of the sales, while the large-farm category represented only 6 percent of the farm numbers but over 52 percent of the sales. Based on 1974 data, a second study has documented increasing concentration in farming, in which 125,000 farms accounted for half of all farm sales in 1974 compared to 205,000 farms a decade earlier (USDA 1979a). It also found that farm programs favored larger producers, with almost one-half the payments going to only 10 percent of the program participants.

Analysis of a survey on land ownership has found that 1 percent of ownership units own about half of all the privately held land in the country. The study found that 40 percent of the U.S. land area is owned by government at all levels, and 60 percent is in private hands. Two-thirds of all privately held land is owned by sole proprietors, and almost three-fourths of that land is in farms and ranches. While many of the larger farms fall into the incorporated family farm category by type of ownership, continuing public concern exists about nonfarm corporate involvement in farm production and foreign ownership of farmland. Nonfamily farm corporations hold about 9.5 percent of all private land but only 2.4 percent of farm and ranch land. Heaviest concentration of nonfamily farm ownership was found in the Pacific region and Southeast (USDA 1979b). Foreign ownership by 26,000 non-U.S. residents was found to be one-tenth of 1 percent of the privately held land and consisted of about 400,000 acres.

In addition to this work, three major task force efforts set forth alternatives for future organization of the food sector (North Central Public Policy Education Committee 1973; National Public Policy Education Committee 1976; Knutson 1978). Later, the USDA (1981) addressed the overall structure issue in agriculture. The first deals primarily with ownership of the productive resources, while the second two are concerned with the issue of maintaining access to the market which, in the long run, may be an equally relevant issue.

The various issues cannot be viewed independently, since there is both an ownership and market dimension to the problem. The choices are to maintain an open market for independent producers, develop a bargained contract market with production contracts negotiated between farmers and processors, or develop an integrated system of producer-owned cooperative processing units in competition with private corporate production processing units (French et al. 1980). The third alternative is the expected outcome if current trends go unchecked (Breimyer 1964).

Given the diversity of U.S. agriculture and viewpoints of these systems, the most likely result will be the persistence of some combination of these alternatives for some time. This points up the need for research to show the economic and social advantages and disadvantages of these alternatives.

Farmers can take action to influence these alternatives. One action would be for producers to form bargaining associations to improve the imbalance of power between farmers and large independent processors or processors integrating via production or other contracts. Such a development would require new legislation providing for recognition and establishment of bargaining rules. Torgerson (1970) has described the legislative process involving the first piece of bargaining legislation. He concluded that the political struggle over control of agriculture has already manifested itself in Congress. Further analysis of the Michigan exclusive-agency bargaining law has involved litigation and decisions validating its constitutionality (Brown 1981; Skinner 1981). Another alternative might involve implementing a forward deliverable contract market to interface between the integrator and the farmer.

Other actions farmers might take involve maintenance of an open-market system. The most promising possibility of maintaining independent processors and family farmers coordinated by an open market involves the use of centralized exchange mechanisms. Along these lines, the USDA is testing four electronic trading systems. NC-117 has established a task force to evaluate these and other electronic exchanges. Some work at evaluating the effectiveness of these systems has been done to show whether a teleauction system has increased prices farmers receive for sheep and lambs (Holder 1979). Earlier work by Holder (1972) shows the operating and pricing efficiency of a computerized market for forward contracts. While these studies show the benefits of centralized trading, they only allude to but do not prove that such open-market exchanges could be expected to result in the persistence of independent processors and family farmers.

The other alternative, and one most likely to prevail if producers are to remain in control, involves organization of integrated producer processing cooperatives in competition with private corporate counterparts. While cooperatives have been in processing for some time, a fundamental question is whether they have achieved an adequate level of market penetration that can outperform the corporate integrator system in a totally integrated agriculture. Similar questions exist relative to the other alternatives. Is anyone interested in maintaining open markets? Is bargaining preferred to corporate and cooperative integration? While qualitative lists of advantages and disadvantages have been developed for each of these alternatives, we have few quantitative or decisive results.

One exception is represented by Carter and Johnston (1978), who echo G. Wunderlich concerning the development of a dual agriculture consisting of large commercial farm units and smaller farms based on subsistence, part time, retirement, and hobby interests. Raup (1970) suggests that such an outcome is not surprising, since our system is the result of an aggregation of policy decisions that impact rural communities as well as farmers. Until a cohesive policy is developed concerning the economic organization of agriculture, trends toward a corporation-dominated vertically integrated system will continue. The alternative organizational systems and linkages with the farm firm are shown in Figure 10.1.

FUTURE RESEARCH NEEDS

While much research has been addressed to the subject of ownership and control within the food industry, it has tended to be highly fragmented and

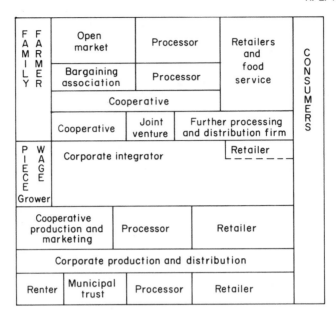

FIG. 10.1. A systems perspective of alternative forms of economic organization of agriculture.

generally lacking in a systems perspective. More often than not, research has been approached from the market structure paradigm, which has not focused on how market channel participants can acquire, gain, and maintain market power in the food system.

Research is therefore acutely needed that:

1. Addresses the uniqueness of the structural makeup of the farm production sector and the manner by which power can be maintained through various forms of collective action in the marketplace to sustain and nourish it.
2. Evaluates new forms of power relationships among market channel participants, their causal relations, and the means of routinizing systems that have private and public impacts on society.
3. Examines the spillover effect of extended vertical relationships on the horizontal competitive atmosphere.
4. Identifies workable rules for negotiating (the reconciliation process among power centers).
5. Explores behavioral aspects of large-scale power centers and the intra- and interorganizational relationships attendant to them.
6. Assesses the ways in which public policy in the form of tax rules, marketing mechanisms, and policies augments power in the ownership and control of the food industry or attempts to regulate or curtail it.

A series of research questions covering these broad topical areas are raised as they apply to ownership and control mechanisms in the food industry. These are identified according to themes used in the foregoing research review. The following will identify broad issues and then discuss practical research that can contribute to an understanding of the issues and to policy prescription if it is required.

Economics of Collective Action. Collective action in agriculture is highly fragmented and layered with a multiplicity of organizations of various degrees of development. Included among these are professional associations (general and single commodity), bargaining associations, cooperatives, breed associations, promotional and advertising associations, and a host of others. Interrelations between these organizations exist in many cases, but analysis of why, how, and at what cost requires further research. Inherent in this is study of the process of organizational maturation and development and elements of decision making as they relate to the farm firm. Means of terminating organizational efforts also require analysis. The notion that we have not found a mechanism for giving organizations a decent burial is all too apparent in the institutionalization and perpetuation of some whose meaningful contribution economically, politically, or socially is questionable.

Research is called for on systems design and rational planning that makes some sense out of the present multiplicity of voluntary organizational efforts in agriculture. The focal point should be on how goals of collective action are determined and how the system is routinized to attain these organizational goals. Based on this research, a plan must then be derived that provides guidance to farmers to better organize their limited and scarce resources through voluntary organizational efforts that maximize goal attainment through improved coordination and at the same time eliminate costly organizational overlaps and duplication of functions, which frequently counteract one another.

Distribution of the benefits and costs of collective action also require research attention. A perennial problem of voluntary organizations is one of internalizing the benefits of group action such that they are contingent upon participation and not available marketwide as a free good to nonjoiners as well as joiners. Restated, it is also a problem of how to get nonjoiners to share in the costs of group action from which they benefit.

In developing this research, careful attention needs to be given to the very complex and fragile nature of voluntary group action as it relates to communication and control within these organizations as well as to the interorganizational linkages necessary for goal attainment. Standard operating procedures internal to these organizations (identified in Chapter 3) and the regulatory environment external to the firms require careful scrutiny in developing organizational design. Attention also needs to be given to the most effective means of achieving goals through economic activity such as cooperative processing and distribution, negotiating over raw product prices, or other contract terms; political action aimed at creating an operating and regulatory climate conducive to collective action; and social organizational attributes that are inherent in the fabric of voluntary organizations and are essential to their internal governing structure. Practical questions that can be researched are grouped as they relate to collective action in agriculture.

Farmers have traditionally used professional associations (general farm and commodity) as a basic organizational type representing the economic and other interests of the farm firm. Only scanty insights are available on the operations of these associations, including their internal organizational structure, revenue sources, and operating functions. A clearer picture is needed on their contributions to farmer welfare and to their role as a source of market power in representing farmers. Questions unique to this organizational form follow.

What is the changing nature and fabric of farmers' professional organizations? Why have farmer movements such as the National Farmers Organizations and the American Agriculture Movement arisen periodically throughout our agricultural history? Were channels for protest not open through existing professional associations? How can those organizations be structured to provide outlets for farmers' grievances? What is the changing functional role of the professional associations as the political strength of farmers declines in numbers? If as Farris (1970) posits, the horizontal common bonds between farmers are disappearing as use of vertical combinations increase, can general farm organizations survive this phenomenon and how do they accommodate to it? What are the alternative means of financing these professional associations? How can benefits from affiliation with professional associations be internalized? What is the interrelationship of cooperatives and professional associations in representing economic and other interests of farmers? Is there a structural and functional relationship? How can these organizational systems be better coordinated? Can a vertical and horizontal dimension to them be identified?

Use of economic organizations, commonly identified as cooperatives, has been widely practiced by farmers as a linkage with the increasingly concentrated food manufacturing and distribution and farm-input sectors. The degree of market penetration by farmers through their cooperatives varies widely among commodities and regions of the country. An important aspect for research is the design of these organizations when they take on a diversified nature and attempt to harmonize various commodity interests. How these organizations can become big and be effective competitors and yet seem small to their members from a communications and control standpoint presents social scientists with a priority research issue. Some of the most fruitful design questions concern the interrelationships of various levels of cooperative activity as national and international linkages are forged. Design of these cooperative systems and maintenance of a legal-economic environment for their operations in achieving farmers' income goals is a key topic for research in the collective action area.

Specific questions to be addressed follow. Do we have an adequate theory of cooperative marketing by farmers and other participants in the food system? How does the performance of cooperatives compare with standard corporations and other business forms? What are the market performance results of more upstream market penetration by farmers through their cooperatives? Are cooperatives playing an increasing role in market access for farmers?

Some of the highest priority research questions in the area of economics of collective action are the following. What are the sources of cooperative market power? If cooperatives possess inherent structural weaknesses that inhibit the exercise of market power (Galbraith 1957), what types of social engineering will make them more effective instruments for achieving farmer market power? What future role is there for pooling and use of marketing agreements? Can the interface with marketing mechanisms such as federal market orders augment cooperative market power? Have existing marketing mechanisms done so? What alternatives—besides direct membership (centralized) cooperatives that restrict membership—are there for cooperatives to control supply? Do full-supply contracts in other subsectors increase marketing efficiency as Boynton (1977) has found in dairying? How can federations and marketing agencies in common be modified to become more effective instruments for attainment of market power? How can cooperatives attain power that accompanies organizational size without sacrificing farmers' entrepreneurial status and control? How is society protected from the exploitive use of market power once it is attained by cooperatives?

Are there any new institutional changes that can provide cooperatives with sounder capital acquisition and financial management strategies? What programs can better address equity redemption issues and maintain ownership by current users? What mix of ownership and various types of debt capital provides farmers with assurances of well-capitalized marketing arms in an inflationary economy? What can be used as a basis of value reporting for financial purposes in single-pool cooperatives when there is no adequate open-market raw product price? How do members remain the primary beneficiaries rather than residual claimants of group action that is designed to benefit them in large-scale organizations?

Vertical Integration and Coordination. A predominant characteristic of food industry organization has been the prevalence of more vertical integration and coordination, with resulting impacts on traditional market channel relationships, concentration of market power, elimination of market pricing mechanisms, and the character of the food industry itself. Research that encompasses the various forms and types of integration and coordination is needed to identify the pervasiveness of this development. Of equal importance is research that identifies the motives for vertical relationships. Earlier it was noted that a lack of theoretical focus exists in treating the vertical integration—coordination phenomenon. Alderson's (1957) approach of functionally examining how firms routinize the system to their advantage deserves priority. Commonly identified causes of integration such as specialization, coordination, and reduction of risk and uncertainty can be usefully explored from the perspective of this approach.

Essential to this work is development of improved understanding of management strategies associated with decisions to vertically integrate and the tactics used to effect such decisions. Of particular importance is the use of cross-subsidization and reciprocity in conglomerate firms and their impacts on market competition. Conglomerate growth in the food manufacturing and distribution sectors requires examination in view of market power relationships that go beyond the industrial organization research approach. Control of information flows and financial resources through holding companies and other mechanisms requires immediate and deeper probing. In particular, knowledge of the role of tax avoidance and its influence is necessary as a possible explanatory variable in emerging firm structure.

Beyond the mechanics of establishing the extent and motives for vertical relationships, research is needed to identify how these structures adapt over time. For instance, use of contractual arrangements in the broiler and turkey industries has often been associated with availability of cheap, underemployed labor resources. If, through organization or other production alternatives, this labor is not available as cheaply, will the industry revert to ownership integration?

Research is also needed to identify the gains derived from integration through ownership and contractual coordination. New institutions associated with vertical coordination (holding companies, joint ventures, franchises, or dealerships) should be assessed to measure their performance and competitive effects on other market channel participants.

Specific questions that require answering in this broad area follow. What explains the tendency for the food industry to gravitate to a more highly concentrated system that is characterized by vertically integrated firms of a diversified form? Is the trend explained more by the separation of management from ownership in corporate firms and associated market power objectives and goals or by system demands for closer coordination? Is the strategy one of integrating to a level that is less competitive? Does the vertically integrated system contain elements that lead to self-destruction or organizational failure? What roles do firm size, tax rules, corporate finance, government marketing mechanisms, contractual coordination, and political involvement play in firm growth and exercise of market power? What functional roles are associated with structure? Are we moving to an industry-by-industry codification (syndicalism) of competitive practices as suggested by Breimyer (1977)?

Why do firms vertically integrate through contracts? Are the advantages of vertical coordination linked to performance, as shown by Boynton (1977) in the dairy industry, or to institutions? How are contracts used as a tool for vertical integration? Is there an evolutionary trend in the type of contracts used and in contract features that can be identified (e.g., in the broiler and turkey industries or others)? If so, what management strategies are associated with such trends? How has the system been routinized to meet firm objectives? Could similar efficiencies be gained through establishment of a forward deliverable contract market?

Where does the initiative originate for joint ventures between food processors and distributors and farmer cooperatives? What management strategies are associated with such arrangements? What have been the performance results of such institutions as a form of vertical integration–coordination? To what extent are cooperatives used as a business form among nonfarm business organizations? Is use of this form growing? What inducements have led to its growth?

What types of system benefits from coordination are found in subsector studies? Are these findings transferable from one commodity sector to another? What change agents are most influential in inducing changes in market relationships given changes in market structure?

Ownership Form. Traditional types of business activity defined as individual proprietorships, partnerships, and corporations have received research attention over the years on an individual basis, mostly from the business schools. The cooperative as a basic business form has been primarily addressed in land-grant universities and specifically in the agricultural economics curriculum. This unique form, however, is coming into wider use in the consumer and other sectors. The attributes of this owner-user form of business and the set of laws, internal governing procedures, and operating characteristics that differentiate it from other types provide a continuing area for fruitful research.

An important aspect of the cooperative is its identification theoretically and practically as an off-farm extension of the farm firm, or the basic membership unit, and the resulting interdependencies and financial relationships that this entails between the two. Analysis of joint maximization of benefits and the role of the cooperative as a form of vertically integrated business activity awaits further discovery.

Perhaps the most insightful aspect of cooperative research involves empirical testing of how cooperatives actually enhance market competition as a dimension of performance. Commonly referred to as the yardstick school as originated by Nourse (1945), this research would examine the impact of voluntary membership, lack of control over member production decisions, and the aggregate effect of the presence of cooperatives as a dimension of market structure.

Extensions of ownership forms other than cooperatives that do not necessarily reflect traditional business forms are in particular need of research. Included among these are the conglomerate enterprise as a new form of raw economic power, the use of limited partnerships as a means to circumvent foreign ownership laws involving our land resources, and other formal or informal means of achieving market presence in our pressure group economy. How specific laws such as those for antitrust affect and deal with these ownership forms and their differential impact is one of our more pressing areas of needed research. Basic identification of data availability to adequately monitor and measure conduct and performance of such firms is essential.

Another research area concerns parastate organizations such as marketing boards and their present or potential role in marketing. The existence of quasi-governmental organizations for dealing with nations where an agency of the state does the trading in export markets, or for attempting to solve domestic pricing or other issues, is in need of critical research.

Specific questions relating to further research on ownership forms are given below in order of importance.

If the food system is increasingly dominated by diversified firms that can cross-subsidize operations and engage in reciprocity in exchange with other firms, how does this affect competition with cooperatives and noncooperatives that are organized essentially along single commodity lines? Are single commodity entities or farmer cooperatives at a competitive disadvantage because they can only engage in member-related services? How do these various ownership forms compare in performance?

What is the source of power for national and multinational conglomerates? Are they corporate "states" unto themselves? What are the dimensions of decision making, communication, and financial power held at the top management level? How do market intelligence and communication systems operate? What is the value of intelligence as gathered internally in terms of accuracy, timeliness, and cost compared with alternative sources? How are financial transactions and accounting systems handled in multinationals? What advantages or liabilities are associated with taxation? What activities do the political affairs divisions embody? What are the advantages and disadvantages of subsidiaries (including foreign-based operations) to such firms? How does the activity of multinationals impact on the performance of international and national markets? Are multinationals more cartelized than is permitted domestically? How does the ability of multinationals to acquire commodities in other countries affect the performance of domestic firms?

What are the market channel participant and system payoffs from more horizontal integration? Is performance improved when producer–first handler transactions are more coordinated rather than less? Why do mergers appear to be the best way to achieve horizontal growth? How does the performance of franchise outlets or dealerships compare to nationally and regionally owned integrated chains? Do cooperative agencies in common lead to superior or inferior performance results?

How do stockholders or member-users exert their voice and control in large-scale organizations? Do technocrats have more say over resource use and the objectives of planning than owners? Do firms grow for the sake of aggrandizement of management or system benefits? What social engineering is required to keep owners as the primary beneficiaries rather than the residual claimants of organized action designed to benefit them?

Who will control agriculture? In what ways do ownership of land and contracting by nonfarm investors, corporations, or land trusts affect the future of family owned and operated agriculture? Should ownership of land by those other than active farm operators be outlawed? What constitutes a family farm? How are supporting institutions such as land-grant universities, the extension service, the USDA, and others affected by the demise of family-farm agriculture? Can the use of cooperatives and professional associations provide the needed market access and clout to permit dispersed farm families to survive? As large-scale commercial farming operations have grown, so has the incidence of organized farmworkers. What is the impact of unionization on the scale and type of ownership integration of farming operations? Does unionization provide an institutional incentive for the growth of corporate agriculture. What are the implications of more farmer cooperative involvement in food processing? How can small-volume producers access markets in an increasingly integrated food system?

If as Shaffer (1968) suggests there is a tendency for control to take on a more political nature, what are the manifestations of it? How has the nature of Commons's (n.d.) "occupational parliament" changed? Are pressure groups more or less effective than in the past? Has political activity by individual firms undermined or bolstered trade association activity? Is the corporate objective control of the state, as Galbraith (1967) suggests? What are the political and economic consequences of such a phenomenon as industry becomes more concentrated? Are property rules, regulations, and legal obligations written by those they are designed to regulate? What explains the political success of public interest lobbies?

Answers to these research questions will provide a much better base of knowledge on how our food system operates and will give guidance to public policy prescriptions to deal with emerging system issues.

HOW TO ORGANIZE FOR RESEARCH ON ORGANIZATION AND CONTROL

Three aspects stand out after reviewing the literature and the scope of research yet to be done. The first is the apparent success of recent efforts to coordinate research and extension undertakings through consortiums of regional committees composed of land-grant university and government contributors. In particular, the NC-117 effort consisting of a nucleus of centrally housed researchers with contributing work from land-grant universities and the federal government represents a unique approach that may provide a model for future intensive efforts. The output has generally been of high quality and coordinated to provide additivity. Since this effort is the research counterpart of extension's "Who will control agriculture?" series, the output, including subsector research and analysis of food distribution, has been of particular significance. Particular focus is needed on linkage of the atomistic production sector with the increasingly concentrated food manufacturing and distribution sectors.

The second aspect of significance is the importance of the scope and findings of the National Commission of Food Marketing (1966) to many of the subsequent research efforts in organization and control of the food industry. The commission provided baseline data that NC-117 updated in part. Most significantly, the findings represented a cross-sectional research effort of massive proportions. While unmatched in its broad coverage, subsequent efforts by the Federal Trade Commission and others to seek appropriations for this activity on an ongoing basis suggest the importance attached to this research by regulators and public policymakers.

The third aspect is the multidisciplinary nature of much of the work that remains to be done. By definition, the study of organization and control involves organized behavior systems. Marketing research must address the systems nature of organized behavior, whether in the form of conglomerates or various forms of group action by farmers. Members of the agricultural economics profession cannot lay claim to exclusive jurisdiction over such studies, even in areas such as vertical coordination where they have done most of the work. The processes of organizational learning, negotiation, transfer pricing, intelligence gathering, capital formation and use, legal maneuvering, and corporate planning and management each have dimensions of social psychology, sociology, political science, and economics involved with them. Means of harmonizing these multidisciplinary inputs and focusing them on organization and control of the food system remains to be accomplished.

What remains is a division of labor concerning a broad and multifaceted research area. Clearly, more subsector research is needed to cover other areas that have not been addressed. In so doing, small groups of two to four researchers may tackle the individual subsectors. The major task that follows is to analyze findings from individual studies on a cross-subsector and multidisciplinary basis. Ultimately, the food system must be analyzed from a total systems perspective that shows interrelationships with domestic food programs, foreign aid initiatives, and economic control programs such as wage and price controls. Each can have a major impact on organization of the food industry. Agricultural economists can be expected to take charge of much of the structure and performance research in economic organization of the food industry.

Control aspects of research will require input from other disciplines, although agricultural economists can maintain project leadership in many instances. Examples are analysis of management strategies and decision making aimed at growth, intelligence acquisition, and political involvement. Research on the process of organizational learning (tracing major decisions to determine how conflicts are resolved within the context of a dynamic political and legal environment) stands as one of the unconquered areas of future work on organizational control within the food industry. Particularly significant is the interface with government facilitating and regulatory mechanisms.

The major barriers to research in the control area are access to data in vertically integrated systems and the development of trust that will permit researchers access to internal decision-making processes. Often the documentation of organizational history is lacking, and it is often highly biased when documented in formally written histories. The researcher is therefore challenged to employ the case study approach on a current basis while tracks are fresh. This is exhausting work but essential to better insights into organizational ownership and control as a dynamic process.

REFERENCES

Alchain, A. A., and H. Demsetz. 1972. Production, information costs, and economic organization. *Am. Econ. Rev.* 62:777–95.

Alderson, Wroe. 1957. *Marketing Behavior and Executive Action.* Chicago: Irwin.

Aresvik, Oddvar. 1955. Comments on the economic nature of the cooperative association. *J. Farm Econ.* 37:142.

Babb, E. M. 1980. Milk Marketing Services Provided by Cooperative and Proprietary Firms. Purdue University Agric. Exp. Sta. Bull. 279, West Lafayette, Indiana.

Babb, Emerson M., D. A. Bessler, and J. W. Pheasant. 1979. Analysis of Over-Order Payments in Federal Milk Marketing Orders. Purdue University Agric. Exp. Sta. Bull. 235, West Lafayette, Indiana.

Bain, Joe S. 1959. *Industrial Organization.* New York: Wiley.

Barnard, Chester. 1938. *The Functions of the Executive.* Cambridge, Mass.: Harvard University Press.
Baumol, William. 1959. *Business Behavior, Value and Growth,* Chap. 6.New York: Macmillan.
Black, Bill. 1978. Marketing agreements. In Agricultural Cooperatives and the Public Interest, p. 235. NC-117, Monogr. 4, University of Wisconsin, Madison.
Blaich, O. P. 1960. Integration in theory with an application to hogs. *J. Farm Econ.* 42:5.
Boulding, Kenneth E. 1968. *The Organizational Revolution,* pp. 50–51. Chicago: Quadrangle Paperbacks.
Boynton, Robert. 1977. Performance of the U.S. Dairy Subsector As Affected by the Vertical Coordination Processes between Cooperatives and Proprietary Handlers. Unpublished Ph.D. diss., Michigan State University, East Lansing.
Breimyer, Harold F. 1964. Future organization and control of U.S. agricultural production and marketing. *J. Farm Econ.* 46:930–44.
_____. 1965. *Individual Freedom and the Economic Organization of Agriculture,* p. 34. Urbana: University of Illinois Press.
_____. 1977. Rules, Roles, Economic Reality and the American Dream. Talk at New Mexico State University, April 21, Las Cruces.
Breimyer, Harold F., ed. 1971. Bargaining in Agriculture. NCR Ext. Publ. 30, University of Missouri, Columbia.
Breimyer, Harold F., and Randall Torgerson. 1971. Farmer Cooperatives: For Whom and by Whom. University of Missouri, Market. Ext. Newsletter, Columbia.
Brown, Gail. 1979. Financial Planning For Cooperatives. Talk to the National Council of Farmer Cooperatives Annual Meeting, Las Vegas, Nevada, January.
Brown, Thomas L. 1981. Opinion of the Court. *Michigan Canners and Freezers Association* v. *The Agricultural Marketing and Bargaining Board and the Michigan Agriculutral Cooperative Marketing Association, Inc.* Nov. 74-16369-CZ, Jan. 30.
Bunje, Ralph B. 1980. Cooperative Farm Bargaining Price Negotiations. Coop. Inf. Rep. 26, USDA, ESCS.
Campbell, Gerald. 1976. Grower–first handler exchange arrangements in the Wisconsin processed vegetable industry. In Coordination and Exchange in the Agricultural Subsector. NC-117, Monogr. 2, University of Wisconsin, Madison.
Campbell, Gerald, and Marvin Hayenga. 1978. Vertical Organization and Coordination in Selected Commodity Sectors. NC-117 Work. Pap. 20, University of Wisconsin, Madison.
Carter, Harold O., and Warren Johnston. 1978. Some forces affecting the changing structure, organization and control of American agriculture. *Am. J. Agric. Econ.* 60:738–47.
Clodius, Robert. 1957. The role of cooperatives in bargaining. *J. Farm Econ.* 39:1271–81.
_____. 1959. Opportunities and limitations in improving the bargaining power of farmers. In E. O. Heady, *Problems and Policies in American Agriculture,* Chap. 17. Ames: Iowa State University Press.
Commons, John R. n.d. *Representative Democracy.* New York: Bureau of Economic Research.
_____. 1950. *The Economics of Collective Action.* New York: Macmillan.
_____. 1957a. *Institutional Economics.* Madison: University of Wisconsin Press.
_____. 1957b. *Legal Foundations of Capitalism.* Madison: University of Wisconsin Press.
Connor, John M. 1980. The U.S. Food and Tobacco Manufacturing Industries. USDA, ESCS, AER-451.
Cook, Michael. 1978. The sources, limits and extent of cooperative market power: Financial laws and institutions. In Agricultural Cooperatives and the Public Interest. NC-117, Monogr. 4, University of Wisconsin, Madison.
Cyert, Richard, and James March. 1963. *A Behavioral Theory of the Firm.* Englewood Cliffs, N.J.: Prentice-Hall.
Edwards, Corwin D. 1955. Conglomerate bigness as a source of power. In National Bureau of Economic Research, *Business Concentration and Price Policy.* Princeton: Princeton University Press.
Eisenstat, Phillip, and Robert T. Masson. 1978. Cooperative horizontal market power and vertical relationships: An overall assessment. In Agricultural Cooperatives and the Public Interest, pp. 281–92. NC-117, Monogr. 4, University of Wisconsin, Madison.

Eisenstat, Phillip, Robert T. Masson, and David Roddy. 1974. An Economic Analysis of the Associated Milk Producers, Inc. Monopoly. Filed with the court in *U.S.* v. *Associated Milk Producers, Inc.* 394 F. Supp. 29(W.D. Mo. 1975), aff'd. 534 F. Supp. 2d 113 (8th Cir. 1976).

Emelianoff, Ivan. 1942. *Economic Theory of Cooperation,* p. 35. Ann Arbor: Edwards.

Emerson, Peter. 1978. Public Policy and the Changing Structure of American Agriculture. Congressional Budget Office.

Farris, Paul L. 1970. The aggregate impact of trends in the farm firm on the economy and on agriculture as an industry. In Emerging and Projected Trends Likely to Influence the Structure of Midwest Agriculture, 1970–85. Monogr. 11, Agricultural Law Center, University of Iowa, Iowa City.

———. 1972. Coordination and the competitive market. In Robert E. Schneidau and Lawrence A. Duewer, eds., *Symposium: Vertical Integration in the Pork Industry.* Westport, Conn.: AVI.

French, Charles E., John C. Moore, Charles A. Kraenzle, and Kenneth F. Harling. 1980. *Survival Strategies for Agricultural Cooperatives.* Ames: Iowa State University Press.

Galbraith, John K. 1957. *American Capitalism: The Theory of Countervailing Powers.* London: Hamish Hamilton.

———. 1967. *The New Industrial State.* Boston: Houghton Mifflin.

———. 1973. Power and the useful economist. *Am. Econ. Rev.* 63: 1–11.

Garoyan, Leon, and Eric Thor. 1978. Observations on the impact of agricultural bargaining cooperatives. In Agricultural Cooperatives and the Public Interest, pp. 135–48. NC-117, Monogr. 4, University of Wisconsin, Madison.

Goldberg, Ray. 1968. *Agribusiness Coordination.* Cambridge, Mass.: Harvard University Press.

———. 1972. Profitable partnerships: Industry and farm cooperatives. *Harvard Bus. Rev.* 50:108–21.

Hadwiger, Don, and Ross Talbot. 1968. *The Policy Processes in American Agriculture.* San Francisco: Chandler.

Handy, C. R., and D. I. Padberg. 1971. A model of competitive behavior in food industries. *Am. J. Agric. Econ.* 53:2.

Heflebower, R. B. 1963. Corporate mergers: Policy and economic analysis. *Q. J. Econ.* 78:537–58.

Helmberger, Peter. 1964. Cooperative enterprise as a structural dimension of farm markets. *J. Farm Econ.* 46:603–17.

Helmberger, Peter, and Sidney Hoos. 1965. *Cooperative Bargaining in Agriculture.* University of California, Div. Agric. Sci., Berkeley.

———. 1972. Cooperative enterprise and organization theory. *J. Farm Econ.* 44:275–90.

Henn, Harry G. 1970. *Handbook of the Law of Corporations and Other Business Enterprises.* St. Paul: West.

Henry, W. R., and R. Rauniker. 1960. Integration in practice—The broiler case. *J. Farm Econ.* 42:5.

Hibbard, Benjamin. 1921. *Marketing Agricultural Products,* p. 187. New York: Appleton, Giannini Foundation.

Hoffman, A. C. 1970. Trends in the food industries and their relationship to agriculture. In Emerging and Projected Trends Likely to Influence the Structure of Midwest Agriculture, 1970–1985. Monogr. 11, Agricultural Law Center, University of Iowa, Iowa City.

Holder, David. 1972. A Computerized Forward Contract Market for Slaughter Hogs. Agric. Econ. Rep. 211, Michigan State University, East Lansing.

———. 1979. Benefits of Sheep and Lamb Teleauction in Virginia and West Virginia. Paper, Southern Agricultural Economics Association Meeting, New Orleans, La., Feb. 5–7.

———. 1980. The rise of economic power: Some consequences and policy implications. *Am. J. Agric. Econ.* 62:866–72.

Holder, David, and Tom Sporleder. 1976. Forward deliverable contract markets. In Marketing Alternatives for Agriculture: Is There a Better Way? National Public Policy Education Committee Publ. 7, Cornell University, Ithaca.

Jacobson, Robert. 1978. Sources, limits, and extent of cooperative market power: Cooperatives and marketing orders. In Agricultural Cooperatives and the Public Interest, pp. 171–86. NC-117, Monogr. 4, University of Wisconsin, Madison.

Johnson, Ralph. 1972. An Economic Evaluation of Alternative Marketing Methods for Fed Cattle. SB 250, University of Nebraska, Lincoln.

Knutson, Ronald. 1971. Cooperative Bargaining Developments in the Dairy Industry, 1960–70. USDA, FCS Res. Rep. 19.

_____. 1978. Who Will Market Your Products? Agric. Ext. Serv., Texas A & M University, College Station.

Krebs, Allen. 1978a. Corporate incest. *The Ag Biz Tiller,* Issue 10.

_____. 1978b. Seed to supermarket. *The Ag Biz Tiller,* Issues 11 and 12.

Kwoka, John. 1975. Pricing under Federal Milk Market Regulation. Paper prepared at North Carolina State University, Raleigh.

Ladd, George W. 1964. *Agricultural Bargaining Power.* Ames: Iowa State University Press.

Lang, Mahlon G. 1978. Structure, conduct and performance in agricultural product markets characterized by collective bargaining. In Agricultural Cooperatives and the Public Interest, pp. 118–34. NC-117, Monogr. 4, University of Wisconsin, Madison.

_____. 1980. Issues in the Design of Farm Bargaining Legislation. Purdue University Dep. Agric. Econ. Sta. Bull. 277, Lafayette, Ind.

Liefmann, Robert. 1932. *Cartels, Concerns and Trusts,* p. 2. London: Methuen.

Logan, S. H. 1969. A conceptual framework for analyzing economies of vertical integration. *J. Farm Econ.* 51:834–48.

Madsen, Howard C., and Earl O. Heady. 1971. Bargaining Power Programs. CARD Rep. 39, Iowa State University, Ames.

Manne, Henry G. 1965. Mergers and the market for corporate control. *J. Polit. Econ.* 73:110–20.

Marion, Bruce. 1976. Vertical coordination and exchange arrangements: Concepts and hypotheses. In Coordination and Exchange in Agricultural Subsectors. NC-117, Monogr. 2, University of Wisconsin, Madison.

Marion, Bruce W., Willard F. Mueller, Ronald W. Cotterill, Frederick E. Geithman, and John R. Schmelzer. 1977. The Profit and Price Performance of Leading Food Chains, 1970–74. Joint Economic Committee of Congress. Washington, D.C.: USGPO.

Masson, Robert, and Phillip Eisenstat. 1978. Capper-Volstead and milk cooperative market power: Some theoretical issues. In Agricultural Cooperatives and the Public Interest, pp. 51–68. NC-117, Monogr. 4, University of Wisconsin, Madison.

Matthews, Stephen, and V. James Rhodes. 1975. The Use of Public Limited Partnership Financing in Agriculture for Income Tax Shelter. NCR Publ. 223, University of Wisconsin, Madison.

Meisner, Joe, and V. James Rhodes. 1974. The Changing Structure of the U.S. Cattle Industry. University of Missouri Spec. Rep. 167, Columbia.

Mighell, Ronald, and William Hoofnagle. 1972. Contract Production and Vertical Integration in Farming, 1960 and 1970. USDA, ERS-479.

Mighell, R. L., and L. A. Jones. 1963. Vertical Coordination in Agriculture. Agric. Econ. Rep. 19, USDA.

Mueller, W. F. 1960. Discussion: Farmer cooperatives. *J. Farm Econ.* 42:503–5.

_____. 1977. Conglomerates: A nonindustry. In W. Adams, ed., *The Structure of American Industry,* 5th ed. New York: Macmillan.

_____. 1978. The control of agricultural processing and distribution. *Am. J. Agric. Econ.* 60:848–55.

_____. 1979. The National Antitrust Commission and Cooperatives. Talk to National Council of Farmer Cooperatives Annual Meeting, Las Vegas, Nevada, January.

Mueller, W. F., and N. R. Collins. 1957. Grower-processor integration in fruit and vegetable marketing. *J. Farm Econ.* 39:1471–883.

National Commission on Food Marketing. 1966. *Food from Farmer to Consumer* and technical reports. Washington, D.C.: USGPO.

National Public Policy Education Committee. 1976. Marketing Alternatives for Agriculture: Is There a Better Way? Publ. 7, Cornell University, Ithaca.

Nicols, John P., and Tom Sporleder. 1976. Producer-first handler exchange arrangements for upland cotton. In Coordination and Exchange in the Agricultural Subsector. NC-117, Monogr. 2, University of Wisconsin, Madison.

North Central Public Policy Education Committee. 1973. Who Will Control U.S. Agriculture?

Series of 6 leaflets, University of Illinois, Champaign-Urbana.

Nourse, E. G. 1922. The outlook for cooperative marketing. *J. Farm Econ.* 4:80–91.

———. 1945. The place of cooperation in our national economy. In *American Cooperation 1942–45*, p. 34. Washington, D.C.: American Institute of Cooperation.

Olson, Mancur, Jr. 1965. *The Logic of Collective Action.* Cambridge, Mass.: Harvard University Press.

Padberg, D.I. 1966. Efficiency and welfare considerations in integrated agriculture. *J. Farm Econ.* 48:1391–1400.

Parker, R.C., and J. M. Connor. 1979. Estimates of consumer loss due to monopoly in the U. S. food-manufacturing industries. *Amer. J. Agric. Econ.* 61:626–39.

Paul, Allen B. 1964. Capital, finance and market structure—Two approaches. In P. L. Farris, ed., *Market Structure Research.* Ames: Iowa State University Press.

———. 1974. The role of competitive market institutions. *Agric. Econ. Res.* 26:41–48.

Phillips, Michael J., Thomas L. Sporleder, James R. Baarda, and Gilbert W. Biggs. 1977. Processed Potato Growers' Associations. USDA, FCS Res. Rep. 35.

Phillips, Richard. 1955. Economic nature of the cooperative association. *J. Farm Econ.* 34:75–76.

Purcell, Wayne D. 1973. An approach to research on vertical coordination: The beef system in Oklahoma. *Am. J. Agric. Econ.* 55:65–68.

Raup, Phillip. 1970. The Impact of Trends in the Farm Firm on Community and Human Welfare. Emerging and Projected Trends Likely to Influence the Structure of Midwest Agriculture, 1970–85. Monogr. 11, Agricultural Law Center, University of Iowa, Iowa City.

Rhodes, V. James. 1969. Changes in Turkey Contracting, 1967–68. University of Missouri Exp. Sta. Bull. 886, Columbia.

Robotka, Frank. 1947. A theory of cooperation. *J. Farm Econ.* 29:94–114.

Rogers, George. 1976. Producer-handler exchange in the poultry and egg industries. In Coordination and Exchange in the Agricultural Subsector. NC-117, Monogr. 4, University of Wisconsin, Madison.

Roy, Ewell Paul. 1970. *Collective Bargaining in Agriculture.* Danville, Ill.: Interstate.

Schneider, Vernon, Clement Ward, and Ramon Lopez. 1978. Corporation Involvement in Agricultural Cooperatives. Texas A & M University Departmental Tech. Rep. 78-2, College Station.

Shaffer, James D. 1968. A Working Paper Concerning Publicly Supported Economic Research in Agricultural Marketing. USDA, ERS.

Shaffer, James D., and Randall E. Torgerson. 1976. Exclusive agency bargaining. In Marketing Alternatives for Agriculture: Is There a Better Way? National Public Policy Education Committee Publ. 7, Cornell University, Ithaca.

Skinner, Robert. 1981. Economic Implications of the Michigan Bargaining Law. Talk to the National Bargaining Conference Annual Meeting. Washington, D.C., January.

Stigler, George. 1951. The division of labor is limited by the extent of the market. *J. Polit. Econ.* 59:185–93.

Thompson, James D. 1967. *Organizations in Action.* New York: McGraw-Hill.

Torgerson, Randall E. 1970. *Producer Power at the Bargaining Table.* Columbia: University of Missouri Press.

———. 1971a. Alternative structural approaches to farm bargaining: A cooperative systems model. *Am. J. Agric. Econ.* 53:823–24.

———. 1971b. *Farm Bargaining.* Oslo: Landbruketsforlag.

———. 1978. An overall assessment of cooperative market power. In Agricultural Cooperatives and the Public Interest, pp. 261–80. NC-117, Monogr. 4, University of Wisconsin, Madison.

Trifon, Raphael. 1959. Guides for speculation about the vertical integration of agriculture with allied industries. *J. Farm Econ.* 41:734–49.

U.S. Department of Agriculture. 1976. The Question of Undue Price Enhancement by Milk Cooperatives. Capper-Volstead Study Committe Rep.

———. 1979a. Status of the Family Farm—Second Annual Report to the Congress. ESCS, AER-434.

———. 1979b. Who Owns the Land?—A Prelimary Report of a U.S. Land Ownership Survey. ESCS-70.

_____. 1981. A Time to Choose: Summary Report on the Structure of Agriculture. Office of the secretary.

U.S. Senate Committee on Agriculture, Nutrition, and Forestry. 1979. Status of the Family Farm. 96th Congress, 1st Session, June 18.

Walsh, Richard. 1968. The Role of Cooperatives in Vertical and Horizontal Integration in Agricultural Production and Marketing. Seminar on Agricultural Organization in the Modern Industrial Economy, sponsored by NCR-20, Chicago.

Ward, Clement. 1977. A Contract Integrated Cooperative Cattle Marketing System. USDA, FCS, Market. Res. Rep. 1078.

Ward, Clement, Vernon Schneider, and Ramon Lopez. 1979. Voting Systems in Agricultural Cooperatives. USDA, ESCS, Coop. Res. Rep. 2.

Williamson, O. E. 1971. The vertical integration of production-marketing failure considerations. *Am. Econ. Rev.* 61:112–23.

Youde, James. 1978. Cooperative Membership Policies and Market Power. In Agricultural Cooperatives and the Public Interest, pp. 219–25. NC-117, Monogr. 4, University of Wisconsin, Madison.

11

PREFERENCE ARTICULATION AND FOOD SYSTEM PERFORMANCE

THIS CHAPTER is a discussion of research involving the relationship between preference articulation and economic performance of the food system. Preference articulation refers to the processes by which choices among alternative mixes of goods and services may be expressed and taken into account within the opportunity set (Commons 1968). Economic performance refers to the total flow of consequences or outcomes of the system that organizes transformation of resources. The research question in its broadest sense is, Is the economic performance of the food system consistent with the preferences of the relevant participants and, if not, what changes in the organization of the system would make it more consistent with what is desired? Presumably, the ultimate objective of most of the research discussed in this book is to contribute to improved performance of the food system. By what criteria should performance be judged? What are the appropriate mechanisms for articulating preferences to achieve desired performance? How can the system be structured to be more responsive to preferences? Whose preferences?

Obviously, this deals with fundamental elements of political economic organization: the mix between market and political processes for providing information about preferences and opportunity sets, the jurisdictional boundaries of decision units for expressing preferences, organization and relationships of decision-making units that respond to current and future preference and opportunity sets in organizing production and distribution, rules and customs determining what is taken into account by decision-making units, and processes that influence preferences.

JAMES D. SHAFFER is Professor of Agricultural Economics, Michigan State University, East Lansing. This chapter benefited from comments and assistance from Compton Chase-Lansdale, A. Schmid, E. Babb, S. Hiemstra, P. Farris, W. F. Mueller, P. Wandschneider, B. Ferres, and C. Cordes.

The topic is too big and too fundamental to be dealt with comprehensively in this short chapter. All the topics of this book could be logically included. Conceptual work is needed to develop a more comprehensive theoretical framework for analysis of preference articulation and economic performance. However, it would be easy to get bogged down in a morass of complex issues. This chapter is concerned with questions, issues, and practical research. It attempts to raise some questions about preference articulation and performance of the food system, which may add a dimension to the research topics discussed and to suggest some research related to specific problems in preference expression and responsiveness of the food system to those preferences.

IMPORTANCE OF QUESTIONS

The problems and issues related to preference articulation are central to evaluation of food system performance and formulation of policies dealing with economic organization. The effectiveness of the food system in identifying and responding to preferences within the production opportunity set is a major dimension of performance. This goal is often assumed to be embedded in a performance goal identified as efficiency, but it is often not dealt with explicitly in evaluating performance. Problems of articulation of preferences are also involved in establishment of performance criteria. What characteristics of performance are preferred, and what trade-offs among alternative performance characteristics are preferred? For example, how effectively are preferences articulated concerning the trade-off between the degree of food safety and prices or between some explicit measures of progressiveness and equity?

The issues of preference articulation involve such basic related questions as: What should be the mix of market and political processes used for preference articulation? What can be done to improve market mechanisms for preference articulation? What can be done to improve political processes for preference articulation?

The last two questions are interdependent. All markets are politically instituted in the sense that the rules of the game governing the market are political. To improve the market mechanism will probably involve political processes (Dahl and Lindblom 1953). Similarly, the response to the first question dealing with the mix of market and political processes depends upon the relative effectiveness of these two processes as instituted. Thus it is useful to think of the effectiveness of the political process as a mechanism for articulating preferences about the rules of the market, property rights, and the like, and as a means for directly expressing preferences in regard to specific outputs. Quite different mechanisms may be appropriate for determining the rules of competition, for example, than for determining the most desirable characteristic of a peach. The political process must be used directly in the first and need be involved only indirectly in the latter.

A major function of the market is to identify preferences and organize responses to them. But the output of a market system is much more than simply the products sold. Preferences for some outputs cannot be effectively articulated through the market; thus the mechanisms of political preference articulation must be used to get desired performance. The institutions for political preference articulation are thus critical in determining the performance of the food system.

EVOLUTION OF CONCEPTUAL APPROACH

Much marketing research has been directed to the problem of improving the effectiveness of the market in identifying and responding to preferences. Until recently, most of this research has been within the context of the neoclassical paradigm. An implicit assumption was that moving the market for food and farm products in the direction of meeting the assumptions and conditions of optimality of the pure competition model would improve its effectiveness in responding to consumer preferences. This directed attention toward a particular set of questions.

Considerable agricultural marketing research has been directed at decreasing market imperfections. Researchers investigated price patterns, asking such questions as: Are price differentials among markets and through time consistent with cost differences? If not, why? Elaborate spatial equilibrium models were developed to investigate the price relationships over space (Martin 1977b). Market imperfections were associated with a lack of information and competitive structures. Thus attention has been directed at improving market information and most recently at the problems of obtaining relevant information in a market situation where open spot markets did not exist or were of minor importance. Contracted vegetables for processing is an example.

A major problem involving the responsiveness of the system to preferences is the lag between production decisions and consumer purchases. Production decisions must anticipate future supply and demand conditions. Thus much marketing research was undertaken to improve information about the outlook (Martin 1977a).

The problem of coordinating supply and demand, i.e., organizing supply to meet demand at prices high enough to assure long-run supplies (or acceptable incomes), has been a significant focus of marketing research, reaching beyond improving outlook information. Research dealing with marketing orders and agreements, marketing cooperatives, forward contracting, and vertical integration has focused upon this issue (Martin 1979a).

Similarly, demand analysis was intended to assist in outlook and the coordination of supply with demand. This research focused on identifying basic relationships between supply, demand and prices, and other factors that influence demand.

Other research inquired more directly into consumer preferences and behavior. For example, several national workshops on food marketing were devoted to research dealing with identification of consumer preferences and relevant product attributes so that products could be designed to be more consistent with consumer preferences (USDA 1951). Studies were also made of consumer purchase patterns, using in-store observations, one-time interviews, and consumer purchase panels. Some attempts were made to better understand consumer purchase decisions and to measure utility. An extensive literature about consumer behavior has been produced by business schools, marketing firms, and advertising agencies.

The search for market imperfections included studies of competitive structure, profits, and marketing costs and margins. This was based in part on the assumption that excess profits or margins were evidence of monopoly and a lack of responsiveness to consumer preferences. At least it assumed that more products could have been supplied within the resource opportunity set.

A related set of research and literature has been identified with the concept of market failure. The basic concept is that a market as instituted may fail to respond to an abstract concept of consumer preferences because important consequences of actions are not taken into account by decison makers. The most discussed examples involve environmental effects that are disregarded in production and consumption decisions. These are often referred to as external effects and might better be called third party effects. The usual suggestion by economists has been to find a rule that will cause decision makers to take the consequences into account. There are a number of examples of market failure research involving the food system. Examples include studies of water pollution from feedlots and processing plants and unemployment and associated consequences resulting from the adoption of technology (Master 1977a). Hightower's (1973) *Hard Tomatoes, Hard Times* is concerned with this problem.

Another related area is identified as public choice. A basic concept in this literature is that some goods and services have a marginal cost of zero for an additional user and that additional users cannot be economically excluded; when these conditions exist, preferences for the good or service will not be accurately reflected in the market. This is involved in a number of issues in food policy. For example, the option demand for future use of farmland has this characteristic and is central to problems of conservation. The issue of providing food reserves involves a similar problem. This is sometimes referred to as a "free-rider" problem and is frequently mentioned in discussions related to food policy and organization of farmers seeking collective goals such as price stability.

While the main thrust of agricultural marketing research, especially during the period from the passage of the Agricultural Marketing and Research Act in 1946 until the late 1960s, has been directed toward perfecting the market within the basic neoclassical paradigm, other views also have influenced research related to preference articulation.

A significant amount of research originated from a business school view of the world. While the emphasis of agricultural economists was on research intended to improve the general performance of the market, the business school orientation emphasized merchandising. Research with this orientation was aimed at expanding the market for specific farm products. Much of the research was very similar to that just discussed but with a modest twist. For example, in-store experiments selling apples in different sized bags were designed to discover means of expanding the purchase of apples, not necessarily to improve the performance of the market. Similarly, research into consumer preference and behavior was intended to help expand sales, perhaps by improving promotion rather than the product. Some public and much private research has been done to test consumer acceptance of products. This information may be used to contribute to the design of products more preferred by consumers or to aid in promotion, market segmentation, and product differentiation. Market research of this orientation may lead to products more consistent with consumer preferences either by adapting products to preferences or preferences to products. The outcome depends upon the characteristics of the market. In any case, this type of market research is a significant factor in the effectiveness of the market in identifying and responding to consumer preferences.

There is another difference between the neoclassical and business school paradigm. It has to do with differences in meaning attached to competition and the burden of proof in respect to performance. The notion of perfect competition in neoclassical economics is competition among sellers who are so numerous and small that no individual seller is able to influence market price. Each seller seeks to market a product that will best respond to consumers' demand as revealed in open markets for undifferentiated products. Promotion and merchandising are unnecessary costs. It is assumed that if the characteristics of the perfect market are obtained, desired performance will result. Thus the role of government is to promote the conditions of perfect competition. The business school paradigm views atomistic competition as chaos; nothing can be planned (Galbraith 1967). It views competition basically as head-to-head strategy to outsell the competitors where their behavior is interdependent. It accepts no necessary relationship between the structural conditions of the market and performance. The burden of proof for government participation in modifying the market is with those who argue that performance could be improved. The emphasis is on performance, not the conditions of competition.

This difference in paradigms was implicit in the views of the majority and minority of the National Commission on Food Marketing (1966); e.g., the majority advocated compulsory grades and standards to improve information and thus facilitate the responsiveness to consumers' preferences. The minority argued that compulsory grades and standards would undermine private brands and take away the seller's incentive to market quality products designed to cater to preferences of consumers. Further, they argued that private brands create an incentive to maintain quality control because a poor product would cause a loss in the value of the brand. The majority proposed expanded consumer information (a move toward more perfect knowledge), while the minority considered it an unnecessary cost. The majority generally believed selling costs to be a waste and an indicator of noncompetitive organization. The minority believed selling costs and product development were evidence of high levels of competition and that consumers benefited. An interesting observation is that after more than fifteen years little research has been undertaken to deal directly with the recommendations on which the commission members disagreed. For example, apparently no one has done research on the question of the relative effectiveness of a system of private brands compared with a system of compulsory grades in responding to consumer preferences.

The field of industrial organization contributed what some refer to as the structure-conduct-performance approach to marketing research. This framework draws attention to the relationship of economic structure to performance of an industry (Scherer 1970). This has been discussed in other chapters. The framework can be expanded to the examination of the relationship between institutions and behavior of participants and in turn can be related to performance. Concepts of economics, sociology, and political science can be integrated into this broad framework.

The concept and measurement of performance becomes critical. Research in industrial organization has concentrated on the dimensions of efficiency, equity, and progressiveness. Marion and Handy (1973) have discussed the dimensions of performance in considerable detail. A number of studies have been undertaken with the intention of evaluating some segment of the food system. The basic idea of evaluation is to compare existing performance (outcomes) with some performance norms. The norms are presumably preferred outcomes; i.e., they are beliefs about preferences. The NC-117 project, mentioned in other chapters, has as its major overall objective the evaluation of performance of the U.S. food system. This program of research includes a number of subsector studies that attempt to relate institutional structure and conduct within the production-distribution system for a class of commodities to a number of dimensions of performance. Special emphasis is placed on the effectiveness of vertical coordination that is related to responsiveness to consumer preferences. Some performance norms are implicit in the analysis but no attempt has been made empirically to relate information about preferences for various outcomes with actual performance.

Agricultural economists doing research in food system organization and performance have tended to be problem-oriented and eclectic in the selection of theoretical concepts. Research has been rich in use of concepts but lacks an integrated framework relating articulation of preferences and economic performance.

The evolving approach to research related to preference articulation and economic performance adopted in this chapter is also theoretically eclectic. It starts with the view that economic performance is multidimensional and research needs to be concerned with the total flow of consequences from any economic subsystem under investigation. It is accepted that the market cannot articulate some preferences and that the responsiveness of the economy to preferences is a function of the way the market is initially instituted. Emphasis is placed on identification of relationships between institutional structure and performance of the system. The integrated theory has not been developed.

ASSESSMENT OF PAST RESEARCH

A considerable amount of research has been produced, contributing to improvement of the market as a mechanism for identifying and responding to consumer preferences, given the general institution of the market. Some research has been produced relating the organization of the market to the responsiveness of the system to preferences. The nature of this research was discussed in a previous section, and evaluation (although perhaps in a somewhat different context) will be included in other chapters.

There has been relatively little research dealing directly with some of the

major questions in regard to preference articulation and economic performance. The relative effectiveness of political and market processes as means for
articulating preferences for food system outputs has not been systematically
researched. Work has been done by theoretical economists and political scientists dealing with the problems of preference articulation through the political
process. Arrow's (1951) work on social choice and Riker and Ordeshook's
(1973) study on positive theory are examples. But agricultural economists have
not evaluated the comparative effectiveness of the two processes as means of
preference articulation for specific attributes of performance. Little has been
done to compare market structures that include different mixes of planning
based upon political expression of preferences and market processes. There
has been little inquiry into the effectiveness of political processes in identifying
desired performance criteria for the food system and the related questions of
how to implement the preferences. Within agricultural economics, marketing
and policy have been considered separate areas of research, even though the
focus of policy has been modification of the outcome of the market. This
separation may have influenced the approach to the question of preference articulation and economic performance, separating the political expression of
preferences from the market expression.

The list could be continued, but this can best be done by turning attention
to the question of research for the future.

FUTURE RESEARCH AND QUESTIONS

Major conceptual, philosophical, and measurement problems are involved in attempting to answer such fundamental questions as: Is the performance of the food system as consistent as possible with preferences, given
the opportunity set? What changes in the way the system is instituted would
make performance more consistent with preferences within the opportunity
set? How are preferences and the performance of the food system associated
with long-run well-being? Making judgments about performance and the effects that changes in the system will have on it requires conclusions or assumptions about the nature of preferences and human nature and the fundamental
role of government and individual rights.

Ultimate decisions concerning the choice of mix between market and political processes for preference articulation must be political. Thus voter rather
than consumer sovereignty is assumed. And while unarticulated preferences
are unknowable by definition, it is assumed that at least the Constitution is an
accurate articulation of preferences. What follows are some relevant topics for
research dealing with preference articulation and performance; the list is not
comprehensive.

What Are Preferences? What do people really want? seems to be the logical
first question for research on preference articulation. A great deal has been
written about the nature of wants and preferences. Yet much conceptual work
remains to be done. We do not know the psychological characteristics of

preferences, which may not exist except as revealed in the processes of articulation.

It is generally assumed that people know what they prefer and that they will like what they get if it is what they prefer. However, uncertainty and ignorance are major facts of life. People frequently do not know what they want or what they are getting when they buy a product, and they do not know the consequences of alternative political and market choices. Frequently, the actual outcome is quite different from the expectation.

The notion of a social welfare function implies that people have a set of preferences in regard to the dimensions of performance and that the problem is to discover the social welfare function. A more realistic view is that the social welfare function does not exist independently of the process of determining it; i.e., it is not discovered but is determined.

Research on the psychological nature of preferences is important but beyond the scope of this chapter. Nonetheless, several assumptions about the nature of preferences is important as background to discussion of suggested research for agricultural economists. It is useful to think of preferences as learned predispositions for responding to situations. These predispositions are related to positive and negative reinforcements in the environment and to physiological characteristics of the individuals. Thus preferences are not fixed. They evolve and are influenced and cannot be taken as given. People seek to improve and change their wants and preferences, and a condition of well-being is dependent upon developing those that are appropriate. Some wants and preferences are more basic than others. One purpose of research and education is to help people understand the consequences of acting on specific preferences. Research on the relationship of food preferences (and consumption) and health is designed to help people develop those preferences consistent with better health. This is critical research for improving the performance of the food system.

Some psychologists suggest a two-self theory. One self is the manager and the other the actor. The managing self is a controller and is concerned with the long run. It seems that it is a psychological fact that individuals (the acting self) have a strong tendency to respond to immediate reinforcers. The successful individual has a managing self that manipulates the environment to provide immediate reinforcers to reward behavior consistent with long-run welfare; e.g., students use self-rewards for studying. In other words, we manipulate our own psychological environments to shape our preferences and behavior. The managing self is socialized; that is, the longer run goals are learned from the social environment created by the society. Designing this environment is the most important function of any society.

It is useful to think of two classes of preferences, personal and social (Thurow 1973). Personal preferences are those that can be expressed within the existing economic structure or environment. They deal with the economic structure and environment, which cannot be effectively articulated by individual action. For example, individuals may prefer an unpolluted environment but may contribute to pollution by their behavior. The fact that they pollute does not indicate a preference for pollution. They simply cannot get a significant reduction in pollution from actions within the system by their behavior. This distinction is important in thinking about the mix of market and political institutions for articulation of preferences. Some preferences simply cannot be effectively articulated by market mechanisms.

Research dealing with food system performance is concerned with the effective articulation of both social and personal preferences. An important aspect of market policy research must be to identify social preferences about performance characteristics of the food system.

Identifying Dimensions of Performance. The political articulation of preferences requires knowledge about the relevant dimensions or characteristics of performance. An important research task for agricultural economists is the definition of important outcomes of the food system and the design of performance indicators. This has been a significant contribution in the past and will be more important in the future. Research needs to be expanded to encompass a more comprehensive set of outcomes. Both conceptual and empirical research is needed to identify outcomes that would be perceived to be important if known. Techniques for measurement need to be improved. Practical empirical research would include surveys of food system participants to find out what they believe to be outcomes and what makes a difference, as well as developing general monitoring indicators. Systematically developing a comprehensive inventory of dimensions of performance may reduce unintended and unexpected outcomes.

Many dimensions of performance are dealt with in other chapters. This discussion will not be comprehensive, but it may be useful to raise some research questions. How should nutrition and food safety be defined and measured? There seems to be considerable confusion with current measures. Consumers are concerned about the reliability of the food supply. In an industrialized food system, reliable food supplies depend upon reliable supplies of many inputs. How should this be monitored? Many participants in the food system express concern about price stability and uncertainty presumably because of the effect it has on planning. What are meaningful measures?

Equity is a very important performance dimension. Considerable data relative to equity, such as the distribution of income, are generated; yet problems remain in both definition and measurement. Similarly, progressiveness or innovativeness is a commonly identified dimension of performance that requires additional research. Policy decisions intended to improve performance of the food system would also have to consider employment and inflation. While aggregate measures of both exist, it is not clear that adequate measures related to the food system have been developed; e.g., there is no adequate definition or measure of underemployment in the food system. Price increases from such factors as poor weather are not distinguished from changes in prices as a result of the money supply. Similarly, many policy trade-offs must be made between environmental effects and other outcomes of the food system. More meaningful measures of environmental quality are needed.

The dimension of performance of most interest to agricultural economists has been market efficiency, which might be defined as producing the mix of products most consistent with consumer preferences as expressed through the market at the least cost, given the opportunity set. Definition and measurement continue to be problems requiring research.

Another important class of performance dimensions is even more difficult to define and measure. These are such important outcomes as freedom, distribution of power, sense of community, justice, values, attitudes about society and government, alienation, and quality of community leadership and life. Agricultural economists have considered some of these outcomes in their discussions of the family farm and farm production controls. Good indicators still do not exist.

All indicators of performance of the food system deserve the research attention of the agricultural economist. Some clearly require contributions of other disciplines. Political articulation of preferences in regard to food system performance would be facilitated by improved descriptions and indicators of its dimensions.

The Trade-off Matrix. Preference articulation has little meaning independent of the opportunity set. A valuable characteristic of a market system is that prices carry much information about preferences and the opportunity set in a condensed, easy-to-understand form. Markets and marketlike processes can contribute information about the opportunity set to the policy process, but nothing comparable to the market has been devised to provide information for decisions about the trade-offs among the many performance dimensions or goals for the food system. Research plays a major role in helping decision makers understand these trade-offs. An understanding of the food system as a system is needed. Another way of viewing the research problem is to ask if a specific policy designed to achieve one goal will have effects not only in respect to that particular goal but also for all the other dimensions of performance. Research will never be able to do this comprehensively, but a systematic display of the nature of the trade-off matrix, empirical evidence on important trade-off relationships, and inferences about others would contribute to the articulation of preferences in the political process. Designing the trade-off matrix and systematically identifying important relationships would be a useful input in planning research. Several specific trade-off issues will be discussed subsequently.

Whose Preferences Count? Obviously, the preferences of different participants in the system carry different weights. Preference articulation has to do with power. Research designed to describe whose preferences count and how they are weighted could be useful in designing improved institutions for preference articulation. The first task would be to develop meaningful classifications of participants. People have different roles and their preferences in respect to each role are relevant, not just their role as consumers. Again, a comprehensive description would be impossible, and even gross description would be difficult. Nonetheless, specific research could yield meaningful information.

Samuels (1972) reminds us that the market is a system of mutual coercion. Each participant in a market theoretically influences the opportunity set of all others. If a rich person buys a large quantity of a product and the price increases, others have to buy less. This is called a pecuniary externality and is usually ignored in economics, but it cannot be. For example, the rich may adopt no conservation and bid fuel from "critical" uses of the poor. The result may be inconsistent with the social preference for justice or fairness of many citizens. A political mechanism would be needed to articulate these preferences. The way the distribution of wealth and rights affects the opportunity set in regard to a few strategic situations in the food system is worthy of research.

Hundreds of laws, administrative rules, and programs determine whose preferences are catered to in the political economy of the food system. Whose preferences count in instituting these rules and programs? Individuals and groups hold strategic positions to influence legislation, administrative rules, and judicial decisions. There is an economics of influence; for some individuals the benefit-cost ratio from resources spent on influencing these decisions is high, for others low. The cost of organizing groups to influence decisions affects the outcome. Information is power (Bartlett 1973). It is profitable for some individuals and groups to produce information and not for others. Some are in a position to control information. Industrial structure affects this ability and also the profitability of producing influence (see Chapter 2).

Nor can it be assumed that preferences are independent of the influence system. Large amounts are spent on public relations to influence public opinion to gain political support for particular positions. Thus distribution of influence is unequal, and in turn some people's preferences receive much greater weight than others. Empirical research substantiates the phenomenon of media impact on political preferences, although the extent is circumscribed in the short run, and the manner of influence is complex (Robinson 1976).

Understanding and improving preference articulation requires good descriptions of how the influence system works. This requires going beyond examination of the process for major legislation. For example, the intent of legislation can be greatly altered by influencing its administration. Monitoring and influencing day-to-day operations of an agency are more expensive than influencing a bill, and this aspect of the economics of influence is more important. Also, since influence is more profitable where the outcomes of a rule are uncertain, research reducing this uncertainty reduces the power of concentrated interests (Bartlett 1973).

The Existing Mix of Market and Political Processes. As mentioned, a basic question is the optimal mix of market and political processes for preference articulation. This is an impossible question to answer. However, research can seek to provide information to improve the mix and to improve the market and political processes. Similarly, it would be impossible to determine the percentage of the output of the food system that is in response to market as contrasted to political processes. All products are in response to both processes; i.e., all markets are politically instituted by rules determining what has to be taken into account by participants, and all public programs respond to information about preferences and opportunity sets obtained through market processes. Nonetheless, research to describe the major features of the mix of the current food system and to identify trends would be a useful input to the policy process in considering institutional changes. Such research would require development of a useful taxonomy. How do we circumscribe the market processes of the food system through political articulation of preferences? How extensive are subsidies, taxes, tax "subsidies," government purchases, specific property rights, rules facilitating collective bargaining, regulation for food and safety, worker protection, subsidized consumer information, price supports and controls, direct government production, credit regulations and sanctioned procedures, rules influencing types of business organization, industrial structure, definitions of theft and extortion, and rules on packaging and labeling? All these devices are attempts to implement preferences through the political process. The more important question is, of course, how effective in the aggregate are these attempts in articulating preferences about performance of the food system?

Market Processes. Research is needed to better understand the problems of market processes in articulating preferences. How effective is the market as a process for articulating preferences in regard to the dimensions of performance for the food system? What are the limits of the market in preference articulation? This is the subject of much of economics and this book.

Economists are fond of the concept of Pareto optimality. There is another equally important concept. This is the situation when a person expressing his or her individual preferences and acting in his or her individual interest causes an aggregate outcome that none prefer. This extreme case is probably as rare in nature as the case of Pareto optimality, but the researcher who can identify either is doing good work. The relationship of micro to macro phenomena has never been dealt with satisfactorily in economics.

Resources are seldom allocated to accurately reflect preferences where the added cost of an additional user is zero, where benefits to a user cannot be captured by a potential provider, or where costs to a third party need not be considered. These are discussed as problems involving free riders, externalities, public goods, joint impact goods, and social traps. They are ubiquitous. The tragedy of the commons is an example of this class of problem. For example, consider the problem of consumers handling fruits and vegetables and destroying them in the process. Research could usefully focus on identifying these types of problems and the means of improving preference articulation where they exist. Note that public action may be neither necessary nor desirable; e.g., packaging the fruits and vegetables to protect them may be more effective than enforcing a rule against handling the products.

Limits to Exit. Preferences are articulated through the market primarily by decisions to buy or not buy a product. But not buying a product usually gives the seller little information about desired quality characteristics. Markets articulate preferences effectively only to the extent that prices summarize the relevant information. Hirschman (1970) provides an extensive development of this theme. He deals with the specific question of quality deterioration. Suppose that a firm is performing well (i.e., providing products consistent with consumer preferences) and then for some reason, quality slips. Owners of the firm do not know what is causing the reduced sales. The seller may go out of business before understanding the problem and recovering. If the firm has some loyal or inert customers, the seller is more likely to be able to recover in time. A more complex situation is one where dissatisfied customers go from firm to firm, always being dissatisfied; but by trading such customers, little discipline is placed on the lazy managements to respond to consumer preferences. Thus Hirschman argues that oligopoly, with its easy alternative, may result in firms less responsive to consumer preferences than monopoly, because if there is no close substitute, the buyer is more likely to respond with what Hirschman calls voice. Consumers are likely not only to complain but to organize to bring some kind of pressure for improved performance. When price does not carry the information alternative, information systems are required. But uninstituted voice may not be enough. It appears that a vocal response to quality deterioration has both a third-party effect and a free-rider problem. All consumers benefit from appropriate complaints about quality, but the benefits to any individual customer may not be enough to bother.

Thus an important topic of research is the improvement of voice mechanisms as means of communicating information about desired quality characteristics and influencing producers to respond to preferences. A particular problem is to ensure representation as well as reliability in information and influence. Public interest groups and consumer testing organizations provide this function. Research on class action suits, a special case of the voice option, would also be appropriate.

Cooperatives mix exit and voice in a unique way and are important in organization of the food system. They should be studied to determine their potential as a superior business organizational form for articulating preferences. Customers of a proprietary firm usually have little incentive to improve its performance; they take their business someplace else. But a cooperative may be able to organize the combined effect of voice and potential exit to influence the performance of the management. Study needs to be made of means for improving the incentive to effectively exercise voice. Judging a cooperative's performance on the same basis as a proprietary firm (i.e., profits) loses the unique advantage of the cooperative; attention needs to be paid to its unique political character.

Scale Economies and Variety. Another class of problems in preference articulation involves economies of scale and the trade-off between cost and variety where two or more systems are incompatible. Consumers usually cannot effectively articulate preferences for options that are not offered; if only one option is offered, it is not necessarily the one that would be preferred by the majority of consumers. For example, do consumers prefer a very limited variety of cheeses at lower prices to a larger variety at somewhat higher prices? If there are economies of scale in production and distribution in total and for each variety, the market process will not necessarily accurately reveal preferences. The lowest price would be associated with offering a single type of cheese. Each additional type of cheese increases the cost of cheese already marketed. The loss of utility caused by higher prices of the more popular varieties of cheese is traded for utility gained by those who get more choices. The outcome will depend upon the structure of the market, standard operating procedures in regard to allocation of overhead, and guesses about market demand. What mechanism would facilitate expression of preferences for alternative systems in regard to the number of varieties of cheese?

Assume a delivery route exists with 100 customers. The cost of serving the route is $100 plus $1 per stop. Total procurement cost is thus $200 or $2 per customer. An innovation is introduced in the market when a store offers the product for a $1 service charge (equal to cost). Ten customers have an opportunity cost for picking up the product at only 90¢, thus they can save 10¢ each by the switch to the store. Now the cost to delivery customers is ($100 + 90) = $190 or $2.11 each, while the cost to store customers is $1.90 each or $1 for the 10. Total procurement cost is $209. But the process continues. Another 10 delivery customers have an opportunity cost for store pickup of $1.00. It is now advantageous for them to switch, saving 11¢ each. Now the cost of delivery is ($100 + 80) = $180 or $2.22 each, while the cost to store customers is ($1.90 × 10) + ($2.00 × 10) = $39. Total procurement cost is $219. For another 10 the opportunity cost of pickup is $1.10 and they can save 12¢ each by switching. Now, the delivery cost goes to ($100 + 70) = $170 or about $2.43 each. Another 10 have an opportunity cost of $1.30 for store pickup, etc. Note only the first 10 store consumers are better off than they would have been without the innovation. If we extrapolate the above cost relationships, the process will continue until each customer, acting in his own short-run interest, will become a store purchaser.

The total real cost of procurement will be increased substantially and only a very small proportion will be better off than at the start. The result is due to the fact that costs external to the accounts of individual buyers existed, and there was no means to shift these costs to the original 10 buyers or to compensate them to stay with the delivery system.

Each buyer looks at the effective set of prices and acts to the greatest advantage. As each one acts, he or she may change the price structure. In sequence this may result in higher prices for almost everyone as shown above. If the 90 cannot negotiate, the higher price is not a payment that accrues to some producer's profits or some consumer's utility but is reduction in consumer's real income. It may be in terms of leisure time foregone; it is not transferred to others but lost altogether.

Emerson Babb (personal communication 1979) made a useful comment on the above example:

> I am troubled by your delivery example. The point you are making is very valid, but this is also the rationale of state milk control agencies for policies which in-

hibit innovation, produce spatial monopoly through restrictive licensing, and resist changes in consumer preferences. I could use your numbers, assume that the two stores and customers are at the identical location, assume a cost of delivery and a cost of pickup, and come up with different results. In the case of $100 fixed costs plus $1 per stop, assuming that the $1 reflects average transporation cost to spatially separated customers, nearby customers are subsidizing more distant customers. This presumably gives rise to the first 10 customers electing to pick the product up at the store, and incentives for more distant customers to follow suit as the process proceeds. Which system would result in lowest total costs basically depends on the aggregate cost of customers picking up their goods versus having them delivered, ignoring asset fixity. If preferences were strictly cost related, I would encourage the system that minimized total cost. Your results hinge on a set of customers paying a larger share of the fixed cost for one system and do not consider just closing that store. The above does not invalidate your point that the wrong choice may be made for "rational" reasons when decisions are made about private goods which involve fixed cost and arbitrary allocations. The real question is which system should have received the investment or have been encouraged to continue. The "market" may not make the best decision, but once the decision is made we are stuck with assets.

This comment raises the important fact that all the dimensions of performance need to be considered in evaluating system alternatives.

On the other hand, an innovation may be introduced with a high fixed cost and low variable cost, which would result in lower total cost to the total group if generally adopted. The lower cost innovation may never be adopted because the critical mass of customers cannot be attracted in time to sustain the enterprise. Some extramarket institution is required to achieve the economies of the new system.

The neighborhood grocery lost out in competition with larger stores serving more extensive trade areas. It is generally assumed that the change reflects the preferences of consumers. However, it can be shown that this situation may be very similar to the delivery case and that the neighborhood is unable to organize to effectively communicate its preference for retaining the neighborhood store. The neighborhood store has some of the characteristics of a public good, and the free-rider problem exists. The convenience store now fills the void created by the loss of the neighborhood store. Why? Did preferences change, did consumers learn, or did it take new management and different pricing practices? Considerable assets were lost in the process.

These examples are an attempt to stimulate diagnostic research into problems of preference articulation. A second step is to devise institutions that address the problem. A third step is to evaluate the probable effects of the alternatives on performance.

Missing Information or Hero. Prior to 1948, canned red tart cherries intended for consumer use were packed almost exclusively in #2 cans. In 1948 the 303 can was introduced. Twelve years later, no tart cherries were packed in #2 cans, the 303 had won out. From the first year it was introduced to the final year, the percentage sold of cases that were 303s, of the two sizes combined, progressed as follows: less than 1%, 1%, 3%, 4%, 13%, 31%, 53%, 69%, 89%, 93%, 94%, 100%. The transition was a process, and the evidence is that the result was not a desirable one. And it appears to be irreversible without a change in market rules. The process seems to have been about as follows. While the 303 can contains about 17% fewer cherries, it is not easily distinguished by consumers from a #2 can. The two sizes appear to contain about the same amount. Thus cherries in 303 cans alongside #2 cans, if priced differently per can, appear to be a bargain. Brands packed in 303 cans won sales from those packed in #2 cans and caused the brands losing market shares to convert to the smaller can. Since there are economies in packing and distributing a single can size, an incentive existed for a single processor to pack in only one size. As more and more brands were packed in 303s, the losses of sales of brands in #2 cans increased, creating more incentive to switch from that size. What was the performance result? Almost all the canned cherries are used by homemakers for pies. It is generally agreed that the quantity of cherries in a 303 can provides a skimpy pie. The great majority of consumers would choose a pie from the contents of a #2 can. The difference in cost of the two pies would be of little significance. The total number of cherry pies baked would not increase but would probably decrease (because the pies produced would be less satisfactory, although the consumer would probably not identify the cause). Thus fewer cherries would be sold, to the disadvantage of the cherry producers. The total cost of processing would be increased because a cost is involved in setting up facilities for a different can size. The initial innovator benefited by a temporary advantage, but the total profits from processing would remain the same or be lower. Thus a temporary gain by a few processors at the expense of others had a detrimental effect on almost all participants in the market and would be judged by almost everyone to have been an undesirable performance.

Where was the missing hero to assume the costs of informing the consumers of the change in product? Did rules against collusion prevent a trade association that would have considered the aggregate consequences? Could trade association committees with consumer representation improve preference articulation? Does requiring unit pricing information on the grocery shelf contribute to the solution of this type of problem?

Vertical Coordination and Planning. Yet another class of problems in preference articulation are those involving vertical coordination. Production decisions are made in anticipation of a future demand. Producers may be many steps removed from consumers. How can improved information about future demand for alternative products of an individual producer be generated? The hog and beef price cycles continue. Prices of broilers continue to fluctuate. Farmers often find that in particular years they produce at prices below their costs of production because of general overproduction, and in other years there are "shortages." There is, of course, the fact of weather variation and uncertainty associated with farm production. Beyond the weather the variations are due largely to problems in coordination rather than shifts in the physical opportunity set or preferences. Neither information on the technical opportunity set nor the preferences are accurately reflected in the market process.

Consider research to test the hypothesis that coordination in the food system could be improved through a system of compulsory forward deliverable contract markets. Such markets are made technically possible by the existence of computer and communication systems. Contract markets could be established at all points in the food system between the farmer and the retailer; e.g., a market for beef contracts could link farmers and processors. Processors could bid for lots of cattle with specific characteristics. Cattle producers would offer cattle based upon their costs. The contract would assure the price. The farmer in turn could have contracts for feed, and the processor could have contracts for the processed beef with wholesalers, retailers, and restaurants. The system would require retailers to anticipate demand and would shift the risk of mistakes in assessing future demand from farmers to retailers. Retailers should be in a better position to assess demand and are in a much better position to deal with the risk of having excess supplies because they can influence demand by advertising and merchandising. This system change would provide for planning production to meet future demands by using the market mechanism. Theoretically, preferences would be transmitted much more accurately than under the current system. However, a political decision is required for implementation of the system. It would be most effective if all a commodity is produced and sold within this market planning system. Since some would oppose such a system, the political problem of determining whose preferences count would have to be resolved.

This is but one example of a possible approach to the problem of coordinating production decisions of many producers with future demand. It is of particular interest because it relies on the market mechanism, taking advantage of its capacity to identify preferences and search the opportunity set. It is an example of an institutional innovation that needs to be evaluated along with alternatives.

Prices and Information. A discussion of market mechanisms for preference articulation might well focus on the role of prices and information. Research on both is appropriate to preference articulation and food system performance. Both are treated in other chapters and thus receive light treatment in this discussion. Also, grades and standards are a means of facilitating the market articulation of preferences. They deserve research attention with emphasis on the design of grades and standards that do in fact increase the information content about product characteristics that make a difference to users. Also, the controversy among members of the National Commission on Food Marketing over private brands versus compulsory grades and standards deserves research attention.

Inflation also deserves special research attention. Inflation involves the micro-macro phenomena where each group attempts to improve its position and in the process reduces production and creates higher prices. As a process it can lead to economic disaster for almost everyone. Inflation also distorts the price signal and thus interferes with the articulation of preferences. Both these phenomena are of great importance to the long-run performance of the food system.

New Technology. The combination of low-cost, high-capacity computer systems and sophisticated communication networks creates new opportunities for preference articulation. The electronic exchange and the forward deliverable contract market depend upon these technologies. The potential may exist to feasibly respond to preferences for much greater variety and specificity. With potential information processing and communication systems, retailers may be better able to respond directly to individual family specifications and product mixes by communicating preferences and even orders from their customers back through the food supply system. The computer information system now makes it possible to order and coordinate the production of a great variety of combinations in mass-produced automobiles. Relevant research would involve evaluating alternative designs of new systems, including the broader implications to food system performances. This new technology also expands the opportunity to collect and process political preferences.

Political Processes. The diagnosis of a problem in preference articulation and in the responsiveness of the market does not necessarily mean that a political process would solve the problem. There are many problems in the design of political institutions to effectively identify preferences and be responsive to them. Some basic problems include determining whose preferences should count on specific issues, registering intensity of preferences, and designing mechanisms to deal with complex trade-offs and problems of bureaucratic implementation. Extensive discussion of these issues is beyond the scope of this chapter, but several areas of potential research are identified.

Research on jurisdictional boundaries and rules of representation is fundamental to preference articulation. A jurisdictional boundary refers to the definition of the decision-making group and the scope of its authority. Rules of representation refer to procedures for expressing and counting votes or other indications of preference within the jurisdictional boundary. Jurisdictional boundary was a major issue in the *Federalist Papers.*

A central question in defining jurisdictional boundaries is, Who will be affected by the decisions of the unit? The problem involves both free riders and unwilling riders. If preferences and circumstances were homogeneous, it would make little difference. But preferences and situations may be very different among groups and areas. An example of a jurisdictional boundary issue is Michigan's attempt to set different standards for processed meat than is established by the federal code. State officials ask why some concept of a national preference for minimum standards should be imposed upon the consumers of Michigan. The issue is complicated by the economies of scale in mass production and merchandising, which is not unusual in issues of jurisdictional boundaries (Bish 1971). Processors argue for national standards in this case. Some Michigan consumers would prefer the national standards if they resulted in lower prices. Michigan's exit from the federal code may impose costs on others because of economies of scale. Whose preferences should count?

Another aspect of jurisdictional boundaries has to do with the extent of authority or responsibility of decision-making units. The behavior of a firm or consumer is constrained by rules and procedures determining what is taken into account just as it is with government agencies. Whose preferences and what effects are taken into account? For example, the modern highway system has had a very substantial effect on the structure of the food processing industry, but this effect and the preferences of those affected were ignored by the transportation agencies. Preference articulation would be different if responsibility for food safety were placed in the USDA as contrasted to the Department of Health and Human Services; e.g., different trade-offs would be considered and different congressional committees would monitor programs. Yet another dimension of jurisdictional boundaries is the structure of authority within the agency. What is the balance in decision making between the center and periphery? This will influence whose preferences are counted and the responsiveness of the agency to differences in preferences and situations.

Rules of representation determine who gets to express a political preference and how it is counted. A problem with direct voting is that it is difficult to express intensity of preference. Consider, for example, a direct vote on the budget of the USDA. A person might have very strong preferences for supporting research to increase yields for soybeans and be completely indifferent to programs on food safety. An individual might vote for both, but such a vote would not reveal preferences. Nor would voting on the total budget solve the problem. The transaction cost could be very great. Direct voting usually is not structured to deal with trade-offs. This, plus the economics of information and analysis, requires representative government. Given such government, the structure and rules of legislatures and administrative agencies become critical to preference articulation. An issue of special importance is the requirement for public notice and hearings on administrative rule making for regulations of the food system. Related research could inquire into the effects of the USDA employing public counsel to represent the interests of groups who would be affected by specific rules, but would not ordinarily be represented in hearings on administrative rules, e.g., small farmers and poor consumers.

Jurisdictional boundaries and rules of representation clearly are central to the effectiveness of a democratic government. These and other problems in structuring the political process are obviously complex, and the basic structure of government is beyond the scope of this book. It is suggested, however, that research needs to deal not only with the structure and rules of the market but also with the structure and rules for making the rules. This is critical to performance of the food system.

Preference Surveys. Surveys of preferences may be used as an input to both market and political processes. But great care must be taken in their use. It is a well-known phenomenon that consumer reports about their preferences are frequently inconsistent with preferences revealed by their market choices. This is often because the context of the survey differs from the context of market choice. There are many product attributes that may be perceived differently by those conducting a survey than by those responding. For example, consumers asked to describe characteristics of a desired product will often omit attributes that would cause them not to buy the product if they are missing. Consumers might not mention the color of a cup of coffee as important but would not drink green-colored coffee. Merchandising firms do elaborate preference surveys and preference testing, but ultimately the product is tested in the market. Research can be useful to identify desirable product attributes that are not being supplied by the market. For example, an industry may fail to provide products that meet nutritional standards desired by consumers. Consumers cannot express preferences in the market for attributes that are not offered. Incentives may be lacking to seek adequate information about desired attributes. This is especially true where the attribute is not easily identified by the buyer, as is the case with nutritional content.

The problem is more difficult and probably more important in obtaining reliable preference information as an input to the political process. Surveys to identify preferences about important political issues from representative samples is especially important because of the potential influence of small vocal and protesting minorities claiming to represent all of a group. Earlier social preferences were defined as those difficult to express through the market. They are also difficult to express in the political process, and the political test of their validity is uncertain. Note the discussion of political commentators about the meaning of an election. No one knows, including the candidates, the content of the message sent by the voters in an election. Thus obtaining additional information about social preferences by surveys is an important potential input for improving performance of the food system. Such surveys are not neutral, and the research is difficult. The same problems of context and attributes inhere in opinion and product preference surveys. Ordinary public opinion polls fall short largely because the questions lack content and respondents lack information. Thus much more elaborate survey procedures are needed, and extreme care must be exercised in wording questions and interpreting results. This makes it a worthy challenge for specialists in the subject matter of the survey.

For example, consider a survey to reveal preferences about regulations to achieve food safety. One approach would be to ask such questions as, Should nitrite in bacon be banned? But the responses reveal little about relevant preferences without information on trade-offs, consequences, means of implementation, and the like. A respondent might favor banning nitrites and oppose banning pesticide residues and be willing to trade one of these goals for the other. Getting preferences in terms of realistic choices rather than a general set of preferences is also necessary. Consider the problem of choosing a meal for a friend based upon a verbal expression of preferences in contrast to simply asking that choices be made from a menu. The problem is to provide an adequate menu.

It should be stressed that preference survey data dealing with the performance of the food system should be used as an input to the political process, but substitution of direct polls for deliberate debate and informed decision making is not advocated here. Additional information for the deliberations deals with the real problems of trade-offs.

Preferences As Product. Performance as outcome has been defined. Preferences are learned and influenced. Thus they may be the most important

product of the political economic system. The dynamics of the evolving political economy is wrapped up in the two-way relationship between preferences
and the organization of the political economy. This general evolutionary
process is beyond the scope of this book, although researchers dealing with
preference articulation and the food system need to recognize this relationship
to keep their research in proper context. We too need to be concerned with the
basic question, What kind of a society do "we" want?

There are several very specific cases of preferences as products that are
appropriate subjects for research by agricultural economists; e.g., To what extent do advertising and promotion influence food preferences and how is this
related to nutrition and health? To what extent does the fact that it is generally
more profitable to advertise processed foods result in heavy advertising, promotion, and retail shelf-space allocation favoring such foods, and how is this
related to costs and health? To what extent do advertising and public relations
influence preferences and behavior in use of agricultural chemicals and other
technical inputs, and how is this related to performance of the food system?
To what extent do advertising and public relations influence preferences and
attitudes toward environmental quality and resource conservation, and how
does this in turn affect performance? How does the way work is structured in
the food system affect attitudes and in turn influence performance? These are
all difficult and important questions. Research is not likely to provide
definitive answers but could provide information contributing to improved
performance.

Institutional Design. The market enterprise system creates great incentives
to explore the opportunity set and identify preferences that can be satisfied
within that constraint. However, the performance of the system always depends upon the way it is instituted, i. e., the structure of competition, rules,
and procedures that determine what is taken into account. At any time the outcome of the market may be inconsistent with potential preferred performance.
The research task is to provide information to assist in the political process to
identify unmet preferences and help design institutions that will move the
system in the direction of desired performance. The trick in institutional
design is to use the market mechanism to the extent possible to search the opportunity set and cater to individual preferences, while meeting collective
preferences about performance of the system. Preference articulation is
enhanced if both a majority and a minority can achieve their preferences. For
example, if a majority prefers to pay the cost of being protected from a food
that causes cancer and others believe the cost is too high, the problem is to
devise a rule to cater effectively to both. This is, of course, oversimplifying,
since the majority may be affected by the effect of cancer in others. Nor does it
argue for individual choice based on an assumption of perfect information.
Some combination of political rules and market processes would articulate
preferences better than a ban on any food with a potential to cause cancer or
ignoring the problem.

In the example of the demise of neighborhood stores (which was hypothesized as being inconsistent with preferences) preference articulation would be
enhanced by an institution making it possible for those benefiting to pay for

maintaining the stores. A national program designed to keep them in business would probably be less responsive to the variety of preferences and situations.

There are many different ways of instituting markets and many ways of using market processes to achieve public objectives. Marketing research can contribute to improved performance of the food system by identifying and evaluating the consequences of alternative institutional arrangements that combine political and market processes.

ORGANIZATION OF RESEARCH

The research suggested here requires a variety of organizational forms. Some awaits conceptual contributions, which come from the individual researcher and the interaction among researchers and other system participants. Thus it is emphasized that organized symposia should be a critical input to research and also noted that other disciplines have a contribution to make; perhaps the initial contribution can be sought through inviting them to a symposium. Other issues are embedded in the overall evaluation of food systems, and efficient research requires a task force approach.

REFERENCES

Arrow, Kenneth J. 1951. *Social Choice and Individual Values.* New Haven: Yale University Press.
Babb, Emerson. 1979. Professor, Department of Agricultural Economics, Purdue University, West Lafayette, Ind. Personal communication.
Bartlett, Randall. 1973. *The Economic Foundations of Political Power.* New York: Macmillan.
Bish, Robert L. 1971. *The Public Economy of Metropolitan Areas.* Chicago: Rand McNally.
Bish, Robert L., and Vincent Ostrom. 1973. *Understanding Urban Government.* Washington, D.C.: American Enterprise Institute.
Commons, John R. 1968. *Legal Foundations of Capitalism.* Madison: University of Wisconsin Press.
Dahl, Robert A., and Charles E. Lindblom. 1953. *Politics, Economics, and Welfare,* pp. 413-26. New York: Harper and Row.
Galbraith, John K. 1967. *The New Industrial State,* pp. 22-27. Boston: Houghton Mifflin.
Hightower, J. 1973. *Hard Tomatoes, Hard Times: The Failure of the Land Grant Complex.* Cambridge, Mass.: Schenkman.
Hirschman, A. O. 1970. *Exit, Voice, and Loyalty.* Cambridge, Mass.: Harvard University Press.
Marion, Bruce W., and Charles R. Handy. 1973. Market Peformance: Concepts and Measures. USDA, AER-244.
Martin, Lee, ed. 1977a. *A Survey of Agricultural Economics Literature,* vol. 1, Minneapolis: University of Minnesota Press.
National Commission on Food Marketing. 1966. *Food from Farmer to Consumer.* Washington, D.C.: USPGO.
Riker, William H., and Peter C. Ordeshook. 1973. *An Introduction to Positive Political Theory.* Englewood Cliffs, N.J.: Prentice-Hall.
Robinson, John P. 1976. The press and the voter. *The Annals.* 427:95-103.
Samuels, W. 1972. Welfare economics, power and property. In G. Wunderlich and W. L. Gibson, Jr., eds., Perspectives of Property, pp. 63-67. Institute for Research on Land and Water Resources, Pennsylvania State University, University Park.
Scherer, Frederic M. 1970. *Industrial Market Structure and Performance.* Chicago: Rand McNally.
Thurow, L. 1973. Toward a definition of economic justice. *The Public Interest* 33:56-80.
U.S. Department of Agriculture. 1951. Market Demand and Product Quality. Report of the marketing research workshop, conducted at Michigan State College, East Lansing, July 13-21.

DANIEL I. PADBERG
RANDALL E. WESTGREN

12

ADAPTABILITY OF CONSUMERS AND MANUFACTURERS TO CHANGES IN CULTURAL PATTERNS AND SOCIOECONOMIC VALUES

MORE ECONOMIC COMPLEXITY and specialization inevitably accompany rising material well-being or level of living. The subsistence household produces many things and buys few. The household itself is complex, but the economic system supporting it is simple and not well developed. In the subsistence household, more knowledge and analysis (per product) is brought to the role of consumer than to the role of producer. The few items purchased are staples of well-known characteristics and extreme importance in the very small total household budget. On the other hand, the household is the producer of many items, mostly for home consumption, each requiring less concentration of knowledge, experience, and analysis.

As we move toward affluence, these roles reverse. The affluent household marshals an impressive array of education, analysis, and experience to be used along with another array of capital-intensive and scientific production aids such as computers for the production of a very few products or services. While the affluent household's role as a producer becomes very specialized, the role as a consumer moves the other way. Affluent households face choices among literally hundreds of thousands of consumer product options. Almost all

DANIEL I. PADBERG is Dean and Director, College of Food and Natural Resources, University of Massachusetts, Amherst; RANDALL E. WESTGREN is Assistant Professor of Food Marketing, Department of Agricultural Economics, University of Illinois, Urbana.

education and preparation is oriented to the producer role. As a consumer, it is not humanly possible to understand the complex details of the product options among which we choose. While our behavior as a producer involves calculated and meticulous responses, our behavior as a consumer uses only the grossest thumb rules in some mosaic of habits and spontaneous impulses.

Consumer purchase behavior in the affluent household involves individual decisions low in importance to the well-being of the household but involving enormous behavioral complexity. The very fact that an individual consumer product purchase is trivial allows the choice process to be affected by matters of style, habit, or whimsical impulse. Another aspect of the complexity of consumer purchase decisions deals with qualitative change in products. The knowledge and analysis of producers can bring product change in almost any direction consumers would like. Finding the "right" or "best" direction for product evolution among products trivial and insignificant to the household is a complex and delicate process.

In microeconomic analysis generally and the study of agricultural marketing specifically, we have tended to prefer the traditional price-oriented economic models to explain behavior of households in the consumer and producer roles. By and large this has been well chosen and successful. Only in the last half-century has the modal U.S. household made the transition from subsistence economic conditions to affluence. An increasing proportion of consumers has discretionary income above basic needs. Still, the price-oriented models are most useful and powerful in explaining the role and behavior of producers. Prices and costs are central economic features of the few (broad) products or services we produce. But the price-oriented traditional models are less effective in describing our role as consumers. The price and value of many consumer products are not large factors in the well-being of the affluent household; they are often meant to provide levity, amusement, or variety. These values are hard to measure and analyze. Impulses may serve the consumer as well as analysis in purchase choices among some of these transitory values. Price-centered analytical models explain such behavior only poorly at best.

Many data trace the economic changes in society that affect consumption patterns. A selected few of these appear in Table 12.1. Inferences that can be drawn from these figures include the following:

TABLE 12.1. Selected indicators of socioeconomic change

Indicator	1950	1955	1960	1965	1970	1975
Average size of household	3.37	3.33	3.33	3.24	3.14	2.94
Median family income (1976 $)	7,850	9,393	10,803	12,552	14,465	14,510
Median family income (current $)	3,319	4,418	5,620	6,957	9,867	13,719
Female labor force participation (% of women 14 years old and older)	31.4	33.5	34.8	36.7	42.6	45.9
Percent of female work force married with husband present	23.8	27.7	30.5	34.7	40.8	44.4
Median school years completed by population	12.0		12.3		12.6	12.8

Source: U.S. Department of Commerce, *Statistical Abstract of the United States*, Bureau of the Census, 1977.

1. As family income increases and family size decreases, more discretionary income is available to the household. This income will be spent on household assets above basic requirements and allows for product experimentation. In the cases where income growth transcends social class, groups of overprivileged consumers relative to their reference groups will develop. The high proportion of discretionary income in these households will be manifested in goods (i.e., intensive leisure and innovative behavior in consumer goods).
2. The increase in female labor force participation creates a new market for consumer goods. The career woman is a new role model to which marketers must address communications and product design. A substantial portion of female labor is as secondary employment for the household. Primarily, this income is used for discretionary spending, with the woman as the purchasing agent for the family. Growing demand for preprocessed foods and convenience items stems from this consumer group.
3. Higher aspirations for future income follows increasing levels of education. If aspirations cannot be met within the confines of employment, they may be manifested in spending patterns to establish life-style and status.

The transition from subsistence to affluence has profound effects on our culture and values as well as economic behavior. This deep and complex topic (culture and values) is most difficult to develop usefully. On the other hand, we can identify some widely held observations:

1. The household and church have declined as shapers and transmitters of values, while education and commerce have increased in importance.
2. Young people from the affluent households have a very different conception of creative and productive work as well as recreation and fulfillment compared to young people growing up in subsistence households.
3. The new economy has emancipated its people from the constraints of the household and the church (more broadly from rules in general) and given greater opportunities for participating in various institutions.
4. The consumer goods industries and the powerful communications system they maintain (particularly television) are intrinsic components of the process in which we choose a life-style as well as our process of implementing those choices.

How does this process of interaction between human aspirations and economic possibilities dictate behavior? Economic incentives give thrust to science and technology. To what extent do economic incentives channel technology as an instrument of human aspiration, and to what extent do both technology and economic incentive shape human aspirations? What conceptual models are useful in understanding these relationships and guiding public policy in a way most related to some concept of the public interest?

Nicosia and Mayer (1976) posit a relationship between the technostructure of the economy, cultural values, and consumption patterns. They suggest that basic cultural values that are inhibited from expression by institutions are "deflected" into consumption; e.g., the value of "getting ahead" is often stifled by bureaucratic pressures in the workplace or by restrictions such as labor union seniority clauses. A person facing this problem will deflect the desire for identification in the workplace by differentiating consumption patterns and attaining status through this unfettered form of self-expression. Because consumption is less regimented and conduct norms are weak, expression through consumption can take myriad forms. This tends to increase the dimensionality of product characteristic bundles demanded by affluent consumers.

The process of qualitative change of consumer products is a very important aspect of competition among consumer product sellers and the interests of the household. This process is very poorly addressed in the traditional price-oriented models; for the most part, it is not addressed at all.

Food industries are of particular importance in these general questions. Food is a big part of the total economy, even in affluence but more so in a subsistence economy. Economic development in the food system is almost imperative if levels of living are to rise in the economy generally. The food economy affects all consumers, while other consumer goods industries may reach only a small proportion. The high volume and levels of specialization in the food industry often enable it to set efficiency norms useful as examples to other industries.

A particularly important aspect of these economic transitions in the food industry relates to two significant consequences for human health. As the food supply becomes more processed, the nutrient values and proportions of basic plant and animal materials are altered. Changes in the nutritional properties of food products and our diet may result. In addition, as the advertising communications from big-business food sellers rise in significance as an influence on our diet, the convenience, taste, texture, color, and other less concrete attributes may receive greater emphasis (in competition strategies of firms and in our choices) than the very important nutritional values. For all these reasons, we need to better understand the process of adaptation and response involving the food products industries and our modern consumer.

CONCEPTUAL IDEAS AND UNDERLYING HYPOTHESES

Our challenge here is to bring the marketing system response mechanism under the microscope for economic analysis. The most significant element that must be accommodated in this analysis is the process through which products change. As we move from subsistence households to those of affluence, consumption patterns are not transformed by use of more subsistence goods; rather, we consume different goods. Therefore, a central feature that must come under analysis is the process of product evolution.

What should constitute the unit of inquiry? The traditional price-oriented models (supply and demand) have firms as a unit of activity, but the coordinating process and the incentives generated come from interactions of groups of firms at the market level. These price-oriented models are great overarching patterns of cooperation that are perhaps unconscious. The firm is not very colorful in these models, acting passively in response to incentives generated at a higher level of aggregation by the market.

To what extent is the process of product discovery and evolution a phenomenon of a market? There is certainly an obvious, intense, and significant aspect of competition among firms in product evolution. This competition among a group of firms gives product evolution a marketlike setting. Yet the development and introduction of a new product is not a market phenomenon. It does not originate in this higher level of aggregation but is an action of individual firms. In an appropriate model of analysis, the individual firm must be studied carefully. Since the behavior of individual firms as well as the pattern of incentives is oriented to the firm level, a useful model must incorporate the capabilities, incentives, and outcomes related to individual firms.

These two imperatives, the need to deal with product evolution and the necessity of treating the behavior of individual firms, make it clear to almost everyone that the traditional models are not very useful in bringing this response mechanism into analysis. If supply and demand analysis is not very useful here, then what pattern of analysis and experience might represent a more enlightened and useful approach? Here the trail becomes weak and hard to follow because clearly very little work has been done in this direction. While we have pursued the conventional microeconomic scenario to great length, often with exotic quantitative instruments, relatively little has been done in this much softer science dealing with consumer behavior and the nonprice competition chosen by individual firms to condition or modify consumer purchases. Since this new direction is oriented to institutional analysis when it relates to the firm and to a pattern of consumer behavior affected by habits and impulses rather than analysis, it has not been well adapted to quantitative models. This may explain the relative dearth of thought and work in this part of our economic system.

DEVELOPMENT OF ANALYTICAL MODELS

The decade of the 1930s seemed to put a great deal of stress on the conventional economic models of analysis. Not only did exotic new models for macroeconomic analysis emerge during this period, but a collection of ideas and work developed around imperfect competition as well. Robinson (1933) and Chamberlin (1933) are the intellectual progenitors of this experience. Perhaps the greatest contribution emerging from this work was the recognition of product differentiation and its impact on producers and consumers. The kinked demand curve was the result of an effort to put this product differentiating behavior into a marketlike setting.

This pattern of development extended the traditional market-oriented theory to recognize the existence of product evolution and its symptoms. It provided the roots for much of the later work. This approach to economic analysis was carried into food products industries by Nicholls (1941) in an impressive and useful book. Although these models were much referred to in the 1940s and 1950s, they did not offer a strong basis for analysis.

A pattern of analysis that seems to have begun in the 1950s eventually became known as industrial organization. Bain (1959) deserves a significant share of the credit for bringing this pattern into being. It incorporated most of the ideas of imperfect competition but came to include a much greater variety and richness of behavior and analysis. Industrial organization had a very central focus on oligopoly. While the structure and behavior of individual firms was not completely bypassed, the greater focus was on linkages between and among firms that might, particularly in highly concentrated oligopolies, condition the behavior and performance of firms and industries toward a more monopolistic character. The greatest shortcoming of industrial organization for the purposes involved in this study is that it was never really used as a setting for the analysis of the response mechanism involving firms and consumers but was used as the intellectual and conceptual roots and underpinnings for antitrust policy. The massive and very careful work by Scherer (1971) carries this use of industrial organization to the highest level of statesmanship as well as scholarship.

While one cannot help but admire this industrial organization tradition and use, there are still disappointments. Our frustration with the antitrust tradition does not suggest it is not a useful part of our food economy. We see no reason to expect it will not continue to be needed and strengthened in future decades. Our concern grows from the view that antitrust as now designed does not deal with the major issues within the interface between consumers and large firms, especially where differentiated products are involved.

Probably the most far-reaching and notable tradition of analysis relating to the response mechanisms between producer and consumer in an affluent society is the work of Galbraith (1958, 1967). He has been such a determined renegade that his work includes many distractions. This may be an important reason why it has received less attention than it probably deserves. One of his great contributions was the willingness to observe that we as consumers act in such a cavalier way that our patterns of consumption are significantly determined by careful and meticulous persuasion (producer sovereignty). With this recognition, he found it appropriate to give an enormous amount of attention to the anatomy of the institutions that emerge as carriers and developers of technology (and as persuaders). The conceptualization of the Galbraithian (affluent, limp, passive) consumer are very useful in bringing the response mechanism between the food industry and the consumer into analysis.

The Galbraithian Firm. The Galbraithian firm is specialized to production and distribution activities that contain scientific complexities. Several reasons are identified as to why this firm must become a very large and powerful institution. A most significant influence grows from the greater risk associated with product experimentation and the high cost associated with consumer advertising and product introduction. Another reason for firms to be large in size grows from the nature of the scientific establishment itself. Scientific inquiry and input is a very indirect process. Expenditures may be incurred for years with no payback. Only large and stable institutions or firms can afford continuous research and provide an isolated and protected setting for it. Special needs for financial planning give incentives for large size. Conglomerates are assembled from firms in product lines that seem to have no relation. Even in these cases, greater financial resources may be marshalled with more efficient use of internally generated capital. On the other hand, some of the Galbraithian overheads, including the work of bench scientists as well as advertising expertise and brand names, may be more successfully used across different products in a conglomerate structure than with more conventional production inputs. We refer to the large diversified food manufacturers that specialize in product development and introduction as Galbraithian firms because their behavior is better described in the Galbraith scenario than conventional microtheory.

What are the economic incentives of these large and powerful institutions? It is argued that the leadership cadre of the large firms grows in the direction of less dependence on and relationship with stockholders and more dependence on and relationship with scientific capabilities and the people who have them. This leads to the suggestion that profit maximization is not as important in these people-oriented firms as in the older capital-oriented firms. This probably is not worth much of an argument because in both cases the firms want to make money, and they see success in that very common definition. What they do with the money may be somewhat different, with the Galbraithian firm showing a high propensity for internal reinvestment at the expense of dividends. The incentive of the technostructure is to have good salaries and benefits for themselves and then to achieve financial flexibility for experimenting with the products of their laboratories.

Another economic incentive of these firms is to avoid head-to-head competition with others in stereotyped product lines. This competitive business tends toward low profits and returns for firms with low overheads. To avoid these conditions, the Galbraithian firm is motivated to be forever in a state of introducing new products. This may be achieved by varying combinations between altering product characteristics and altering consumers' perceptions of products. In more old-fashioned parlance, one might say the Galbraithian firm is seeking a stream of innovative profits. Perhaps one might say that such a firm would like to be a temporary monopolist. This pattern of competitive strategy complements the capabilities of the firm as well as existent antitrust laws and probably leads to the highest expectable and respectable rates of earnings.

To what extent are these firms accountable to the general public? Are there specific or general laws that condition and channel their enormous power in directions consistent with public well-being? They are certainly affected by such general laws as for food safety. To the extent that they have better laboratories than the Food and Drug Administration and employ more and better scientists, they may have outgrown public surveillance somewhat. In their role of conditioning consumers' tastes, preferences, and diet they seem to have outgrown public surveillance entirely. Certainly the antitrust laws scarcely affect them. Yet there may be some inherent tendencies or patterns of incentives, with the likely result of bringing their behavior patterns into some general consistency with public well-being.

They must have some general interest in and incentive for maintaining a good name with the public, simply because they are large institutions and could be vulnerable to new policy made by an aroused public. But more specifically, there may be other mechanisms of greater importance. If these firms are more sensitive to science-oriented people than investors in choosing their internal policy, we must consider who these people are, where they come from, and what kind of social conscience they have. These science-oriented people generally are recent graduates from public universities. At the same time that they learned their science, they acquired at least some humanities and must have been exposed to debate concerning various concepts of the public good.

What patterns of motivation do firms have in recruiting and selecting these people? Increasingly, in the past few years we see large Galbraithian firms making vivid public declarations of alignment to current and popular public concerns. Texaco avows great concern about the nation's energy posture. Food companies take on a noticeable sensitivity to human nutrition. These large companies are increasingly making a public case for their commitment to and alignment with popular manifestations of the public good. This halo effect is undoubtedly an instrument in acquiring the best people for employees, selling products, and perhaps favorably affecting public policy concerning themselves. We may very quickly question the motives and sincerity of these manifestations in advertisements. There is no question that advertising claims associated with products give an incentive for improving them, and it is possible that institutional advertisements that make commitments to the public good give incentives for aligning performance in the same direction. Inasmuch as such behavior is probably a profit increasing choice, it does not seem improbable.

We do not know how important this loose link between Galbraithian firms and the public interest may be. It seems quite likely that it is a much more important influence and force in our economy than it might have been in the 1960s. It also seems likely that as graduates from public institutions of higher education from the 1960s and 1970s move into responsible positions in Galbraithian firms, this alignment may be strengthened. These firms are relatively new in our economic history. How they will eventually fit into our culture and public policy patterns is not well understood at this time. It certainly merits attention, however.

Adaptations within the Food Industries. The major adaptation to the food industry involves the linkage between the manufacturer and large retailing firms. Large food manufacturers tend to specialize in product development and introduction of new products. Most of their competitive policy and strategy relates to the final consumer through advertising rather than promoting to the trade. Promotion to the final consumer is a process that is very expensive and only affordable by the very large. Smaller manufacturers usually have more conventional products not so easily promoted to the consumer in the first place and usually find it more appropriate to choose promotional strategies oriented to the trade (i.e., wholesalers and retailers).

The Galbraithian firm is motivated to make its food products relevant to consumer ideas and concerns of the moment. If consumers are interested in nutrition and are being nourished with vitamins and proteins, nutrition is built in and emphasized in the food products. If consumers are calorie conscious, then calorie consciousness is emphasized in the food products. Where consumers are interested in sex, products are made sexy. To the extent that economy becomes the focus, you find out how much Campbell's soup you can buy for a dollar. The large Galbraithian firms probably have the most developed scientific talents relevant to probing, measuring, and understanding human motivations and subliminal desires. They are motivated to use and develop this knowledge as an element in development and experimentation with food products and images that can be given to them.

The process of experimentation with product characteristics and images that takes place within the Galbraithian firm is a very expensive one. The nature of the distribution channel requires an army of hundreds of salespeople working in concert with advertising campaigns and other promotional activities to introduce a product to the supermarket trade. Among ideas generated and tested, those actually being adopted represent a very small percentage. Nevertheless, with 200 firms of large size, aggressively managed, the number of new products introduced within the space of year is impressive. If these new offerings catch the fancy of the consumer, they can be very profitable to the manufacturer. Each food product purchased is small enough in relation to the total budget of the affluent household that there is quite a bit of latitude to price above production cost. Also, in the case of new products, quite a time period exists in which there is either no competition or competition only in products of dissimilar characteristics, allowing some pricing latitude.

The role of the large food distributor, particularly the largest retail chains, includes a couple of functions of particular importance in food industry response to consumer preferences (Handy and Padberg 1971). The supermarket setting offers an excellent opportunity for stimulating impulse purchases. Probably the major feature of merchandising strategy within supermarkets involves the special exposure of a selected list of products that is varied from week to week. This is another way that the professional in the food system takes over chores once done by the household. By selectively promoting different products from week to week, the supermarket operator is helping the household plan supper (Shaffer 1960). Food distributors have an enormous array of products available from which they can choose their weekly features. The special exposure given these products is available at very low cost. While quite a different process from advertising, it has a significant effect on bringing the special characteristics of these food products to the attention of consumers.

Another role that the large distributor performs in the process of product evolution pertains to economy products. Most of the products developed by the Galbraithian firms are unconventional, special, and expensive. This leaves a bit of a competitive vacuum in the product space pertaining to economy. Merchandising simpler, standardized goods to the consumer is particularly well adapted to the large distributor, the largest few of which operate from 12-30 manufacturing plants in their own organizations. In addition, they are able to contract for the production of many other plants under their specifications. This gives the opportunity to offer an enormous array of private-label products. The nature of these is appropriately linked to the economy image. They tend to be conventional and undifferentiated and are oriented to the market for more stable and staplelike products. They are designed to serve everyday needs rather than subliminal desires.

Large food chains are often able to make these products available at an economy price by controlling the distribution system from manufacturing to sale to consumers. Their sales costs are minimal. They do not advertise. Their exposure to consumers is affected by merchandising strategies within their stores. They are able to harmonize production with sales levels at their stores and thereby reduce costs of finding markets. Many smaller manufacturers find it very difficult to get their products into supermarkets under their own labels and are well served by the availability of these private-label programs. From a consumer's perspective, this action and initiative on the part of major distributors provides an economy offering to balance against the exotic products pushed by the large Galbraithian firms.

Through this complex industrial structure and sophisticated pattern of competitive strategies, a wide variety of product features and characteristics are offered to the food consumer. Perhaps the variety of industrial structure types and competitive strategies operating effectively in the food industry is greater than in most other industries. It does not seem illogical that one of the largest and most established industries in the economy might support a very diverse and specialized pattern of industrial structure and competitive behavior. Through this complex, the industry engages in a process of experimentation either with measuring, understanding, and influencing human wants and needs or with possibilities for alterations and improvements and changes within the nature of food products.

It should be observed, however, that the combined variety and richness of alternatives described here is the province of manufactured food products. Something over a third of the food sales is in fresh produce and meat. Within these products, we do not find the presence of the large Galbraithian manufacturer. As a result, the handling and presentation of these products rests primarily with the supermarket operator. The scope for developing and altering fresh food products is certainly more limited than with manufactured products. For all these reasons, fresh produce and meat have been the scene of much less product experimentation, have undergone less transformation and evolution, and represent an area of the food system in which price tends to be of greater importance than nonprice competition.

Consumer Interaction. The consumer's role in this pattern of interaction is a very passive one. Consumers are overwhelmed by the number and complexity of alternatives offered to them. The same person who brings impressive sophistication to the production of some product or service may act naively in behavior and reaction to the thousands of sophisticated products faced in the market. No one likes this vision of consumer behavior. We inherit a vision of an analytical consumer from past realities and traditional models. Our transition in the last three or so decades to an affluent society and a sea of consumer products makes this vision antiquated. Considerable attention has been given to theorizing Rube Goldberg devices in which the peculiar and indeterminate behavior of consumers might seem rational. For the most part, this effort has been of little value in structural or predictive description of this behavior.

Once we recognize that inevitably we must deal with a passive consumer, however, there may be new hypotheses that will be useful in bringing this interaction pattern into analysis (Padberg and Westgren 1979). It is most interesting to look at the passive consumer's reaction to new product introductions. The conventional hypothesis is that consumers prefer bold and startling innovation. It is not infrequent, particularly among experts, to see very critical views toward products advertised as new and different, with only very modest and even hardly noticeable physical changes. This mentality well suits the subsistence setting in which the household bought only a very few consumer products, was expert in appraising their quality, and was able to assess the full implications of the innovation (i.e., the better mousetrap). The use and meaning of these products in the subsistence household was almost entirely one of functionality. They were not for show or image among one's friends. It was possible and desirable for the consumer to make the full leap from the present product to a much superior one and make it in one bold maneuver.

As we look at the analogy in our present affluence, almost none of these conditions hold. Consumers are not able to assess the full implications of a bold innovation. They are cautious about choosing and using products radically different from their reference groups. While it is hard to visualize how a boldly different product would fit into a complex pattern of living, it is much easier to assess a product change that is only modestly different from our experience. Consumers probably have a noticeable and realistic preference for approaching product evolution in small steps rather than the massive ones that would certainly be possible within today's available science and technology. The large Galbraithian firms studying consumer action and reaction probably have a much better understanding of this matter than the experts who critique and appraise the process of product evolution and competition.

Another aspect of the consumer reaction within the process of product evolution is the clear preference for and appreciation of third-party judgments on product changes. Consumer frustration and anxiety about highly manufactured food products, including concern about additives, has led to programs requiring information on labels. Within the pattern of rationale used in legislating these requirements, we always revert to our concept and vision of the analytical consumer who needs more information. However, the proliferation of information surpasses the human capacity to process it (Scammon 1977). Efforts to assess the importance of such labels to consumers indicate that they have a strong preference for information on labels but do not use it (McCullough and Padberg 1971; Lenahan et al. 1973; Jacoby et al. 1977). Labels are rarely read. Their primary purpose is more like an insurance policy. Consumers want to know, even though they do not understand the big multisyllable ingredients, that a third party (the government) is watching and that anything manufacturers do must be open and reported (Padberg 1977).

IMPORTANT RESEARCH CONTRIBUTIONS

Research into the response mechanisms of firms and consumers to the socioeconomic patterns discussed above has lagged along with the perception of these changes. This situation has been observed in surveys of agricultural marketing literature (Breimyer 1973; Helmberger et al. 1981), with some notable exceptions. Business marketing reseachers, psychologists, and business economists, however, have pursued the issues inherent in this research topic across broad fronts. While the sheer force of numbers of these research efforts implies some progress in effectively addressing this complex topic, the commodity orientation of agricultural marketing research leaves our profession relatively undeveloped in this direction.

A simple taxonomy of research effort will facilitate this brief review. Three thrusts may be identified for historical and future efforts at describing the relationship of the market consisting of Galbraithian firms and consumers characterized by a complex purchasing behavior: the competitive interface among the manufacturers of food products, among purveyors of these products and their attendant services, and between these two groups in the market; the broad research area of consumer behavior, specifically purchasing behavior for food (characterized as the polar case of low value, frequently purchased nondurable goods); and the intermediation between seller and consumer. Holbrook and Howard (1978) define the separation between the consumer and firm as the market space and the marketing activities of the individual firm as the instruments to close this spatial gap.

Breimyer (1973) classifies the cadre of agriculture marketers pursuing the organization and behavior of the food industry from the orientation of the consumer as the "market development" school. With license this genre of research can be be said to include some aspects of market structure–performance research in addition to the narrower definition given. Helmberger et al. (1981) review the history and contributions of market structure–performance research, including food manufacturing (cf. FTC 1969) and food retailing (Mueller and Garoian 1961; Marion et al. 1977). While this approach has been criticized as insufficient to show the market effects on firm profits (Bass et al. 1978), the heavy emphasis on profits as a performance variable has caused the most dissatisfaction among other researchers. Bloom (1977) suggests that without a measure of consumer satisfaction caused by brand proliferation, product design, and communication, industrial organization research is limited in serving the needs of antitrust legislation. Helmberger et al. (1981) posit the need for measures of progressiveness in establishing the performance of food industries in the market. With a broadened scope of performance measures, the industrial organization contribution to Galbraithian firm behavior will be enhanced.

Staff analyses from the Federal Trade Commission (1969) and the National Commission on Food Marketing (1966) have provided much data used in market structure–performance research as well as some descriptive works on the competitive behavior of food sector firms. Padberg (1968) examines the structure and nonprice competitive behavior of the food retailing industry. This analysis is extended to the interface between the food manufacturing industry and the food distribution industry, where the use of price competition and distribution policy are found to be a result of the size of the firms participating in exchange (Handy and Padberg 1971).

Metwally (1976) has compiled a unique study of the theoretical and empirical nature of nonprice competition. Using firm theory to derive optimal levels of advertising, promotion, and product competition under oligopolistic market conditions, he tests the optimality of existing advertising levels in New Zealand and makes several cross-country comparisons of firm and industry marketing strategy. While the need for research into nonprice competition has been addressed in agricultural economics literature (Moore and Hussey 1965; Tongue 1965; Beem and Oxenfeldt 1966), the nature of the data necessary for such analysis has hindered this effort. A study by Wildt and Bass (1973) is a significant departure from this. They specify a simultaneous system of equations that determine the market share, media advertising expenditure, and local retail price advertising expense for three disguised firms in a consumer nondurables market. Exogenous variables include new product activity, price relatives, seasonal dummies, and lagged endogenous variables. Their results are illustrative of the testable nature of some of the hypotheses posited about the nature of competition under the new scenario of food industry competitive behavior (Padberg and Westgren 1979).

The area of consumer behavior is replete with contributions to the theory of buyer behavior. Several texts (Howard and Sheth 1969; Hansen 1972; Engel et al. 1973; Howard and Ostlund 1973; Woodside et al. 1977) describe the development of buyer behavior theory based upon a complex multistage process of perception, cognition, and evoked behavior. This body of theory does not presume the restrictive assumptions of ordered preferences and perfect knowledge that economic utility theory presupposes. One of the significant empirical results of this research area is identification of an information overload facing consumers (Jacoby et al. 1975; Scammon 1977). The multiplicity of products, product attributes, and communication and distribution matrices confound the information processing capacity of the consumer. This forces habituation and the tendency toward incremental moves between product characteristic bundles, particularly with the introduction of new products.

One of the revealing areas of effort in buyer behavior is the evidence on shopping behavior for food products. Shaffer (1960) examined the in-store choices of shoppers compared with their intentions. He found that aggregate purchases of total food and major food groups did not differ greatly from intentions, although individuals made significant switches from intentions when confronted with in-store promotions and displays. The merchandising techniques in the store have been shown to contribute to formation of brand preference for milk products (Padberg et al. 1970). For a review of shopping behavior research see Granbois (1978) and Holbrook and Howard (1978). Research includes examination of shopping behavior among income, ethnic, socioeconomic, and age groups. These studies are the cross-section analog to the dynamic economy described here and reflect observations on the increasing complexity of the shopping task, mobility of the affluent shopper, and purchase differences across the subsistence-affluence dichotomy.

Scholarly research into the resolution of the marketing space has proliferated during the 1970s. This can be attributed to use of marketing mix instruments by industry marketers in establishment of new products and identification of markets for existing products in optimal promotion schemes, and in applicability of quantitative techniques for analysis. A common thread in this work is identification of market segments by specifying exclusive groups of consumers by demographic and psychographic (life-style, beliefs, self-image) characteristics (Engel et al. 1972). Some researchers criticize the predictive capability of these models for individual purchases, but Bass et al. (1968) correctly observe that these models have predictive value for these segments and that group determinance of behavior is relevant to the market-process.

After the characteristics of the target market segments are identified, the marketing firm can create the bundles of product characteristics that approach the ideals of the segment and take advantage of the blanks in the preference map for those not served by competing brands (Engel et al. 1972). This strategy accounts for the proliferation of brands in several food groups (e.g., breakfast cereals, beer, coffee, convenience foods). Segment analysis also provides information for designing appropriate advertising messages and media, distribution and outlet strategy, and price positioning.

NEEDS FOR FUTURE RESEARCH

Several general objectives may be appropriate as we contemplate research possibilities. One would be to develop a research program that would improve the understanding and information within the academic establishment. This proposition may seem of low priority and excessively general. Yet if a democracy is to govern itself with wisdom and efficiency, it is important that its primary patterns of interaction and behavior be understood not only by the experts but by the general public. We may occasionally succeed in legislating the right policy for the wrong reasons (the case of nutritional labeling), but even here we get strong negative reaction from requiring industry to apply labels that consumers do not read. A better understanding of the basic parameters of industry and consumer interaction is needed to enable the public to guide complex and powerful industrial behavior.

Another general objective for research might be to provide advice regarding industry behavior to enable its improvement. If the educational establishment contains within it a better understanding of the interaction pattern between industry and consumers, it may be useful in training of new entrants for the business world and in consulting with and advising ongoing business operations. This does not seem to be the most powerful incentive to justify public research. We judge the present state of affairs to be one in which industry has made enormous investments in understanding and executing an operational pattern of behavior, while the academic establishment has just begun addressing these research topics.

Another general objective might be to design and conduct research specifically oriented toward improving policy and our ability to make good public decisions. More than just developing an awareness on the part of the general public as discussed above, the experts must be able to conceptualize the overall patterns of behavior and develop policies sensitive to guiding this powerful industrial process in the public interest. This approach to and use for research seems to be paramount.

As this pattern of research is developed, we need a better overview of the whole pattern of response mechanism(s). Strategies and incentives of each of the separate actors needs to be better understood. Constraints and limits, including competitive effects and policy potentials, must be identified.

Very early in the process we must also approach the matter of normative choices. What characteristics of a product development process and industry-consumer interaction pattern identify them as a good combination compared to other characteristics that might identify an antisocial situation? In general, social scientists have spent very little time with norms. Perhaps economists are the least sensitive to this question because subsistence conditions have made it easy to identify efficiency as the greatest normative property. As we move into affluence, only the most insensitive economists can argue that efficiency still is the most important normative property of a system of behavior. When the pattern of behavior alters our diet (and our health as a result), the efficiency with which this diet of poor nutritional quality is delivered cannot be the most important topic of analysis.

Pareto optimality and many of the hallowed scenarios of welfare economics have little connection with public welfare, where the focus of competition turns on qualitative change in products rather than their quantitative price and availability. Making progress in this area will be likely to require the work of a liberally trained political economist more than use of the skills of econometric analysis so highly developed within our profession.

The public accountability of large firms is a topic of considerable importance in this research program. Accountability to the public was primarily a personal function in the small business world of two generations past. The application of science, which has freed us from a handful of staples and enabled an enormous variety of convenient products, has also brought the inevitability of the large firm. What linkages and mechanisms are most useful in relating that large and powerful firm to the public interest? Since our turn-of-the-century antitrust laws deal hardly at all with these questions, this matter remains an important focus for potential research. What sort of policy might be the most effective in recognizing the public interest and giving the public an ability to guide and channel the activities of these large firms?

The required public disclosure of basic objective information may be a much more powerful instrument in this regard than we have understood. It is impressive to see the effects of requiring automobile manufacturers to disclose estimated mileage data. Once this factual information is disclosed, it cannot be held out of public communication and advertising in support of this product. By choosing elements of particular public interest, a society may be able to force that aspect of the product into an important focus in the process of competition.

Nutritional labeling has had quite a similar experience and effect in the food industry. Large food manufacturers, like any large consumer goods firm, absolutely require a good name in their dealings with the public. Requiring labels or other disclosures of information about product characteristics of particular concern to the public may inevitably focus their attention on providing better products in those dimensions. This seems a much softer and less positive control mechanism than the type of remedies available in conventional antitrust policy. The truth may be just the opposite. We observe that our society is very reluctant to impose the very stringent and positive control measures made available to antitrust regulation. So while these remedies seem strong, they may in fact be weak because of a reluctance to use them. On the other hand, positive labeling may be a much more important force in public guidance of activities of these large firms.

SUMMARY

The great part of marketing research oriented toward the food system has focused on sales by farmers and the intermediate handlers of food commodities. The concepts of positive economics serve well as a general guide to this pattern of analysis. The study of behavior patterns associated with marketing of food products to final consumers is much less developed and represents only a tiny share of the total effort of our profession. It is not surprising that we find less bold achievements in this area in view of the fact that we have invested much less.

At the same time, the possibilities of altering our food supply and our consumption preferences are seen by the public as being sufficiently important that a stronger emphasis of research in product marketing seems our destiny for the future. This work will probably involve different conceptual models, particularly as they relate to the behavior of our passive consumer. It may well be possible to find ways through research to improve the American diet and health through careful guidance of the activities of the food manufacturer rather than depending on the self-interest and analytical behavior of the consumer. The choice of human nutrition as a focus of a federally supported program of grants for research signals the rising priority of these questions (U.S. Congress 1977). It is certainly our hope that our profession can be a useful instrument in society's effort to make the best use of our food raw materials, science and technology, and emerging economic structure.

REFERENCES

Bain, J. S. 1959. *Industrial Organization.* New York: Wiley.

Bass, F. M., P. Cattin, and D. R. Wittink. 1978. Firm effects and industry effects in the analysis of market structure and profitability. *J. Market. Res.* 15:3–10.

Bass, F. M., D. J. Tigert, and R. T. Lonsdale. 1968. Market segmentation: Group versus individual behavior. *J. Market. Res.* 5:264–70.

Beem, E. R., and A. R. Oxenfeldt. 1966. A diversity theory for market processes in food retailing. *J. Farm Econ.* 48:69–95.

Bloom, P. N. 1977. Potential contributions of consumer research to antitrust decision making. In H. K. Hunt, ed., Advances in Consumer Research, vol. 5, pp. 138–44. Ann Arbor, Mich.: Association for Consumer Research.

Breimyer, H. F. 1973. The economics of agricultural marketing: A survey. *Rev. Market. Agric. Econ.* 41:115–66.

Chamberlin, E. H. 1933. *Theory of Monopolistic Competition.* Cambridge, Mass.: Harvard University Press.

Engel, J. F., H. F. Fiorillo, and M. A. Cayley, eds. 1972. *Market Segmentation: Concepts and Applications.* New York: Holt, Rinehart, and Winston.

Engel, J. F., D. T. Kollat, and R. D. Blackwell. 1973. *Consumer Behavior.* New York: Holt, Rinehart, and Winston.

Federal Trade Commission. 1969. The Influence of Market Structure on Profit Performance of Food Manufacturing Companies. Superintendent of Documents, Washington, D.C.

Galbraith, J. K. 1958. *The Affluent Society.* Boston: Houghton Mifflin.
_____. 1967. *The New Industrial State.* Boston: Houghton Mifflin.
Granbois, D. H. 1978. Shopping behavior and preferences. In Selected Aspects of Consumer Behavior. National Science Foundation, Superintendent of Documents, Washington, D.C.
Handy, C., and D. I. Padberg. 1971. A model of competitive behavior in the food industries. *Am. J. Agric. Econ.* 53:182–90.
Hansen, F. 1972. *Consumer Choice Behavior.* New York: Free Press.
Helmberger, P. G., G. R. Campbell, and W. D. Dobson. 1981. Organization and performance of agricultural markets. In L. R. Martin, ed., *A Survey of Agricultural Economics Literature,* vol. 3. Minneapolis: University of Minnesota Press.
Holbrook, M. B., and J. A. Howard. 1978. Frequently purchased nondurable goods and services. In Selected Aspects of Consumer Behavior. National Science Foundation, Superintendent of Documents, Washington, D.C.
Howard, J., and L. E. Ostlund. 1973. *Buyer Behavior: Theoretical and Empirical Foundations.* New York: Knopf.
Howard, J. A., and J. N. Sheth. 1969. *The Theory of Buyer Behavior.* New York: Wiley.
Jacoby, J., R. W. Chestnut, and J. Silberman. 1977. Consumer use and comprehension of nutrient information. *J. Consum. Res.* 4:119–24.
Jacoby, J., D. E. Speller, and C. E. Kohn. 1975. Brand choice as a function of information load. *J. Consum. Res.* 1:63–69.
Lenahan, R. J., J. A. Thomas, D. A. Taylor, D. L. Call, and D. I. Padberg. 1972. Consumer Reaction to nutritional information. *Search* 2:1–26.
_____. 1973. Consumer reaction to nutritional labels on food products. *J. Consum. Aff.* 7:1–12.
Marion, B. W., W. F. Mueller, R. W. Cotterill, F. E. Geithman, and J. R. Schmelzer. 1977. The Profit and Price Performance of Leading Food Chains, 1970–1974. Joint Economic Committee, 95th Congr., 1st Sess. Committee Print.
McCullough, T. D., and D. I. Padberg. 1971. Unit pricing in supermarkets: Alternatives, costs, and consumer reaction. *Search* 1:1–27.
Metwally, M. M. 1976. *Price and Non-price Competition.* New York: Asia Publishing.
Moore, H. L., and G. Hussey. 1965. Economic implications of market orientations: *J. Farm Econ.* 47:421–27.
Mueller, W. F., and L. Garoian. 1961. *Changes in the Market Structure of Grocery Retailing.* Madison: University of Wisconsin Press.
National Commission on Food Marketing. 1966. Studies of Organization and Competition in Grocery Manufacturing. Tech. Study 6, Washington, D.C.
Nicholls, W. H. 1941. *Imperfect Competition within Agricultural Industries.* Ames: Iowa State College Press.
Nicosia, F., and R. N. Mayer. 1976. Toward a sociology of consumption. *J. Consum. Res.* 3:65–75.
Padberg, D. I. 1968. *Economics of Food Retailing.* Ithaca: Cornell University Press.
_____. 1975. Emerging effectiveness of competition and the need for consumer protection. *Am. J. Agric. Econ.* 57:196–205.
_____. 1976. Governmental policy issues relating to food marketing and food distribution. In Marketing Strategies in Food Distribution, pp. 1–26. Proceedings of the 1975 World Food Marketing Conference, American Marketing Association, Chicago.
_____. 1977. Nonuse benefits of mandatory consumer information programs. *J. Consum. Policy* 1:5–14.
Padberg, D. I., and R. E. Westgren. 1979. Product competition and consumer behavior in the food industries. *Am. J. Agric. Econ.* 61:620–25.
Padberg, D. I., F. E. Walker, and K. W. Kepner. 1970. Measuring consumer brand preference. In R. G. Murdick, ed., *Mathematical Models in Marketing,* Chap 14. Scranton, Pa.: Intext.
Robinson, J. 1933. *The Economics of Imperfect Competition.* London: Macmillan.
Scammon, D. 1977. Information load and consumers. *J. Consum. Res.* 4:148.
Scherer, F. M. 1971. *Industrial Market Structure and Economic Performance.* Chicago: Rand McNally.
Shaffer, J. D. 1960. The influence of impulse buying or in-the-store decisions on consumers' food purchases. *J. Farm Econ.* 42:317–24.

Tongue, W. W. 1965. Competition in the affluent society. *J. Farm Econ.* 47:433–44.

U.S. Department of Commerce. 1977. *Statistical Abstract of the United States.* Bureau of the Census.

U.S. Congress. 1977. Food and Agriculture Act of 1977. Conf. Rep. 95–599.

Wildt, A. R., and F. M. Bass. 1973. Multifirm Analysis of Competitive Decision Variables. Pap. 422, Institute for Research in the Behavioral, Economic, and Management Studies, Krannert Graduate School of Management, Purdue University, West Lafayette, Ind.

Woodside, A., J. N. Sheth, and P. Bennett. 1977. *Consumer and Industrial Buyer Behavior.* Amsterdam: North-Holland.

WILLIAM T. BOEHM
ROBERT J. LENAHAN

13

HEALTH AND SAFETY REGULATIONS AND FOOD SYSTEM PERFORMANCE

A MAJOR RESTRUCTURING of social priorities has been occurring over the past several decades. Increasingly, concerns regarding the quality of life have taken on more importance than earlier concerns for quantity. Health, safety, and environmental quality have become generally recognized as primary social goals.

Few policy issues have generated more public controversy than those resulting from the desire to accomplish the broader social goals through the regulatory process. The debate itself is understandable. Changes in the rules of exchange almost always imply changes in distribution of income and in economic opportunity. Policy options being considered in the health and safety area are no different.

Agricultural producers and the food system, more generally, have not been unaffected by these changed social priorities. Reports by the Senate Select Committee on Nutrition and Human Needs highlight the available scientific evidence linking diet and health and recommend rather specific changes in dietary patterns, including a reduction in animal fat intake (U.S. Congress 1977). Agricultural producers and food processing firms have become subject to work safety and environmental quality regulations. The Food and Drug Administration (FDA), U.S. Department of Health and Human Services, has proposed to ban or severely restrict use of certain chemicals and drugs once

WILLIAM T. BOEHM is Director of Economic Research, The Kroger Company. He was Senior Staff Economist, Council of Economic Advisors, Office of the President, when the chapter was completed. ROBERT J. LENAHAN is Economist, Office of Pesticide Programs, U.S. Environmental Protection Agency. He was Agricultural Economist with the Economics, Statistics, and Cooperative Service, USDA, when the chapter was completed. The views expressed are the authors' own and do not reflect official views of organizations with which they have been associated.

thought to be essential for improved feed efficiency and protection of animal health.

Simultaneously, regulatory actions to enhance the economic viability of the farm sector have come under increasing attack. Federal and state milk marketing orders are being studied for their social welfare implications (Buxton 1977; MacAvoy 1977; U.S. Congress 1978; Ippolito and Masson 1978). The once widely accepted antitrust exemptions granted to farmer cooperatives are being tested in the courts (U.S. v. Associated Milk Producers 1972; U.S. v. Dairymen, Inc. 1973; U.S. v. Mid-America Dairymen Association 1973). A new analytical group has been formed in the USDA to help enforce the undue price enhancement provisions of the Capper-Volstead Act (USDA 1979a).

The impact of such actions on agricultural producers, food marketing firms, and consumers will doubtless be significant. But in many cases, the rigorous economic research and policy analysis needed to help formulate the options have not been available during the policy debate. Some even say that the present institutional system encourages decision making based on incomplete analysis (Food Safety Council 1978).

Emphasis is given in this chapter to regulatory initiatives involving the health and safety of food. First, a conceptual foundation for analyzing regulatory initiatives is developed. Next, the statutory authorities that exercise broad controls and influence allowable market actions in the area of food safety and quality are reviewed. The final section focuses on the need for improved economic research and policy analysis. A plea is made for a more balanced treatment of the benefits and costs of proposals that seek to alter existing rules of exchange. A framework for a more comprehensive examination of the issues is discussed and a program for future research is presented.

FRAMEWORK FOR REGULATORY POLICY

Regulation, in its broadest sense, involves interpretation of any rule that governs or otherwise influences behavior. The rule can be written, spoken, or assumed. In this context, every society is regulated. From earliest days it has been recognized that maintaining social order requires a set of property rights and contract laws outlining the rules of the game with respect to owning, using, buying, and selling property. Given such a set of rules and an initial distribution of resources, trading among economic agents theoretically continues (in a simple and static world) until no further trades will benefit some members of society without harming at least one (Ferguson 1972).

In reality, however, the world is not simple, and neither the rules of exchange nor the distribution of resources is static. Preferences change as well as the available technology and stock of natural resources. Since changing the rules influences the distribution of wealth, there is constant pressure to alter rules to affect the relative wealth position of individuals and groups.

Proposals for rule changes, therefore, always evolve from a dissatisfaction with the result of transactions and activities that take place in the existing market setting. The problem is not that side effects exist, but that the benefits they confer or the costs they impose are often not reflected in the prices and costs that guide private decisions (Schultze 1977). Indeed, it is a truism that whenever the existing system is perceived as generating results that are unacceptable to established or potential political action groups, the pressure for rule changes intensifies.

The fact that economic behavior (i.e., trading) takes place in a social setting argues strongly against a simplistic view of regulation. Regulatory policy analysis undertaken as an exercise to help determine whether society needs more or less regulation is not very useful. The useful information will come from studies that deal with questions of how best to regulate, including suggested procedures for dealing with determinations that ultimately attach relative importance to the winners and losers. In developing such a framework for regulatory policy, it is important to address the question of market intervention. That is, when is it more efficient to use market incentives to achieve the desired result? When should government mandates be involved?

Most cases of mandated obedience (thou shalt or shalt not) evolve from the principle of relative risk. In high-risk health areas, for example, it is generally presumed better to prevent consumer or worker harm than to compensate for it later. Information costs, frequency of purchase, and relatively low costs per purchase make it more difficult (sometimes impossible) for consumers (or workers) to make what might otherwise be rational decisions. The problem arises because they either are not aware of or do not accurately calculate the hazards that accompany private decisions in the marketplace. The consumer may not know (indeed, may have no way of knowing) that a food item contains a dye possessing a health threat. The worker may not know that the flour milling company does not have an effective system for controlling dust and is therefore susceptible to an explosion.

The ignorance of the consumer in these transactions may arise because of fraud, where the employer or food manufacturer is aware of the risks but chooses to hide them and/or perhaps not fully investigate them; myopia on the part of workers or consumers who refuse to realistically assess the dangers even when they have full knowledge; and/or prohibitively high information costs that make it difficult or impossible for the worker or consumer to collect the necessary information—for example, how can consumers effectively evaluate the health impact of a new food preservative?

An often suggested alternative to mandated obedience, including product sale restrictions, is for government to provide the information needed to make informed decisons. With such a policy, it is argued, the consumer is free to decide from among the available alternatives. However, in the cases just discussed, the level of risk and the technical nature of the information necessary for making fully informed decisions often makes that alternative unrealistic. Not only is the cost of the search sometimes prohibitive, but calculating the associated risks, direct and indirect, can be impossible. Even under the best of circumstances, not all parties are (or probably can never be) fully informed.

There is also the related queston of consumption externalities. No smoking sections on airplanes have little to do with the health and safety of the smoker; indeed, they make the smoker worse off. Their stated purpose, and their foundation in the law, is to protect the rights of the nonsmoker. That argument can be logically extended to support the simple prohibition of product sales. Since death, injury, or disease impose both economic and emotional costs on others, some product purchase decisons and some seemingly private actions (i.e., riding a motorcycle without a helmet) do impose externalities. Whether and/or how to best internalize such externalities are questions that will continue to be settled in what is largely a legal and political social setting.

On the producer's side, there is usually little incentive to make such social determinations. For example, if a single firm furnishes the information to consumers on highly technical matters and in areas where there is a great deal of uncertainty regarding potential harm, private costs increase. If this cost increase is passed on to consumers in the form of higher prices or if consumers act on the information, the quantity demanded for that company's output will probably fall. This process is amplified in relation to the complexity of the information. The more information necessary and the more serious the potential harm, the higher the private costs will be.

There are, however, significant economies to providing such information. In some cases where the potential for harm is thought to be great, regulatory agencies have been assigned responsiblity for gathering, sorting, and disseminating information. In other cases, government mandates require that all firms provide the information.

The real question, therefore, is not whether we regulate, but how. To what degree and in what areas is it least costly (relative to the benefits) to rely only on private decisions and market incentives to regulate? When is it preferable for society to mandate obedience to achieve the desired outcomes? These are the questions economists must be prepared to help answer. Given this backdrop, regulation research becomes policy research involving an identification of the issues and available policy options and an assessment of the resource allocation consequences of the alternatives. It is a research environment familiar to many economists.

The view of the regulatory process just described is based on the proposition that rules are always necessary for the conduct of market exchange, even in purely competitive systems. Within this framework, discussions about the old and new regulations take on less significance (Weaver 1978), and searches for the optimum amount of regulation lose most of their meaning, as do accounts of increased budgets for the so-called federal regulatory bodies (Hildreth 1978). Indeed, within this framework all government bodies are regulators and all rules influence resource allocation.

EVOLUTION OF FEDERAL FOOD HEALTH AND SAFETY STANDARDS

The debate over how best to regulate the safety of the national food supply is not a contemporary phenomenon. Quite the contrary, public debate about food safety and its implied consequence for the public health predate most public actions dealing with farm commodity price and income support (U.S. Congress 1979). The reasons for public action in the area of food safety relate primarily to the naturally perishable nature of most foods and their ability to carry diseases that cripple and/or kill. As a result, legislative actions in the area of food safety and health have sought to assure that raw products are wholesome and safe for use in processing and consumption; production, processing, and commercial consumption environments are safe and sanitary; and chemical substances and other additives used for functional purposes are not hazardous to health.

Food Inspection. Federal food inspection began in 1890 with enactment of the Meat Inspection Act, which was aimed mainly at promoting exports of U.S. meat products to other countries and contained very limited provisions for meat inspection to protect consumers in the United States. When publication of *The Jungle* (Sinclair 1906) exposed the unsanitary conditions prevalent in many meat packing plants, the resulting public outcry culminated with enactment of the Federal Meat Inspection Act of 1906. This remained the major piece of legislation regulating meat inspection for more than 60 years. Poultry inspection, however, was voluntary until the passage of Poultry Products Inspection Act of 1957. Responsibility for enforcing these regulations rests with the USDA.

The acts were designed to protect the public from unwholesome, adulterated, or misbranded meat and poultry products. They require mandatory antemortem, postmortem, and processing inspection of all meat, poultry, and derivative products prepared for commerce. Federal inspection is also required in states that do not maintain inspection systems at least equal to the federal programs.

Under the authority of these acts, the USDA conducts inspection in approximately 7300 plants with a field inspection force of approximately 9200. On average, 456,000 head of livestock, 13 million head of poultry, 287 million pounds of domestic product, and 7 million pounds of imported product are inspected daily.

While visual inspection during slaughter and processing is effective for locating signs of disease and other forms of adulteraton, residues of various additives that may also adulterate the product are not so readily detected. Thus the USDA conducts a residue monitoring program through which random samples of meat and poultry carcasses are submitted for laboratory analysis. Various other federal agencies also inspect food products; e.g., the Department of Commerce provides a voluntary inspection service for plants preparing fish products. The other federal agency with major responsibility for food inspection is the FDA.

Methods of inspection differ among the various regulatory agencies. The USDA requires that each processing plant be registered and have official inspectors on the premises; the inspection is continuous. The FDA checks food establishments at infrequent intervals, depending on the risk associated with the specific commodity. A high-risk establishment such as a manufacturer of milk products may be inspected twice a year, while low-risk establishments such as a soft-drink bottler may be inspected once every few years.

Whether or not one system is better from a public welfare point of view is a researchable question. Surely, both systems could be operated more efficiently. Federal expenditures for meat and poultry inspection currently amount to more than $250 million. A study by Booz, Allen and Hamilton, Inc. (1977) stated that recommended program changes could save the USDA $210 million (in 1976 dollars) by 1985, with no loss in consumer protection. The study recognized that differences in the kind and degree of risk exist among the various processed products and meat and poultry. However, it also indicated that certain aspects of the FDA approach (e.g., industry quality control) could be adapted by the USDA. Some of these changes are under way. In 1979 the USDA announced its intentions to use the concept of voluntary quality control more extensively.

Food Production and Processing Statutes. The initial federal effort to control health and safety conditions in food processing occurred with passage of the Federal Food and Drug Act in 1906. The act was administered by the Bureau of Chemistry in the USDA and was the result of a 27-year congressional debate on the issue. Passage of the Sherley Amendments in 1912 made violators subject to prosecution. Even though the law required that government prove potential harm, more than 6000 violators were brought to trial during the first 10 years of the law.

New legislation, the Federal Food, Drug and Cosmetic Act (FFDCA), was passed in 1938. The FFDCA was designed to protect the public from dangerous and unwholesome products. A then new agency within the USDA, the FDA, was made responsible for all administrative decisions relating to its enactment. In 1940 the FDA was removed from the USDA and placed in the Federal Security Agency, a precursor of the Department of Health and Human Services. This move represented a significant shift in congressional attitude. The food and drug regulatory function was separated from other food-related programs and became aligned with an agency with a health responsibility.

The original statute included prohibitions against "poisonous or deleterious substances in food which may render it injurious to health," but did not include provisions for the premarket testing of substances such as food additives, animal drugs, or color additives. Subsequent amendments prohibit addition of such articles to food unless they have been shown to be safe by appropriate tests.

The Food Additives Amendment (P.L. 95-929) was enacted in 1958. It was designed to protect consumers by requiring premarket testing of all food additives. Included in this amendment is the first so-called Delaney clause. This clause, added to the bill as a committee amendment, states "that no additive shall be deemed safe if found to induce cancer when ingested by man or animal, or if it is found, after tests that are appropriate for the evaluation of the safety of food additives to induce cancer in man or animal. . . ."

In 1960 the Color Additive Amendment (P.L. 86-618) was added to the FFDCA. This amendment requires that the safety of color additives be demonstrated before approval for use is granted. Another Delaney clause was added to the act. Its language is virtually identical to that of the Food Additives Amendment. The safety and testing procedures regarding color additives parallel those required for food additives. The sponsor of such a substance has the burden of establishing its safety, and the FDA is expressly precluded from permitting the use of any color additive that has been found to have induced cancer in humans or animals.

The various provisions of the FFDCA governing premarket approval of drugs intended for use in animals were consolidated under the Animal Drug Amendments of 1968 (P.L. 90-299). Under these amendments, approvals for use of chemical substances for treatment and prevention of animal disease and/or as growth promoters may be granted after a two-part evaluation by the FDA. First, there must be a determination that the drug is safe and effective for use in animals. Second, the data must be reviewed to assure the safety of potential residues that might occur in the foods derived from such animals. These amendments also include a reiteration of the Delaney clause. In this case, however, a carcinogenic animal drug may be approved for use provided the substance will not adversely affect the animals involved and that no residues will be found in the edible portions.

The absolute nature of the various Delaney clauses and subsequent improvements in detection technology have made compliance increasingly more difficult. At the time of enactment, the test methods used for the detection of carcinogenic substances were valid to about 20 parts per million. Advances in testing procedures have now brought these analytical methods to quantitative accuracies of about a few parts per billion.

Pursuant to the original 1936 act, the FDA also exercised regulatory authority over the amount of pesticide residues in or on foods. This area was regulated primarily through a system of unofficial and information tolerances until the enactment of the Pesticide Chemical Amendment in 1954 (P.L. 85-791). This amendment defined a category of poisonous substances known as pesticide chemicals and authorized their use in or on raw agricultural commodities unless they were unsafe within the meaning of the newly enacted amendment.

The Environmental Protection Agency (EPA), established in 1978, establishes tolerances for pesticide chemicals under the FFDCA. The EPA also has registration responsibility. The Federal Environmental Pesticide Control Act of 1972 was substantially amended to provide for more complete regulation of pesticides and gave the EPA additional powers to prohibit misuse of registered pesticides. These statutes establish a regulatory system for pesticide chemicals that is essentially the same as that established for food additives, color additives, and new animal drugs except that there is no reiteration of the Delaney clause.

RESEARCH AGENDA

Economic research and policy studies that make explicit the relative costs and benefits of alternative sets of property rights and rules of product exchange have become more important in recent years. This is partly so because the decision process today is more open than previously. Organized citizen groups have become quite active, increasingly turning to government action to get rule changes. But in addition there is a more general recognition that the public sector responsibility for food and agriculture goes beyond administration of regulatory programs to enhance commodity prices and farm income. Public actions in the areas of food safety and quality, nutritional balance, and nutritional information and education are becoming as important as programs that ensure adequate supplies of food and an economically viable farm sector.

The issues tend to be emotional because they imply explicit changes in resource allocation. In the absence of objective analyses there is a high probability that they will be resolved in the courts with arguments based less on informed judgment than on emotional appeal. While the availability of economic research does not negate the legal requirement for action, estimates of the resource adjustments implied by various alternatives have much to contribute to improvements in the decision process, particularly if the results are effectively communicated (Foreman 1978).

Herman Talmadge (1978), chairman of the Senate Committee on Agriculture, Nutrition, and Forestry, expressed his concern for such research: "Many members of my Committee are concerned that regulatory agencies are making decisions that will have major implications for the economy without full knowledge or consideration of these impacts. While no members of the Congress would ever encourage the use of products that endanger the health and safety of our citizens, they are also concerned that we not endanger the economic welfare of those citizens in the pursuit of a minor benefit or to eliminate a theoretical risk." The needed economic research falls into three areas: issue identification, consequences of policy alternatives, and evaluations of existing systems.

Issue Identification. In many cases the policy debate related to food safety and health issues could be improved significantly by a clear, concise discussion of the issue. While the entire scientific community has a role to play in this area, economists who understand the political and legal systems have a potentially significant role. Often the available options are constrained by statute, regardless of how unreasonable they appear. In other instances, the political setting is such that only incremental changes are possible, even if a total reform would appear to be the best option.

Participants in policy debates cannot be relied upon for objective issue identification. Their role as advocates often mandates an overstatement. The responsibility for such analysis then belongs in the public domain.

A proposed rule change dealing with meat and poultry product labeling provides an example (*Federal Register* 1977). The proposal was, in essence, to change the definition of tare in product weight determination. Labeled product weight would be made equal to total package weight minus the weight of all packaging materials and free and absorbed liquids. Consumers could better evaluate product value if they knew the drained weight of the product in the package, it was argued.

The response to this proposal by advocate groups was predictable. The meat industry argued against the proposal, saying that it was inflationary since reported price per pound would increase. They also said it was impractical since they could not control the amount of liquid that drains from meat products. Consumer groups, on the other hand, argued in favor of the proposal, saying that the industry does exercise significant control over the amount of free liquid and that the proposed change would improve efficiency rather than increase costs. They also argued that consumers had been paying "chicken prices for water" (USDA 1978a).

After a year of debate, the Government Accounting Office (GAO) conducted an investigation of the proposed regulatory action and the available evidence. The GAO concluded that there were no data to help evaluate whether the proposal was needed. No one, to their knowledge, had identified the existence of an economic problem (USGAO 1978). The issues had not been sorted out.

A subsequent analysis by USDA economists indicated that the effects of the proposed rule change were grossly misunderstood by both consumers and producers. Consumers, for example, could not expect the reported price per pound of product to remain unchanged if free liquids were excluded from labeled product weight as a result of a drained-weight rule. The price per pound on the label would increase and would increase most for products with relatively more free liquid. However, the per package cost to consumers for usable product would remain unchanged. Contrary to contentions of many producers, actual producer costs would not increase because of the change in the definition of tare. The reported weight and price per pound would simply be adjusted to reflect the change. There would be no overfill, and processing costs per drained-weight pound would be unaffected.

The main benefit of a drained-weight system would appear to be an increase in the accuracy of information available in the marketplace for all meat and poultry products. As with unit pricing, consumers would be able to make better per pound price comparisons, and producers would not be competitively disadvantaged by those who use techniques that result in a higher proportion of free liquid to usable product (immersion chilled versus air chilled poultry).

The point here is not to assess whether the net weight labeling proposal should have been adopted. Indeed, it is not clear that economic research can resolve this issue. But economists, with a framework from economic theory, did play an important role in placing the issue in perspective. A subsequent USDA proposal in this area reflected the importance of economic analysis in placing boundaries on the discussion and helping to keep the debate focused. This is an important role for economists in a democratic system.

Consequences of Alternatives. A recurring theme throughout this chapter has been that the real policy question is not whether we regulate but how. Stating the question in that way makes explicit the need for research on the resource allocation consequences of regulatory alternatives.

One area receiving attention relates to proposals that would affect more complete compliance with the animal drug and food additive provisions of the FFDCA. Studies assessing the impact of restricting use of animal growth hormones, antibiotics, and food additives are examples.

The continued subtherapeutic use of some chemicals is being questioned because of scientific evidence linking resistant strains of certain organisms to the chronic intake of antibiotics. The development of such resistance is of concern because it makes the drugs potentially less effective in dealing with human health problems. Uses of other drugs are being proposed for restrictions since they have been found to produce tumors in laboratory animals.

Such proposals tend to generate emotional responses from the farm and agribusiness community. Widespread use of these chemical substances is closely identified with changes in the structure of livestock and poultry production. The ability to suppress and control many infectious animal diseases has often been cited as facilitating development of large-scale confinement feeding operations. Although these animal production systems require large investments of capital, many argue that they have resulted in significant decreases in operating costs, especially for labor. Proposals that would prohibit their use appear on the surface to threaten economic disaster for some producers.

The available research evidence, however, suggests that the economic system would probably be quite resilient to a more restrictive policy on animal drug use (Allen and Burbee 1971; Dworkin 1976; Mann and Paulsen 1976; Gilliam et al. 1977; USDA 1978b). This is particularly the case if only the subtherapeutic uses are restricted. Costs of production and consumer food prices would increase initially; but because the farm-level demand for most livestock products is inelastic, initial decrease in output would cause farm prices to increase more than enough to offset the higher production costs. As a result, net revenue to the farm sector would increase and farmers would be encouraged to expand output.

There could be structural dimensions to such a change in policy. But even here, the results are not clear-cut. Increased risks associated with feeding poultry and livestock in confinement production systems could make such systems less viable. On the other hand, dramatic changes in management practices could occur in anticipation of or following enactment of such rules, thus negating their impacts. These are issues needing more research.

The same can be said for proposals that would restrict use of some food additives. Certain of these substances have been used for years to protect the naturally perishable food supply and increase its shelf life. At the most basic level, such additives increase the marginal productivity of producing food out of raw agricultural products. Whether proposals that would prohibit their further use improve or adversely affect farmers depends on the change in output price relative to the change in marginal productivity. It is conceivable that such rule changes could increase the demand for some farm products.

An immediate ban on use of nitrites to cure bacon, for example, would affect hog and cattle producers in different ways (Brandt et al. 1978; CAST 1978; Madsen 1978; Meat Packers Council of Canada 1978; USDA 1978c). Hog prices would fall, reflecting the fact that large quantities of diverted bellies would need to be rendered into lard. Food prices, though, would probably increase because of the higher costs for processing pork and the increased demand for substitute meat products. Initially, some hog producers would probably go out of business. However, cattle producers would probably benefit from such a ban. Because of the increased demand for beef products, cattle producers' receipts could be expected to increase. The economic adjustments required by an immediate ban on all nitrite use would be similar to those required by a selective ban such as for bacon but would be more pronounced. Markets would immediately have to be found for products cured with sodium nitrite. While the quantity of uncured meat would increase (a price depressing effect), consumer demand for such products would also increase (a price increasing effect).

While producer advocacy groups have strongly opposed such regulatory changes, the available research indicates that the proposals would likely have little noticeable effect on resource use in the aggregate. Indeed, since the demand for most agricultural products is inelastic, the short-run effect of a unit cost increasing action is to increase industry revenue.

The interesting story that unfolds from these studies is that consumers (those generally supporting the changes) rather than farmers (who are typically opposed to the changes) end up shouldering most of the impact. Food prices tend to rise, but aggregate farm income either increases or remains about unchanged. But do consumers really end up worse off? If these actions have the effect of improving the nation's health, perhaps that means higher food costs but lower costs for medical care.

Economic research in the food regulatory area is important, but such research must initiate as well as respond to policy problems. Over the years, agricultural economics research has played an important role in shaping the policy options. Parity price support programs, land diversion schemes, and food distribution programs were all influenced in important ways by researchers who had a vision of the need for alternatives and then did something about it (Penn and Brown 1977; Sharples and Krenz 1977). Future research in the area of food safety and quality should build on that tradition.

Evaluating Existing Systems. The broad social objectives in the health and safety area tend to be specified in the statutes. In almost every case, responsibility for administration of action programs is lodged with the federal bureaucracy. Once they are in place, there is an important continuing need to assess program performance relative to the legislated objective. Two questions need to be addressed: Is the present system effective, i.e., does it work? Is it efficient, i.e., are the programs being administered in a way that is cost effective?

The effectiveness question is probably the most difficult to research. Seldom is there a benchmark. Most will agree, for example, that the food supply should be safe and that the public at large should be protected from the deliberate or even unintentional sale of foods that will cause death or disease. Implementing this protection is a program question, and alternative programs produce different results at different costs.

Quite often there is little or no opportunity for controlled experimentation, but that does not negate the importance of or the need for research in this area. Decisions will be made, and social scientists (economists in particular) have a responsibility to provide information about the relative benefits and costs associated with present programs.

Regulatory effectiveness studies must be balanced to be credible. It is not enough to measure administrative costs. Benefits must also be considered explicitly. Proposals that would remove regulation from the market must also be evaluated while considering the costs associated with imperfect regulatory bodies. Such research must recognize that social intervention often becomes a race between the ingenuity of the regulatee and the loophole closing regulator, with a continuing expansion in the volume of regulations as the outcome (Schultze 1977). The conceptual framework offered by Posner (1975) in his article discussing the costs of monopolizing is important and relevant to the conduct of realistic program effectiveness regulation research.

The subtle aspect of this research is that it involves an appraisal of overall system performance. To what extent is the entire regulatory system working to effectively reflect preferences of the society (and its individuals)? What are the relationships between technical change (or the lack of it) and the set of rules governing behavior? There is little doubt, for example, that government policies have influenced what we eat. Commodity programs have made the production of some agricultural products more profitable and sometimes encouraged production beyond market needs. Some agricultural producers have doubtless benefited from these programs, and consumers have had abundant food supplies. But do these regulatory efforts result collectively in a food supply that promotes the national health (U.S. Congress 1977)? Do national health care costs exceed the benefits of cheaper food? What is the extent of the trade-off between food prices and medical care costs? These are all legitimate program effectiveness questions for the applied food and agricultural economist to research.

Regulatory program efficiency studies are probably easier to do than the effectiveness studies but are not less important. Efficiency, of course, is measured by the degree to which the present organizational structure achieves a given level of services at the least cost. Questions about administrative costs relative to the task being performed and appropriate sampling for a given level of accuracy are important aspects of the research needed in this area.

The obvious difference in inspection philosophy between the USDA and FDA is an excellent example. The FDA approach is less labor using and is based more on sampling theory. High-risk operations are visited most often, but even then there is no inspection of each food item. USDA procedure now requires the actual physical inspection of each carcass. Their system is labor using.

A study of benefits and costs of the Federal Meat and Poultry Inspection Program indicated that it has been reasonably cost effective (USDA 1979b). Measuring just a few of the apparent health benefits and considering the economic importance of the meat export trade generates a benefit-cost ratio of about 6:1. But that analysis does not indicate whether the benefit of another dollar spent on current programs would be greater than, equal to, or less than the expenditure. Nor does it indicate whether all parts of the inspection program are equally cost effective. Indeed, the analysis suggests that some rather fundamental changes in inspection procedures could generate a benefit-cost ratio exceeding that of the present system. Simply put, it seems the system has not kept up with the changed need for meat and poultry inspection.

The research needed to test this hypothesis seems obvious enough. What would alternative systems cost? What would be the probability of a reduction or improvement in the safety of the food system? How might the present system be changed to improve the delivery of services and perhaps reduce costs? In an age when inflation pressures are foremost in the minds of policymakers and most citizens are calling for reductions in government spending, it would seem that economic research addressing the simple question of how efficiently government services are being performed could have high payoff.

SUMMARY

Most research on food system regulatory topics has been piecemeal. It has been undertaken to respond to specific questions about programs, either proposed or in force. Generally speaking, the research has been done from a rather narrow producer perspective. Only the costs of regulations have been calculated in most cases. There has been little explicit recognition of the benefits received. The benefit calculation, some have argued, is vague at best and perhaps not even relevant. Obviously, if publicly supported economic research is going to help improve the collective decision process in the future, it will need to be done from a broader public good perspective.

Public research institutions have undertaken research to specifically measure benefits in some areas (Conference 1978; National Highway Traffic Safety Administration 1978), at least to learn more about how to deal with benefits. Those with training in the agricultural sciences can potentially benefit a great deal from such efforts.

The reality of the situation is that the change in policy focus from farm to food is likely to continue. The public will become more involved in shaping the food policy agenda. If research responds to the needs of this broadened constituency, it will contribute to improved decision making. Our perspective will help in an important way to determine how far we are able to push back this frontier in agricultural marketing research.

REFERENCES

Allen, G., and C. Burbee. 1972. Economic Consequences of the Restricted Use of Antibiotics at Subtherapeutic Levels in Broiler and Turkey Production. Unpublished staff paper, USDA, ERS.

Booz, Allen and Hamilton, Inc. 1977. Study of Federal Meat and Poultry Inspection Systems. USDA.

Brandt, J. A., M. D. Judge, and M. B. Sands. 1978. An Analysis of the Impact of a Nitrite Ban in Bacon Curing. Purdue University Agric. Exp. Sta. Bull. 200, West Lafayette, Ind.

Buxton, Boyd M. 1977. Welfare implications of alternative classified pricing policies for milk. *Am. J. Agric. Econ.* 59:525–29.

Conference on the Benefits of Governmental Health and Safety Regulations. 1978. Sponsored by the Public Interest Economics Foundation and the National Science Foundation, Berkeley Springs, W.Va., October 12–13.

Council for Agricultural Science and Technology (CAST). 1978. Nitrite in Meat Curing: Risks and Benefits, Report no. 74, Iowa State University, Ames.

Dworkin, Fay H. 1976. Some Economic Consequences of Restricting the Subtherapeutic Use of Tetracyclines in Feedlot Cattle and Swine. Food and Drug Administration, OPE Study 33.

Federal Register. Dec. 2, 1977. 42(232):61279–82.

Ferguson, C. E. 1972. *Microeconomic Theory.* Homewood, Ill.: Irwin.

Food Safety Council. 1978. Highlights of report of Social and Economic Committee, annual meeting, December 14.

Foreman, Carol Tucker. 1978. Consumers and food policy in North America. *Am. J. Agric. Econ.* 60:777–81.

Gilliam, H., R. Martin, W. Bursch, and R. Smith. 1977. Economic Consequences of Banning the Use of Antibiotics at Subtherapeutic Levels in Livestock Production. Texas A & M University, Dep. Agric. Econ. DTR 73-2, College Station.

Hildreth, R. J. 1978. Public Policy and Regulation. Paper presented at the Northeast Agricultural Economics Council annual meeting, June.

Ippolito, Richard A., and Robert T. Masson. 1978. The social cost of federal regulation of milk. *J. Law Econ.* 21:33–65.

MacAvoy, Paul W., ed. 1977. *Federal Milk Marketing Orders and Price Support.* Washington, D.C.: American Enterprise Institute for Public Policy Research.

Madsen, H. C. 1978. Impact of the Loss of Nitrites on Animal Agriculture. Meat Industry Research Conference, University of Chicago, March.

Mann, T., and A. Paulsen. 1976. Economic impact of restricting feed additives in livestock and poultry production. *Am. J. Agric. Econ.* 58:47–53.

Meat Packers Council of Canada. 1978. Economic Consequences of a Ban on Nitrites. Islington, Ontario, Canada. March.

National Highway Traffic Safety Administration. 1978. The Contributions of Automobile Regulations. Preliminary report, Washington, D.C., June.

Penn, J. B., and W. H. Brown. 1977. Target price and loan rate concepts for agricultural commodities. In Agricultural–Food Policy Review: Proceedings of Five Food Policy Seminars, pp. 69–75. USDA, ESCS. AFPR-1.

Posner, Richard A. 1975. The social costs of monopoly and regulations. *J. Polit. Econ.* 83:807–27.

Schultze, Charles L. 1977. *Public Use of Private Interest.* Washington, D.C.: Brookings Institution.

Sharples, Jerry A., and Ronald Krenz. 1977. Cost of Production: A Replacement for Parity? In Agricultural–Food Policy Review: Proceedings of Five Food Policy Seminars, pp. 62–68. USDA, ESCS, AFPR-1.

Sinclair, Upton. 1906. *The Jungle.* New York: Doubleday, Page.

Talmadge, Herman. 1978. Letter to Secretary of Agriculture Bob Bergland, January.

U.S. Congress. 1977. Senate Select Committee on Nutrition and Human Needs. Dietary Goals for the United States.

_____. 1978. Senate Committee on Governmental Affairs. Study on Federal Regulation. Appendix to Volume VI: Framework for Regulation. Committee Print, 95th Congr., 2nd Sess.

_____. 1979. Senate Committee on Agriculture, Nutrition, and Forestry. *Food Safety: Where Are We?* Washington, D.C.: USGPO.

U.S. Department of Agriculture. 1978a. Public Hearing on Net Weight Labeling, Washington, D.C., February 9.

_____. 1978b. Economic Effects of a Prohibition on the Use of Selected Animal Drugs. ESCS, AER-414.

_____. 1978c. Nitrites in Bacon: A Summary Analysis of a Ban on the Use of Nitrite in Curing Bacon. National Economic Analysis Division, ESCS-44.

_____. 1979a. Undue Price Enhancement by Agricultural Cooperatives, Criteria, Monitoring and Enforcement. Capper-Volstead Study Committee Report.

_____. 1979b. The Benefits and Costs of Federal Meat and Poultry Inspection. ESCS Staff Report.

U.S. General Accounting Office. 1978. Proposed Changes in Meat and Poultry Net Weight Labeling Regulations Based on Insufficient Data. Report by the Comptroller General of the United States.

U.S. v. Associated Milk Producers, Inc. 1972. No. SA72Ca49, U.S. District Court, Texas.

U.S. v. Dairymen, Inc. 1973. No. 7634A, U.S. District Court, Kentucky.

U.S. v. Mid-America Dairymen Association. 1973. No. 73CV681W-3, U.S. District Court, Missouri.

Weaver, Paul H. 1978. Regulation: Social policy and class conflict. *The Public Interest* 50:45–63.

14

MARKETING IMPACTS OF THE DOMESTIC FOOD ASSISTANCE PROGRAMS

ON MAY 6, 1969, the president of the United States in a speech to the Congress made a historic commitment to end hunger. He said, "That hunger and malnutrition should persist in a land such as ours is embarrassing and intolerable. . . . The moment is at hand to put an end to hunger in America itself for all time" (Nixon 1969). In August of that year the Food and Nutrition Service was established in the USDA to centralize the operation and expand the size of the domestic food assistance programs. In December, the White House Conference on Food, Nutrition, and Health was held. These and subsequent events have led to substantial increases in size and scope of the various food programs. The primary objectives of the programs shifted from demand expansion to solving the problems of hunger and malnutrition in carrying out the presidential commitment. Even so, this demand expansion has benefited the commercial marketing system, since it has increasingly been used as the vehicle for food delivery.

Most of the U.S. food assistance programs were designed with the dual objectives of assisting participants and making effective use of our agricultural abundance. In the 1930s the Food Distribution Program was established to provide direct food assistance to families and institutions and to provide a useful outlet for surplus food products acquired under farm price and income programs. That program was phased down as the Food Stamp Program expanded geographically. Food stamps were required nationwide by the Agriculture and Consumer Protection Act of 1973 (P.L. 93-86). The Food Stamp Program was considered more responsive to participant needs; it allowed free se-

STEPHEN J. HIEMSTRA is Director, Economic Analysis Staff, Office of Policy, Planning, and Evaluation, Food and Nutrition Service, USDA. This chapter is based on information through fiscal year 1977, prior to implementation of the Food Stamp Act of 1977, which eliminated the purchase requirements for the program.

lection of foods of the same type purchased commercially by the nonpoor. However, direct commodity support was continued for schools, certain other child nutrition programs, specialized family distribution programs, institutions, and foreign donations. When the Food Distribution Program to families was virtually eliminated, there were no food surpluses. Food shortages and rising prices were the policy concerns.

The National School Lunch Program, authorized by the National School Lunch Act of 1946 (P.L. 79-396), also had its roots in commodity support to the schools in the 1930s. Since then, greatly expanded support of school food service programs has mainly been in the form of cash. Commodity support has continued at roughly stable dollar levels, but it has declined to about one-fifth the total value of the federal support.

The legislative objectives of the child nutrition programs continue to point to the dual concerns to safeguard the health and well-being of the nation's children, and to encourage the domestic consumption of agricultural and other foods (Child Nutrition Act of 1966—P.L. 89-642). Shifts in funding levels clearly indicate a shift in importance of the two objectives in favor of improving nutritional levels of children. The Special Supplemental Food Program for Women, Infants, and Children (WIC Program), authorized in 1972 by P.L.92-433 as a revision of the Child Nutrition Act, expanded rapidly and has clear-cut nutritional objectives. It provides assistance to selected groups deemed to be especially vulnerable nutritionally. This program lacks reference in its objectives to stimulate demand for agricultural products.

Shifting of priorities away from producer interests determining consumer food use in favor of consumer choice advanced when the purchase requirement for participation in the Food Stamp Program was eliminated. Increased program participation was traded for reduced food and nutritional impacts per dollar of assistance. Another big step would be to abandon food stamps entirely and provide the consumer with the sovereignty that comes with cash transfers.

SIZE AND MARKET SHARE OF THE PROGRAMS

Federal Costs. In recent years, food programs have grown significantly. Federal costs in fiscal year 1978 totaled some $9 billion, compared with $8.6 billion in fiscal year 1977, and less than half that five years earlier. Funding has increased sharply for both child and family food programs. For more detailed and historical information, see U.S. Congress (1975) and Hiemstra (1978a). Federal costs for the Food Stamp Program totaled about $5.5 billion annually in fiscal years 1976–78. Declining participation from the peak of nearly 19 million persons in 1976 to 15 million in 1978 offset increased costs per person. But this level of cost was more than double the $2 billion of 1973.

The Food Stamp Act of 1977 (P.L. 95-113, Title XIII of the Food and Agriculture Act of 1977) eliminated the purchase requirement of the program, simplified eligibility requirements, and lowered program eligibility to the poverty level. The net impacts will be to increase participation considerably. But per person benefits are expected to drop, aside from the effects of price escalators. The new law for the first time put a ceiling on food stamp spending at about $6.2 billion annually for fiscal years 1979–81. Food price inflation in fiscal year 1979 required that ceiling to be modified.

The federal cost of school and other child food service programs has increased at about the same rate as the Food Stamp Program. Total cost in fiscal year 1977 was $2.7 billion compared with $1.2 billion in 1972. Increased cost has been due largely to increasing numbers of free and reduced price meals served in the National School Lunch Program and to significant expansion in some of the newer programs such as the School Breakfast Program (P.L. 92-32), Child Care Feeding Program (P.L. 95-627), and Summer Food Service Program for Children (P.L. 94-105). Future costs of the National School Lunch Program probably will rise at about the same rate as the Bureau of Labor Statistics index of Food Away from Home to which the escalator on benefits per meal are tied. Unit costs of the School Breakfast Program and other child food service programs probably will also rise at this rate. Costs will be further increased by the expected 5–10 percent growth rate in participation expected. Increases in level of participation have been a factor in the rising costs of several food programs. In addition, the escalators on benefit levels that apply to all the major food programs have added significantly to total costs because benefits are tied directly to rising food prices.

Total Dollar Value of Food Spending. Federal input into several of the USDA food assistance programs is supplemented by participants, state and local governments, or private donations. Increases in gross spending for food exceed the $8 billion in federal program costs.

Under the Food Stamp Program, total spending consists of the face value of the food stamps, which through 1978 included both the federal input plus the participant's purchase requirements. With the purchase requirement eliminated, the face value of the stamps will drop to the level of the federal input. In fiscal year 1977, $8.3 billion in food stamps were issued. Of that total, $5 billion represented bonus value to participants, and $3.3 billion was the size of the purchase requirement.

In the case of the school food service programs, many participants contribute toward the cost of the meals or milk provided. In fiscal year 1977, the $2.7 billion in federal cost under the National School Lunch Program resulted in about $4.5 billion in total value of food served. Under the School Breakfast Program and other child food service programs, most of the meals are served free. The Special Milk Program is an exception. Federal costs cover about two-thirds and children's payments one-third of the total value of milk served.

Monies provided by the Elderly Feeding Program are added to those provided under Title VII of the Older Americans Act for elderly programs of various kinds. In addition, elderly participants are invited to contribute, if able, some share of the cost of the meals they receive. As a result, the value of food served under this program was probably greater than the USDA cost.

In summary, of the estimated $13.8 billion in the total value of food served under these programs in fiscal year 1977, federal expenditures were $8.1 billion (Table 14.1).

TABLE 14.1. Estimated value of food purchased or served through USDA food assistance programs compared with federal costs, fiscal year 1977

Program	Federal cost[a]	Total value of food purchased or served
	($mil)	($mil)
Food Stamp Program	5,058	8,340
National School Lunch Program	2,214	4,457
School Breakfast Program	146	163
Child Care Food Program	126	126
Summer Food Service Program	130	130
Special Milk Program	153	222
WIC Program	212	212
Elderly Feeding Program	19	76
Other food distribution programs	43	43
Total food assistance	8,101	13,769

[a]Fifty states, Puerto Rico, and U.S. territories. These figures do not include federal operating expenses, sharing of state operating costs, and other costs of administration, which totaled about $465 million in fiscal year 1977.

Food Marketing Share. The importance of the food programs may be viewed from several perspectives. Food marketed through these programs may be viewed as a share of the U.S. total farm value of all food sold or as a share of food sales at other levels of the marketing system. Further, the programs utilize more of certain food commodities than others.

The $13.8 billion in total dollar value of food purchased or served through the domestic USDA food programs amounted to perhaps $12.0 billion spent on foods from U.S. farms. The balance was spent on imports, fish, and other nonfarm foods. Farmers received about $4.1 billion of the $12 billion in retail sales, which is about 7 percent of the $57 billion total value of food produced on U.S. farms in 1977. The $12 billion in food program spending for U.S. farm products consists of about $7.2 billion for food used at home and about $4.8 billion for food served in schools and at other away-from-home locations.

Complete elimination of the food programs would reduce food spending by less than federal program costs because approximately one-half the $8 billion in federal spending merely substitutes for food that would be purchased anyway (Reese et al. 1974; West et al. 1978). Based on average marketing bill data for farmers' returns relative to consumer food spending, the farm value of the $4 billion in lost consumer food spending probably would total about $1.3 billion. Net additional demand for food resulting from the food programs accounted for about 2.3 percent of the U.S. total farm value of $57 billion in 1977.

The marginal impacts on prices and incomes of farmers from such a drop in demand for farm products could be substantial for some commodities in view of the inelasticity of supply and demand for farm products. The size of these impacts would vary considerably among commodities, the length of the program phase-out period, and the extent to which the drop in farm value affects quantity demanded. The short-run impacts could indeed be substantial; whereas supply could adjust to lower demand over time.

ASSESSMENT OF PAST RESEARCH

Hypotheses, Assumptions, and Program Objectives. A fundamental premise underlying the design of the Food Stamp Program is that expanding food expenditures of participants is positively related to improved nutrition. In turn, increases in food expenditures are expected to benefit farmers. Both assumptions are correct only to a limited degree.

Inappropriate evaluation criteria often are employed in food program evaluation (Hiemstra 1978c). Many assessments are input (federal government costs) rather than output (utility to participants or marketing system) oriented. Benefits may either exceed or fall short of government costs, and incidence may vary. Benefits may exceed costs because of economies of scale in food procurement, cost sharing requirements, or inducements that provide an entirely new market for food. Benefits may be lower than costs because of costs of administration, diseconomies in procurement, and the forcing of participants to accept foods that ignore personal preferences. Some studies have tried to document the size of the utility of a dollar of food stamp bonus (Clarkson 1975; Peskin 1976).

Policy objectives concerned with accomplishing certain end results often predetermine the degree of market efficiency attainable among alternative programs. Some programs such as food stamps, the WIC Program, and most child nutrition programs by their design use and thereby benefit the marketing system much more than others. The direct food distribution programs tend to compete with the marketing system by purchasing at an early stage in the marketing process, and then government or program operators perform some of the marketing functions.

Direct food distribution has quite different impacts on various farmers and parts of the marketing system because it allows policymakers the opportunity to concentrate purchases among selected products with precise specifications of type or quality and at specified periods of time and place. Exercising choice among commodities purchased allows efficiency in use of federal inputs to either increase farm prices or hold down program costs, depending upon the slopes of supply and demand curves and policy objectives.

Program benefits accrue quite differently to farmers versus users of food provided by some programs such as schools. If farm prices are maximized by a particular choice of procurement pattern, schools may be worse off because of a fixed dollar level of food support measured in terms of government costs. No allocation procedures have been developed to prorate costs of joint benefits between farm programs and food programs. School food service is charged the full cost of many foods with relatively low utility for them because of its assistance in supporting prices received by farmers. Consequently, proper program evaluation must first specify the importance of benefits among competing beneficiaries.

Data Problems. Evaluations of impacts of the food programs are often hampered by lack of available data. Only data needed for program operations and accountability are regularly reported. Data are not routinely collected to assess impacts of the programs on the food marketing system. Special studies have been based primarily on available secondary data. Routinely available data are limited to participation of persons or meals served and federal government costs. Costs of administration borne by the federal government are usually available in total but often not by program. State and local costs of administration are sometimes available because of cost sharing.

Data problems in assessing food marketing impacts include (1) availability of quantities of food served, provided, or subsidized only through special surveys, other than for direct food distribution programs; (2) reporting of program costs only in terms of costs to the federal government regardless of level of purchase in the marketing system, specification of products, location, or time of purchase; (3) availability of national totals only rather than local or regional components, annual rather than shorter time periods, and total expenditures or use rather than commodity breakdowns; and (4) availability of national marketing bill data that provide only rough indicators of relative program impacts on various segments of the market rather than data that would provide an assessment of marginal relationships that could be associated with a program change.

Data problems are even more limiting in attempts to assess the nutritional impacts of the programs on participants. Secondary data are scarcer and less useful because of the precise requirements for analysis. The 10-State Survey and the Health and Nutrition Examination Survey are the main sources of biochemical data, both of which identify food program participants. Dietary data from the USDA's 1977-78 Nationwide Food Consumption Survey will also be analyzed when available. But all three sources have shortcomings (Hiemstra 1978c).

Important Research Areas. For a summary of the major program evaluation activities supported by the Food and Nutrition Service, USDA, see Hiemstra (1976) for studies completed during 1970-75 and Hiemstra and Braden (1979) for studies completed during 1974-78.

NET ADDITIONAL FOOD SPENDING. In assessing the importance of the program to the food marketing system, the most important factor is the marginal propensity to consume or spend (MPC) from a dollar's worth of food program benefits. MPC is also related positively, although indirectly, to nutritional benefits to participants.

Study results have shown MPC for food stamp benefits to vary from about 0.3 to 0.5 (West and Price 1976; Neenan and David 1977; Choudhury 1978; Lane 1978; West et al. 1978). Earlier studies suggested that MPC probably was between 0.5 to 0.65 (Reese et al. 1974). Liberalization of benefits probably has lessened the size of MPC over time. Other early studies by the Agricultural Marketing Service (AMS), USDA (1962), and Benus et al. (1976) showed results of 0.8-0.9. The Benus study was probably the most sophisticated of the group, but it used data that were probably the least reliable. The AMS study did not allow for the value of the directly distributed food that was discontinued.

All these studies are of limited value in national policymaking. Most relate to small geographic areas or samples. Most were dependent upon cross-sectional comparisons of participants and nonparticipants judged, usually with limited data, to be eligible for participation. An implicit assumption in cross-sectional studies is that the samples are homogeneous with respect to response rates, thereby allowing the researcher to use statistical models with specified statistical properties. In fact, participants tend to be quite different from eligible nonparticipants in average age, household size, assets owned, and other characteristics that may be associated with food needs and preferences (USDA 1977). As a result, assumptions of homogeneity are violated, and it should not be surprising if small program differences or none at all are measured. West et al. (1978) found Food Stamp Program participants significantly different from nonparticipants in psychological need as a factor affecting food expenditures.

Some of the statistical models used to evaluate the food expenditure impacts of the food programs have been poorly specified. Food price and expenditure impacts are expected to be overpowered by general market demand shifters. Serial correlation and other statistical problems may also outweigh the impact of program variables in aggregate single-equation analysis. Simultaneous equations or the partitioning of markets that are subsequently reweighted may be necessary to obtain a better picture. Nonsampling errors tend to be completely ignored, even though they may be larger than sampling errors. Use of nonprofessional interviewers, errors and omissions in reports, poorly designed questionnaires, and related problems may bias the results as much as lack of random sampling and sample size.

Past research suggests that net additional food spending induced by programs is quite variable from one situation to another. This factor makes it tenuous to generalize on the basis of pilot studies. However, there are some central tendencies of results despite substantial differences in methodology.

For the child nutrition programs, one study of the National School Lunch Program showed that receipt of free lunches by children reduced household food spending by about 60 percent of the value of the meals when valued at the price paid by students (West and Price 1976). This amount translates to about 30 percent of the federal cost (MPC of 0.3). Unfortunately, this study has not been replicated or verified by other research. Use of this relationship alone is not sufficient in assessing the net additional food use attributable to the child nutrition programs. If it were not for federal assistance, many of these programs simply would not exist. To the extent that these added programs attract state inputs and local contributions, the federal input may result in a multiplier effect. This multiplier may be in the range of 1.8, thereby bringing the MPC of the program up to about 0.5.

There have been few studies conducted on the marketing impacts of the direct food distribution program. Those that have been done suggest that the marginal propensity to increase food use relative to costs is lower than for stamps, at least before elimination of the purchase requirement (Lane 1978). This conclusion is hard to accept by people who regard direct purchasing and distribution as tangible activities that must be having corresponding impacts;

but they tend to ignore that donated food may be substituted for purchased food in the absence of the program.

This analysis is complicated by the purchase of different kinds of products than might be purchased in the absence of the program. Food costing is complicated by the joint benefits derived from such food. Studies that measured food use by food distribution participants who later participated in the Food Stamp Program suffered because they did not allow for the value of the direct food benefits. There have been no such studies conducted for the WIC Program.

NET IMPACTS ON THE FOOD INDUSTRIES. The net economic effects of the Food Stamp Program and the National School Lunch Program on various industries have been assessed using input-output (I-O) techniques. These procedures allow researchers to follow initial changes in demand stimulated by outside sources such as federal spending for the food programs. Ripple effects of initial spending are measured as dollars work their way through the market.

Some studies assumed that federal food program spending was paid with tax dollars. Therefore, taxpaying consumers were assumed to reduce their demand for food, and this was subtracted from the increases experienced by program recipients. In this sense, the resulting impacts on the various industries are net of any tax-induced reductions in demand. For the Food Stamp Program and National School Lunch Program combined, the net impacts on business receipts are given in Table 14.2. These figures show net beneficial impacts on business receipts totaling $1.3 billion for agriculture, forestry, and fisheries, and $2.5 billion for the food manufacturing industries, after allowing for the negative impacts of higher taxes. Wholesale and retail trade also show a combined positive impact of $1.3 billion, even though trade of non-

TABLE 14.2   Net impacts on business receipts by the Food Stamp and National School Lunch programs

Business	Receipts
	($mil)
Agriculture, forestry, and fisheries	1,334
Food manufacturing	
Meat and poultry products	836
Dairy products	523
Grain mill products	269
Bakery products	134
Canned and preserved foods	476
Other foods and beverages	251
Total food manufacturing	2,489
Wholesale trade	542
Retail trade	754
All other industries	−1,805
Net change in business receipts	$3,314

Note: Data for the Food Stamp Program relate to fiscal year 1976 (Nelson and Perrin 1976a; later tabulations for 1976 supplied to the Congressional Budget Office, March 1977) and for the National School Lunch Program to fiscal year 1974 (Nelson and Perrin 1976b). The school lunch data include impacts of cash contributions and direct commodity support combined.

food items showed a decline. Impacts of the Food Stamp Program are strongest at retail, whereas impacts of the National School Lunch Program are strongest at wholesale. The net impact of all food and nonfood industries combined, after allowing for a cut in spending for nonfood industries, is an increase of $3.3 in business receipts. This total is in response to federal program inputs of $5.3 billion in food stamp benefits and $1.4 billion in school lunch benefits and a tax increase of an equal amount of $6.7 billion. Among individual food manufacturing industries, meat and poultry products show the largest combined impacts. However, for the school lunch program alone, dairy products have the largest increase in receipts.

Results of such studies must be interpreted in light of many data problems and restrictive assumptions. Underlying coefficients generally are several years old, but the food programs have experienced rapid growth and change. Marginal propensities to spend for food by detailed commodity groups would be preferable to the average propensities used in I-O studies.

FOODS USED THROUGH THE PROGRAMS. The programs impact quite differently upon different sectors of the food business. Such differences have marketing and production implications for producers and nutritional implications for participants. But such differences are not well documented and analyzed.

Food distribution programs are known to concentrate purchases mainly on commodities that are in need of price support from the standpoint of farmers. Food distributed in schools has been heavily weighted to processed dairy products, meat, fats and oils, and processed fruits and vegetables. Grain products were important items in earlier years of large surplus supply. More recently, less than 40 percent of the total dollars spent have gone for crop products and more than 60 percent for livestock products. The foods affected are known for these programs, but the impacts of these purchases on market prices are not and have been only crudely analyzed, using average rather than marginal relationships.

About 40 percent of the dollar value of all foods used in schools consists of dairy products, according to Van Dress and Putnam (1982). Meats are second, followed by fruits and vegetables. These findings, which represent average use by the nation's schools, are only crude approximations of the food impacts of the school food service programs. It is not possible except by analysis yet to be done, to determine the extent to which these programs actually increase consumption of specified foods and expand the markets for farm products. One such study of the Special Milk Program had only limited success (Robinson 1978).

Food stamp participants are free to buy any kind or type of food they desire, either foreign or domestic. As a result, the type of foods purchased must be obtained by survey or analysis. The 1973–74 Consumer Expenditure Survey of the Bureau of Labor Statistics (diary data) showed the composition of food expenditure for a nationwide sample of food stamp participant and non-participant households (Table 14.3). These aggregates cover up differences among individual foods between participant groups. For example, among the meats, participants purchase larger proportions of pork and poultry but rela-

TABLE 14.3.   Food expenditure by food stamp participants and nonparticipants

Commodity	Participants	Nonparticipants
	(%)	
Meats, poultry, and fish	36.5	36.3
Eggs	3.5	2.6
Dairy products	13.8	13.7
Fruits	5.5	6.5
Vegetables	8.2	7.7
Cereal and bakery products	12.9	11.9
Sugar and sweets	2.7	3.0
Fats and oils	3.3	3.0
Nonalcoholic beverages	7.1	7.2
Other foods	6.5	8.1
Total food at home	100.0	100.0

Source: West (1978).

tively less of beef; more fresh whole milk but less cheese among dairy products; and more cereals but less bakery products among the cereal and bakery products. Among the other foods, prepared foods were less and snacks much less important for participants than for nonparticipants. Again, these data have only recently become available, and they represent average rather than marginal program impacts. They need to be analyzed to assess meaningful impacts upon food markets and nutrition.

An analysis of food sales data in three selected areas of Puerto Rico before the Food Stamp Program started there and one year later showed a marked increase in total sales. Food sales were up by 43 percent in actual dollars and 31 percent after adjustment for retail price increases, despite a drop over the period of 1 percent in real disposable personal income (Choudhury 1977, 1978). Food sales showed a significant decline in quantities sold of starchy vegetables, rice, dried beans, poultry, pork, codfish, and sugar, all of which have been important foods in Puerto Rico. There were sizable quantity increases in sales for beef, fresh milk, dairy products, and other foods that are relatively demand elastic. Unfortunately, similar data are not available for the United States. The study in Puerto Rico was limited in size and included participant versus nonparticipant purchases in three pilot areas. Despite these and other problems, the study provides one of the few insights into the longitudinal impacts of the program on food use.

DIETARY AND NUTRITIONAL IMPACTS. Changes in food expenditures and nutritional intake or in biochemical or anthropometric measures of nutritional status apparently are loosely associated. But the nature of this relationship remains vague. Given an increase in food buying power, households tend to buy more of the highly desired foods, which may also be more nutritious. Research is badly needed to verify or refute these conclusions in light of the billions of dollars being spent on programs designed with the understanding that these relationships are known and positive.

Madden and Yoder (1972), in their assessment of the dietary impacts of the Food Distribution Program and the Food Stamp Program in two counties of Pennsylvania, found essentially no measurable impacts of the Food Distribution Program. They found some measurable impacts of the Food Stamp Program under certain conditions, although they were small. They also tested for dietary impacts associated with income but found no significant association. Lane (1978), in her study of dietary impacts of the Food Stamp Program and Food Distribution Program in Kern County, California, found quite significant dietary impacts of the Food Stamp Program but not much for the Food Distribution Program. Food stamp recipients had significantly higher levels of intake of protein, calcium, vitamin A, and riboflavin compared with nonparticipants. Both studies were useful but lacking in sample size and breadth. They cannot be replicated because they were based on longitudinal comparisons, before and after program implementation, but now the program is nationwide in coverage.

Some nutritional evaluations of the school food service programs have been conducted. Perhaps the most comprehensive was completed in 1976 for the state of Washington by Hard and Price (1975) at Washington State University. Their findings included measurement of the percentage of contributions made by food consumed at school relative to total daily energy and nutrient intakes of participating children. Iron was found to be the lowest, and calcium, vitamin A, riboflavin, protein, and phosphorus the highest. Selected physical characteristics related to nutritional status were examined along with dietary and biochemical measurements. Comparisons were made of children eligible for free lunches with those at higher income levels for three racial groups. Sample selection was a problem because of the difficulty in finding eligible nonparticipating students. The sample also related only to one time and to a single state. It suffered from the usual problems of cross-sectional analysis, but replication of the study would be useful in other areas and at a later time.

A useful study completed by Hoagland (1978) statistically analyzed some of the data collected earlier in the Health and Nutrition Examination Survey (HANES) conducted by the National Center for Health Statistics. Nutritional status data for students were compared for those participating and not participating in the school food service programs.

A major problem in assessing absolute levels of improvement in dietary and nutritional impacts of the various food programs is that the evaluation criteria are extremely crude. Recommended dietary allowances, often used to assess dietary status, are set at two standard deviations above the expected value of the mean of nutrient needs of the population. That means that dietary levels are expected to be too high for 97 percent of the population through the use of these criteria. For that reason, most studies analyze group means rather than assess intake against nutritional standards of need.

Biochemical and anthropometric measurements appear to be a more direct approach to nutritional assessment than dietary measurements, but they have their own problems. Evaluation criteria for these measurements also are ill-defined, partly because of the sensitivity of the tests to proper administration and analysis. Relatively few biochemical studies have been conducted in attempting to evaluate the food programs. The major one, the medical evaluation of the WIC Program conducted by Edozien (1976) that was mandated by the enabling legislation, has been a source of controversy. Nutrition researchers do not agree on the quality of the research and usefulness of the findings.

Everyone seems to be in favor of nutrition education, but little research has documented and analyzed nutritional benefits. Most evaluation studies of nutrition education measure only the learning of nutrition facts rather than testing for behavioral change. In 1976, the USDA funded a significant behavioral research effort ($300,000), but the results were disappointing (Applied Management Sciences 1977). The researchers indicated that more time was needed to obtain benefits that are measurable. In 1973, however, the Office of Equal Opportunity funded a small-scale behavioral study in North Carolina conducted by Head (1974). It showed small but positive measurable results.

FRONTIERS FOR FUTURE RESEARCH

Research and evaluation of programs has not kept pace with their growth in size and importance. This assessment applies with respect to impacts on participants as well as on agriculture and the food marketing industry. Needed research falls generally in the following areas.

Rates of Participation. Research is needed in developing short- and long-term statistical models to understand the factors associated with varying rates of participation and to use in developing program projections consistent with other macrovariables and in relating participation rates to the size of the eligible universe for the various food programs.

Data problems are severe in estimating the universe of some of the food programs; eligibility criteria do not usually coincide with available secondary data. Microsimulation models often have been used in deriving participation rates. These are particularly helpful in assessing participation and cost of alternative program options. Considerable improvement is needed in these models, the basic data used, and application to the various food programs. The current models are largely static but should be made dynamic in design. They should also be broadened to allow for variations in frequency and types of households participating.

Factors associated with participation have been studied superficially in many surveys that ask why respondents do or do not participate. But few surveys have been designed to obtain hard data from random samples for empirical analysis and building of improved microsimulation and macroprojection models. Elasticities are needed that relate degrees of participation to levels of program benefits. Few data have been available to allow such analysis. Generation of appropriate data bases is expensive and has not had support from policymakers.

Macroprojection models have been developed by the USDA for the Food Stamp Program. Such models need to be modified to incorporate elimination of the purchase requirement. Similar models have not been successful in projecting costs of and participation in the child nutrition programs, food distribution programs, or the WIC Program. Obviously, more research is needed on factors (principally noneconomic) that are associated with participation.

Dollar Value of Food Used. The size of the marginal propensity to consume or spend associated with various program benefits has been researched, but refinement and extension to other programs is needed. National panel data detailing household food expenditures and various sources of income would be useful as a basis for conducting such research related to all food programs directly serving households. These data would allow each household to be its own control in assessing impacts of the various programs in which it participated. Turnover of participants appears to be sufficiently large so that tracking all eligible households in a sample would allow before and after comparisons for most by obtaining quarterly data for one or two years. Such a study was designed, but funds were not made available for needed primary data generation. Instead, funds were used to support the Science and Education Administration's 1977–78 household food consumption survey, which provided related data, although cross-sectional rather than longitudinal.

In assessing school food service programs, household surveys are not sufficient. The techniques used in the Washington study of the child nutrition programs are more appropriate. In that study, a sample of students was drawn and surveyed with respect to food consumed at school. This survey was followed by visits to the homes of sample students to obtain household expenditure data. Statistical analysis was used in approximating the net additional value of food used attributable to the program. Replication of such a study in other parts of the country and extensions of this methodology to other nonhousehold programs is warranted.

Quantities of Food Used and Nutritional Evaluation. Food quantity data must be obtained for marginal analysis to assess food programs. Quantity data could be obtained along with food expenditure data from a panel survey of households described earlier. Similarly, quantity data could be obtained in school and related household samples of student food use. Quantity data are useful for two purposes: to add into totals and compare with marketing data in assessing various market impacts (discussed later) and to compute nutrients in the foods used for assessing nutritional benefits of programs.

Net additional quantities and nutrients attributable to the program need to be derived in the evaluation. Simply measuring nutrients delivered by a program without subtracting the amounts that would have been used in its absence grossly overstates any program's actual worth. Nutritional assessments of the food programs using quantities consumed begin with this data problem. As noted earlier, they are complicated further by lack of precise nutritional criteria by which to evaluate food programs. Considerable variability in needs among individuals complicate development of criteria. Nutritional evaluations using biochemical measurements face the same problem. Before-and-after measurements or cross-sectional comparisons are usually used in lieu of comparisons with actual needs.

Despite the problems and heavy costs involved, nutritional evaluation of the programs should be conducted because of the need to justify the high costs. Even so, priority probably should be given to assessment of nutrients in the quantities of foods used. Such studies are more cost effective than biochemical measurements and analysis. Further, some programs such as the

Food Stamp Program are explicitly required to operate through the commercial marketing system. They were designed on the premise that increases in quantities of food provided would lead to improved nutrition (as well as stimulate the demand for food).

Whenever possible within cost constraints, however, nutritional surveillance (biochemical) data should be collected and analyzed statistically to evaluate the food programs. Data on pregnant women and infants on the WIC Program, obtained from routine testing, already are sent to the Center for Disease Control, Health and Human Services, for evaluation. This program perhaps could be broadened to cover all household members receiving other food program benefits. The data could be supplemented by requesting medical examinations for a sample of new participants, followed by periodic examinations over a two- or three-year period.

Evaluation of Nutrition Education. The 1977 National School Lunch Act (P.L. 95-166) provided about $26 million annually to conduct nutrition education in the schools in fiscal years 1977 and 1978. Grants were made to the states for this purpose. The Child Nutrition Amendments of 1978 require states to use 16 percent of their administrative funds for nutrition education in the WIC Program. The Food Stamp Program has lagged in requiring nutrition education, but it has many proponents (Madden and Yoder 1972).

For these reasons, a sound behavioral evaluation study should be conducted to measure the usefulness of nutrition education activities of various types. According to the findings of Applied Management Sciences (1977) such a study should cover a long period. Perhaps a school food service's educational benefits could best be evaluated by cross-sectional comparisons of its graduates with those of schools that had no such programs. Similarly, family educational benefits should be measured in combination with and without food stamp benefits to test the usefulness and design that such activities could take.

Income Transfer. It is clear that all the food programs implicitly transfer sizable amounts of nonfood income. The amount varies inversely with the size of the MPC out of program benefits. Such data are available by subtraction from research analyzing the food benefits of the various programs. Analysis that places such implicit income transfers into better perspective would address several policy concerns. These include assessment of impacts of elimination of the purchase requirement (EPR) from the Food Stamp Program, the desirability of welfare reform in lieu of food stamps, and new measures of the degree of elimination of poverty.

EPR under the Food Stamp Program is a step in the direction of implicit nonfood income transfers as opposed to food support. Congress has required an annual assessment of EPR on program costs and participation, but study of the food versus nonfood impacts unfortunately is not required (P.L. 95-113). Research on such impacts may be useful in guiding related policy decisions. Such impacts should assess both food and nutrition effects on participants and secondary effects on the market for food.

Although not expressed as an objective of any of the food assistance programs, some observers prefer to assess the usefulness of the Food Stamp Program in terms of an implicit income transfer rather than as food assistance (Nathan 1978). Such thinking supports the policy position of eliminating this program in favor of comprehensive welfare reform. Support for using the Food Stamp Program as a vehicle for getting welfare reform is obtained by realization that this program has many of the technical merits that good income transfer programs are expected to have. These include relatively low "negative tax rate," no significant "notches" in benefit schedules, national availability, uniform benefit levels from state to state, and eligibility on the basis of income rather than being limited to specified categories of people. Research in this area will be spearheaded by the provision of cash in lieu of food stamps for the elderly in pilot areas, mandated by P.L. 95-113. Such research is likely to grow in importance and scope as welfare reform issues are debated into the 1980s.

The number of people under the poverty level as officially measured is not reduced at all by the food programs because food benefits provide only in-kind income. The Bureau of the Census defines poverty solely in terms of money income. Better data and more research on the economic value of in-kind benefits are needed to make a more complete appraisal of the magnitude of poverty that remains in this country.

Horizontal and Vertical Equity. Considerable program analysis has been directed at determining the kinds of people served by the Food Stamp Program, their incomes, geographic location, and other socioeconomc characteristics. Ways to administer the program more efficiently and equitably also have been given priority for analysis. Surveys of participants have been conducted annually to determine their characteristics, and microsimulation models have extrapolated the data to other times and program alternatives. Several legislative changes have been made based on such information, and continuing information is required. P.L. 95-113 requires an annual evaluation of the Food Stamp Program that must include such analysis.

P.L. 95-113 altered the Food Stamp Program to make ineligible the higher income of those previously eligible for the program and to increase participation of those at lower incomes. Such changes were made with inadequate measures of the economic factors relating benefits to participation, but the program changes themselves will provide the data for future analysis. Disincentives to work related to income transfers of various sizes are of considerable interest in designing benefit schedules and acceptable levels of trade-offs between benefits and income.

Some useful research has been done on these questions as they relate to the Food Stamp Program and general welfare programs, but more is needed. Much less is known of the economic impacts of program design for the child nutrition and other food programs.

Impacts on Food and Farm Businesses. Importance of the food quantity impacts of the programs, by commodity group, need to be related to the size of the various food industries. Analysis should relate these program impacts to total farm ouput, prices, and income. Impacts should also be compared with other farm policy programs to assess their relative efficiency in providing income and price support to times of weak demand.

Crude measurements of such impacts have been made, but refinements are needed (Nelson and Perrin 1976a,b; Boehm and Nelson 1978; Hiemstra 1978b). Input-output methods appear relevant as a research tool in such analysis. However, fundamental coefficients for such work are limited and usually available only with considerable time lag. State-by-state impacts would be useful, but Texas is the only state known to have the coefficients necessary for this analysis (Nelson and Perrin 1977). Nelson and Perrin's I-O studies could profitably be replicated for some of the other food programs and the results aggregated for all. They could be greatly improved by use of later I-O coefficients, more recent marginal propensities to spend for food, and the latest data on food used by commodity group from the Food Use Survey and the West (1978) study.

Secondary impacts of the programs on food prices and employment have been attempted by use of single-equation least-squares models and by comparative analysis, but no comprehensive analysis has been conducted (Choudhury 1977, 1978). Such impacts have been the subject of public debate when food price changes are large. Thus careful research on this subject would appear warranted. Aggregative statistical models may not be sufficient to assess the impacts.

Relating food program impacts to marketing bill data give a rough indication of average magnitudes. But marginal impacts may be quite different from averages because of inelasticities of supply and demand for farm and food products, particularly in the short run. Fundamental improvements in the marketing bill data and computation of short-run elasticities of supply and demand are needed for comparative purposes.

Assessment of the impacts of cash in lieu of commodities (CLC) support of school food service has been a major policy concern for several years. Food stamps in lieu of commodities for families was settled only as recently as the 1973 farm bill (P.L. 93-86). Research on CLC has been limited in part because of the lack of quality data and fundamental economic research (Hiemstra 1979).

The CLC issue has many ramifications, including (1) relative size of price elasticities of demand and supply in the short run for limited specifications of products and small geographical areas versus annual U.S. elasticities for broader product groups that would apply to local school purchases versus USDA purchases; (2) nature of the economies and diseconomies of scale that exist in local school procurement versus USDA national procurement; (3) economic costs and their allocation for foods acquired for surplus removal and price support purposes and used for school food service; (4) relative benefits to agriculture of the ability to administratively shift short-run purchases of food among selected commodity groups, measured against the possible dislocation and costs forced on schools and other food users by such decisions; (5) importance of local flexibility in the purchase and serving of food that meets local tastes and preferences; and (6) impacts on food use and nutrition of children that may be associated with any impacts that CLC may have on participation rates in school food programs and plate waste. (See Hiemstra 1979 for objectives and study procedures for the 1978–79 USDA study of CLC).

REFERENCES

Applied Management Sciences. 1977. Evaluation of a comprehensive nutrition education curriculum. In Nutrition Education and Training Status Reports, Completed Projects, pp. 55–56. USDA, FNS.

Benus, J., J. Kmenta, and H. Shapiro. 1976. The dynamics of household budget allocation to food expenditures. Rev. Econ. Stat. 58:129–38.

Boehm, William, and Paul E. Nelson, Jr. 1978. Food expenditure consequences of welfare reform. In Agricultural–Food Policy Review: Proceedings of Five Food Policy Seminars, pp. 45–50. USDA, ESCS, AFPR-2.

Choudhury, Parimal. 1977. The Impact of the Food Stamp Program in Puerto Rico. Department of Social Services, Commonwealth of Puerto Rico.

_____. 1978. The Food Stamp Program and Unemployment in Puerto Rico. Department of Social Services, Commonwealth of Puerto Rico.

Clarkson, Kenneth W. 1975. Food Stamps and Nutrition. American Enterprise Institute for Public Policy Research, Washington, D.C.

Edozien, Joseph C. 1976. Medical Evaluation of the Special Supplemental Food Program for Women, Infants, and Children. School of Public Health, University of North Carolina, Chapel Hill.

French, Ben C. 1977. The analysis of productive efficiency in agricultural marketing: Models, methods, and progress. In A Survey of Agricultural Economics Literature, vol. 1, p. 95. Minneapolis: University of Minnesota Press.

Hard, Margaret M., and David W. Price. 1975. Evaluation of School Lunch and School Breakfast Programs in the State of Washington, parts 1 and 2. (Final report September 1976. Available on loan from the Food and Nutrition Information Center, National Agricultural Library, Beltsville, Md.)

Head, Mary K. 1974. A nutrition education program at three grade levels. J. Nutr. Educ. 6:54–59.

Hiemstra, Stephen J. 1976. Program Evaluation Status Reports, Completed Studies. USDA, FNS.

_____. 1978a. Child Nutrition Review, part 1. USDA, FNS.

_____. 1978b. The Impact of USDA Food Programs on Agriculture. Paper presented at annual meeting of the Arkansas State Farm Bureau, Little Rock, June 27–28.

_____. 1978c. Research Needs in Evaluating USDA Domestic Food Assistance Programs. Paper presented at graduate seminar, Kansas State University, Manhattan, November 29.

_____. 1979. Cash in lieu of commodities in school food service programs. School Food Serv. Res. Rev. 3:29–32.

Hiemstra, Stephen J., and Johnny D. Braden. 1979. Program Evaluation Status Reports, Completed Studies. USDA, FNS.

Hoagland, G. William. 1978. The Impact of Federal Child Nutrition Programs on the Nutritional Status of Children. Paper presented at the Southern Economics Association meetings, Washington, D.C., November 8–10.

Lane, Sylvia. 1978. Food distribution and Food Stamp Program effects on food consumption and nutritional "achievement" of low income persons in Kern County, California. Am. J. Agric. Econ. 60:108–16.

Madden, J. P., and M. D. Yoder. 1972. Program Evaluation: Food Stamps and Commodity Distribution in Rural Areas of Central Pennsylvania. Pennsylvania State University Bull. 780, University Park.

Nathan, Richard P. 1978. Public assistance programs and food purchasing. In Agricultural–Food Policy Review: Proceedings of Five Food Policy Seminars, pp. 25–29. USDA, ESCS, AFPR-2.

Neenan, P. H., and C. G. David. 1977. Impact of the Food Stamp Program on low income house-hold food consumption in rural Florida. *South. J. Agric. Econ.*9:89–97.

Nelson, Paul E., Jr., and John Perrin. 1976a. Economic Effects of the U.S. Food Stamp Program, Calendar Year 1972 and Fiscal Year 1974. USDA, AER-331.

_____. 1976b. Economic Effects of Federal Contributions to the U.S. School Lunch Program, Calendar Year 1972 and Fiscal Year 1974. USDA, AER-350.

_____. 1977. Food Stamp Program impact on resource use: Texas compared with the U.S., fiscal year 1974. *South. J. Agric. Econ.* 9:81–87.

Nixon, Richard M. 1969. Message of the President, May 6. Reprinted in White House Conference on Food, Nutrition, and Health, pp. 1–4, Final Report.

Peskin, Janice. 1976. In-kind income and measurement of poverty. In The Measure of Poverty, U.S. Department of Health, Education, and Welfare Tech. Pap. 7.

Reese, Robert B., J. G. Feaster, and G. B. Perkins. 1974. Bonus Food Stamps and Cash Income Supplements. USDA, ERS, MRR-1034.

Robinson, John S. 1978. Special Milk Program Evaluation and National School Lunch Program Survey. USDA, FNS-167.

Rutman, Leonard, ed. 1977. *Evaluation Research Methods: A Basic Guide,* Chap. 5. Beverly Hills, Calif: Sage Publications.

U.S. Congress. 1975. Food Stamp Program—A Report in Accordance with Senate Resolution 58. USDA, FNS. 94th Congress, 1st session.

U.S. Department of Agriculture. 1962. The Food Stamp Program: An Initial Evaluation of the Pilot Program. AMS-472.

_____. 1977. Characteristics of Food Stamp Households, September 1976. FNS-168.

Van Dress, Michael G., and Judith L. Putnam. 1982. Changing food mix in the nation's schools. *Nat. Food Rev.* NFR-18, Spring: 16–20.

West, Donald A. 1978. Food expenditures by food stamp participants and nonparticipants. *Nat. Food Rev.* NFR-3, June: 26–27, 46–48, and unpublished manuscript.

West, Donald A., and David W. Price. 1976. The effects of income, assets, food programs, and household size on food consumption. *Am. J. Agric. Econ.* 58:725–30.

West, Donald A., David W. Price, and Dorothy Z. Price. 1978. Impacts of the Food Stamp Program on value of food consumed and nutrient intake. *West. J. Agric. Econ.* 3:131–44.

VERNON L. SORENSON
E. WESLEY F. PETERSON

15

INTERNATIONAL MARKETING OF AGRICULTURAL COMMODITIES AND INTERRELATIONS WITH THE DOMESTIC SYSTEM

AGRICULTURAL ECONOMISTS have increasingly recognized the need for a worldwide approach to research on the economic and policy problems of food and agriculture. Interdependence and uncertainty have become dominant themes that lead to concern with a wide range of international issues. These issues must be dealt with in any attempt to understand the problems that the United States and other nations face in developing effective policies for food, agriculture, and agricultural trade. Further, these areas need to be of increasing concern to agricultural economists in their role as educators, both in the classroom and in extension education. To do this effectively, research is needed.

Viewed from the U.S. perspective, a number of reasons exist for engaging in this kind of research. American agriculture is highly dependent on world markets. With 20 percent of the value of domestic agricultural production going overseas, a discussion of the U.S. food and agricultural marketing system is incomplete without reference to the international components. Like the internal marketing situation, the global context is dynamic. Changes in foreign demand, production, government policies, power relationships, and such stochastic factors as weather continually work to alter the set of constraints and opportunities facing U.S. agriculture. Since many of these factors are outside the control of U.S. decision makers, it is important to understand their impact on international markets and their consequences for U.S. agriculture.

VERNON L. SORENSON is Professor of Agricultural Economics, Michigan State University, East Lansing; **E. WESLEY F. PETERSON** is Assistant Professor, Institut de Gestion Internationale Agro-Alimentaire (IGIA), France.

Additionally, the United States is committed to a number of international policy objectives related to trade, development, and improvement of the world economic order. This involves problems of international commercial policy, international monetary policy, and the special trade problems facing less developed countries.

In the 1970s the postwar changes in international trading relations have been brought more clearly into focus. The position of the United States as leader in world economic affairs, challenged in the 1960s with the formation of the European Communities (EC) and the emergence of the United Nations Committee on Trade and Development (UNCTAD), was further eroded as a result of the actions of the Organization of Petroleum Exporting Countries (OPEC) and deterioration in the value of the dollar. Problems of inflation and unemployment increased.

In this context, the quantity and value of U.S. agricultural exports have risen sharply. International trade is no longer a marginal activity for U.S. agriculture. The linkages between domestic and international markets are reflected in such areas as price formation, achieving effective market coordination, achieving adequate market stability, defining sources of market growth, and a number of other areas. International market developments influence U.S. agriculture, but U.S. economic conditions and policy also have significant impacts on international markets. This linkage and the dynamism involved present an important challenge to those involved in operation, regulation, and research in agricultural marketing.

Although U.S. political and economic dominance has diminished since the immediate postwar period, American agriculture retains an inherent strength that can be exploited on international markets. Future research will need to provide a basis for U.S. agricultural policy, recognizing that in the context of international interdependence and government intervention, effective performance of the domestic marketing system requires a carefully conceived international marketing strategy to complement internal programs. Thus the challenge to agricultural marketing research is broad. The traditional criteria of market performance, including efficiency and allocative and distributional questions, are all of significance; but in addition, international markets are subject to conflict in goals among nations and to problems associated with gains and losses imposed on individual groups within nations as well as such national interest questions as economic stability and balance of payments.

Before turning to a discussion of the conceptual framework for marketing research and some specific themes around which it might be organized, we will examine some of the relevant dimensions of international markets and the accomplishments of past research.

DIMENSIONS OF INTERNATIONAL MARKETS

The value of international agricultural trade has increased rapidly during the post-World War II period from a total of approximately $27 billion in 1951-55 to over $150 billion in 1977. The greatest share of this growth has been among industrial countries as a result of economic growth and increasing consumption of livestock products, particularly in land-shy importing regions such as Western Europe and Japan. In addition, some communist countries and individual less developed countries (LDCs) that have achieved major steps toward industrialization have become major importers of livestock and, more important, feed material to support livestock production. Along with a steady increase in the import needs of a number of LDCs based on population pressures, these developments have resulted in a continuous expansion of agricultural trade.

The United States has participated fully in this expansion with exports of about 16 percent of the value of all agricultural commodities moving in world trade. U.S. exports grew to include (1) more than 50 percent of U.S. production of wheat, soybeans and products, rice, dry edible peas, almonds, cattle hides, mink pelts, and hops; (2) 33–50 percent of U.S. production of cotton, tobacco, prunes, and tallow; (3) about 25 percent of the grain sorghums, corn, lemons and limes, and dry edible beans; and (4) from 10 to 20 percent of a wide variety of meats, some grains, dry milk, and a number of fruits and vegetables plus smaller amounts of dry whole milk, flaxseed, chickens, and turkeys. These products are distributed on a worldwide basis, although Europe and Japan represent the major market areas.

The United States also imported approximately $14 billion of food and agricultural products in 1977. We are fully dependent on a variety of tropical products, and supplement U.S. production on such important domestic items as sugar, tobacco, many fruits and vegetables, low-grade beef, high-quality cheeses, and a limited number of grain products. U.S. interest in effectively operating international market systems thus stems from the importance of both imports and exports, which cover many products and represent dealings with many countries.

Much of the process of international marketing is similar to that occurring at the domestic level. There is a need to develop buying and selling arrangements, financial mechanisms, product specifications, and delivery times, much as for domestic transactions. Differences do exist however. Some of these are technical in that new product specifications must be developed to fit the desires of individual countries with differing tastes, habits, and merchandising practices. Product labeling, specification, and packaging all can become important in this respect. Another important difference is the need to operate through foreign exchange markets. Given the situation of fluctuating foreign exchange rates and the relatively longer delivery time involved in most foreign shipments, an additional element of risk and uncertainty is injected into the process. In general, however, these kinds of functional differences do not present insurmountable obstacles. The most important differences between the functioning of domestic and international markets are associated with institutional arrangements and the constraints and guidelines established by individual governments or by governments collectively.

While the private sector, in the form of multinational corporations, dominates trading relationships in some commodities such as grains, pineapples, and bananas, a major characteristic of the international trading system is the direct intervention of governments. These interventions involve price shifting practices through tariffs, variable levies, fees, or export subsidies as well as a large number of nonprice-related barriers and constraints such as quotas, sanitary regulations, packaging requirements, and product specification regulations. These constraints can be implemented directly through rules established by governments or through international agreements and practices developed to handle international commercial relationships. However they are instituted, these practices have the effect of changing price relationships and creating trade distributions that do not necessarily reflect underlying economic forces.

A surprisingly large proportion of international transactions in agricultural commodities is handled through state trading. This, of course, includes all transactions with communist countries, but in addition, many other countries handle part or all their international transactions through direct state-controlled mechanisms. Most major grain exporters other than the United States maintain monopoly control through marketing boards. Japan provides an extreme example of state intervention, where strong administrative discipline rules lead to joint government-industry determination of trade patterns. The quantities imported and the differential between cost, insurance, and freight (CIF) import price and the price at which commodities are made available to users are administratively determined.

In addition to the wide range of unilateral action by individual governments to manipulate and control international flow of commodities, increasing emphasis is being placed on international market agreements as mechanisms for handling trade. In the multilateral trade negotiations, commodity agreements are being discussed for wheat, dairy products, livestock and meat products, and coarse grains. An international sugar agreement has been promulgated. The LDCs are suggesting a number of agreements associated with a common fund that would be developed to handle financial requirements. If all the agreements being proposed or under discussion are culminated, the bulk of all international trade in agricultural commodities will be subject to some form of international commodity agreement. In addition, a number of countries are interested in establishing bilateral agreements, wherein governments would guarantee purchases and supplies and hence create more regular international trading patterns. This would represent an addition to existing intervention for those major exporting countries that handle exports through a single marketing board that has established traditional customers. Government intervention through rules and regulations, actual government trading mechanisms, and parastatal organizations closely controlled by govenments thus create a major distinction in international market mechanisms from that existing internally within the United States. This has significant implications for the kinds of marketing problems that arise and the research needed.

ASSESSMENT OF PAST RESEARCH

An impressive amount of research related to international trade in agricultural products has been carried out, and the literature on this subject is extensive; only a sample has been selected for discussion. These examples of international agricultural trade research have been grouped into six categories, which overlap considerably but should provide a framework for assessing achievements of past research as well as the difficulties encountered.

Tariffs and Protection. Agricultural economists tend to view the problem of trade policy in classical economic terms and thus support the search for free international markets. Since all nations employ some form of protection, it has been natural to focus much empirical research on the effects of tariffs, nontariff barriers, and other forms of protection. The orientation of much of this work has been toward measuring welfare losses caused by protection or welfare gains that could be achieved under a regime of completely free trade. For example, a study done at the USDA found that the effects of trade liberalization would be generally beneficial to U.S. agriculture (Vermeer et al. 1975). The approaches used in these studies range from sophisticated models to what are in effect "best guesses." As with any attempt to estimate welfare gains and losses, the static, neoclassical economic base generally obscures the dynamic adjustment costs. The result is that the models will almost always lead to the conclusion that the world would be better off with free trade. An exception to this generalization is a study by Dean and Collins (1966), who found that the EC would obtain a net welfare gain from higher common tariffs on oranges.

A major difficulty in the study of protection is the problem of measurement. The concept of effective tariffs has been used to assess the impact of overall tariff policies on the level of protection. Efforts have been made to include nontariff barriers and domestic farm policies in the effective protection rate. Wipf (1971), for example, has developed an effective protection measure of this nature, which indicates that the effective protection in many agricultural sectors is very high even when nominal tariffs are relatively low. As Wipf notes, measurement of effective protection often relies on the use of input-output data, which imposes restrictive assumptions concerning input substitutability, that affect the results. Although there are many studies on the effects of protection and the best ways to measure protection levels, much conceptual and empirical research remains to be done in this area.

Industrialized Countries. A second research category can be defined to include the work done on agricultural trade among industrialized countries. A substantial amount of descriptive research has been carried out on individual countries and regional groupings. Descriptive studies (including information on volume and direction of trade flows, commodity composition of trade, and the policy and ideological orientation of governments) are crucial in developing analyses of international markets. Since these studies are relatively straightforward and the data are generally reliable, this type of research has generally been useful. In addition to descriptive research there have been many attempts to delineate the forces that determine trade patterns. Zarembka (1966) estimated production functions and derived production possibility curves for the United States and northern Europe. He found that the United States has a comparative advantage in agriculture because of its relative capital intensity.

One of the most important changes of the past 25 years for international trade scholars was the formation of the European common market. This event provided researchers with the raw material for a great many studies related to trade among industrialized countries. The opportunity for research is expanded each time new countries join the EC. Many of these studies deal with the impact on world trade of the EC common agricultural policy and its adoption by new members. In particular, several excellent studies were produced when Britain joined the EC (Ferris et al. 1971; Warley 1967). As with the tariff and effective protection research mentioned above, these studies employ a variety of models and projection techniques. In addition, however, a body of theory exists related to the effects of economic integration on trade flows between members and nonmembers. Young (1975) used this customs union theory to estimate the extent of trade creation and trade diversion in the EC dairy industry.

In general, studies related to the EC and those based on the application of theory have provided insights into the nature of trade among industrialized countries and contributed to the formulation of policy. As time passes, it will be possible to better assess the accuracy of the projections, particularly those relating to the formation of the EC.

Another important area dealt with in studies of industrialized countries concerns the potential for food marketing in other countries. The USDA has done extensive research on supply and demand in foreign markets. One study (USDA 1972) found that there was great potential for U.S. food exporters to take advantage of the growing West German market for more highly processed food. Many of these studies are very similar to the marketing research traditionally done by U.S. agricultural economists.

Less Developed Countries. A third area of research on international agricultural marketing relates to LDCs, or more precisely, the impact of trade on development. Much interesting work has been done on the question of whether trade benefits or retards development. Broadly speaking, there are two schools of thought. The more traditional researchers find that the international trade theory description of gains from trade is accurate. Alternatively, many scholars, particularly those from the Third World, argue that there are few benefits to poor countries from trade. Stripped of its rhetoric, this position is based on the assertion that international institutions, largely determined by the wealthy industrialized nations, in combination with the facts of colonial history result in a dependence syndrome. Dependence, it is argued, negates the gains from trade as portrayed in neoclassical theory, since it limits the power and scope of action of the dependent nations. As a result, rich nations are able to establish trading rules in their favor so that trade becomes another form of exploitation. These views have been extensively developed by Raul Prebisch, Samir Amin, and others and have been the philosophical base of much of the work done at UNCTAD.

While the "trade-as-exploitation" school has been prominent at the UN and in many LDCs, it has remained somewhat abstract; the bulk of empirical research has been based on more traditional economic theory. For example, Valdés (1973) uses neoclassical methods to analyze Chile's trade in agricultural products, concluding that the commercial policies of Chile led to negative effective protection and trade deficits. From the U.S. point of view, the USDA has studied the potential for market development in LDCs as well as the demand prospects for agricultural exports by them (USDA 1970). Other writers have assessed the impact on development of food aid programs such as the Agricultural Trade Development and Assistance Act (P.L.-480), as well as the contribution to economic growth and development of such policy tools as preferential tariffs. Clearly, the range of issues related to agricultural trade and development is quite broad, reflecting the central position of agriculture and trade in most LDC economies. The major impediments to research in this area seem to be the lack of reliable data and the complexity of development problems. This complexity is reflected in the very different philosophical approaches between the two schools of thought on the role that trade plays in development.

International Models. A fourth important category of research is the construction of international trade models. This is not distinct from the others but merits separate treatment because the techniques used are of interest. Bawden (1966) developed a spatial price equilibrium model including policy variables that alter the traditional free trade assumptions. Abel and Waugh (1966) employed transition matrices to measure changes in the level and composition of trade over time. Many of the international trade models are designed to generate projections of world production, consumption, prices, and trade of key commodities. The USDA, for example, has used a mathematical model focusing on grain, oilseeds, and livestock to project changes in important world export markets (Rojko and Schwartz 1976; USDA 1978). At Michigan State University a model of the U.S. agricultural sector has been built, which includes an international demand component. These modeling efforts have been developed in two perspectives. The more general approach seeks to deal with trade in a general equilibrium framework by measuring the worldwide trade flows that will occur under specified supply, demand, and policy conditions. These models are useful in defining worldwide equilibrium adjustments at some future time. The second approach is partial equilibrium and deals with patterns of changing demand and supply, but without seeking to define a complete pattern of trade flows or general equilibrium adjustments. Partial equilibrium analysis is an effective way of dealing with the interaction of international variables and U.S. supply and demand conditions in assessing future prospects for U.S. agriculture.

Commodity Studies. Another research category that cuts across the others includes the many studies of particular commodities. These often serve as vehicles for examining a specific issue in international agricultural marketing. A study by Epp (1968) focuses on the grain-livestock economy of Europe to evaluate the impact of integration on EC grain and livestock markets. Another theme in the research on particular commodities is development and testing of industrial organization models of international markets. Sirhan and Johnson (1971) use a market-share model to estimate export and import demand elasticities for cotton. Freebairn (1968) characterizes the hard wheat market as a cooperative duopoly with Canada acting as the price leader. He sees the soft and durum wheat markets, on the other hand, as oligopolistic with imperfect collusion. Use of industrial organization models and game theory, replacing firms with nations, appears to be a promising approach to research on international markets.

In addition to the research mentioned above, numerous supply and demand projections have been done for particular commodities. Johnson (1977) lists more than 60 studies of supply and demand for specific commodities or groups of commodities in various nations. Research of this nature is important in evaluating the potential for market growth and development in foreign countries.

Interaction of Domestic and International Policy. The final category includes research on national or regional policies and interaction of domestic agricultural and trade policies. Josling (1976) notes that domestic farm policies are often incompatible with a properly functioning world food system. Using empirical evidence from the world wheat market, he shows that pursuit of domestic price stability has often led to destabilized world markets. Many of the studies already mentioned deal with the impact of national policies on agricultural trade. One further policy area where a limited amount of research has been done concerns the effects of monetary policies on agricultural trade. Studies by Kost (1976) and Vellianitis-Fidas (1976) have found that agricultural trade is not very sensitive to changes in the exchange rate. Conversely, Dobbins and Smeal (1977) note that other researchers discovered a high degree of sensitivity of agricultural trade to exchange rate variation. While much has been written on the effects of tariff policies and domestic farm policies on trade, empirical research on the implications of changes in the monetary variables has been limited.

In summary much excellent research on international agricultural trade has been done. Research accomplishments, however, have been somewhat circumscribed by the complexity of the subject. Some of the most effective research has been quite limited in scope, focusing on the supply and demand conditions in specific international markets. Development of more advanced analytical techniques and an expanding data base and a gradual expansion of conceptual insights should provide the basis for extending our analysis of the international trading network. Although the accomplishments of past research have been substantial, much work remains to be done.

THE CONCEPTUAL FOUNDATIONS OF INTERNATIONAL AGRICULTURAL MARKETING RESEARCH

Evaluation of the domestic marketing system has traditionally been done by measuring the extent to which the market functions or fails to function within the framework of certain performance criteria. Such criteria as efficiency, growth, full employment, and equity have been given high priority. Much research effort has been devoted to identifying the sets of attributes or variables that influence economic performance, specifying how these factors are linked to designated market outcomes and determining how they can be manipulated to induce the desired results. It has been recognized that performance goals are not always fully consistent with each other, and compromise has been necessary. The process of ascertaining appropriate means of compromise and weights to be given to each goal has been the focus of further research.

A performance orientation is also appropriate for evaluation of the international marketing system and its links to domestic agriculture. The same broad criteria can be used in work in an international context, although the political and economic diversity brought into the analysis necessitates, in some cases, modification of the way the criteria are specified.

Efficiency. The efficiency goal clearly retains importance when applied internationally, especially as gaps develop in worldwide resource availabilities. In a general sense, the meaning of marketing efficiency is the same whether applied to a domestic economy or to the international economy. Markets should be organized to operate on a least-cost basis in performing functions such as storage, processing, distribution, and financing. Also, the information and price system should result in allocation of resources to production such that the greatest output is achieved at least cost with available technology. As with the domestic marketing system, improvements in international market efficiency can conflict with other performance goals such as equity or economic security. In addition, differences among countries in customs and practices, available technologies, language, political systems, and levels of education and economic development lead to rigidities and barriers that further impede movement toward greater productive and allocative efficiency. The international differences lead to different sets of performance goals and different weightings of these goals with the result that improvements in international efficiency inevitably conflict with the desires and aspirations of some country or group of countries.

Economic Growth. The related objectives of growth, expansion, and progressiveness can be applied to international and domestic markets. However, where growth should take place and how its benefits are distributed become questions with a strong component of national interest and over which conflict may arise. Variations in resource endowments and the differences among nations already noted determine the extent to which growth can occur. This is not to argue that these kinds of differences do not exist regionally within individual countries. What is different, however, is the difficulty of transferring technology, knowledge, and other factors required for growth within countries as compared with international transfers where different economic, political, and social systems come into play.

Full employment and maintenance of low inflation rates are two objectives tightly linked to growth. The issue here centers on the discovery of means to foster these two goals jointly in the international arena to complement domestic efforts. As stated by Bergsten et al. (1975):

> When the postwar economic system was constructed, the attention of virtually all countries was riveted on the fear of unemployment. . . .
> In the 1970s, however, inflation has emerged in virtually all countries as an

economic, and hence, political problem at least as severe as unemployment, if not more so. Indeed, in recent years numerous countries have been seeking to insulate their economies against imported inflation and even export their inflation to others by upvaluing their exchange rates, unilaterally liberalizing their import controls, and instituting export controls.

The goals of growth, full employment, and low inflation as well as the trade-offs among them have thus taken on great importance in most countries. They influence the approach that nations take to both domestic and international policy issues and hence have a major impact on the functioning of markets.

Equity. The equity issue is an emotional one domestically and is no less difficult on the world level. The clamor of the LDCs for a new international order illustrates its importance in questions of international market organization. A basic argument is that for a number of reasons related to differences in the structure of economies and markets, LDCs are disadvantaged and this is reflected in deteriorating terms of trade and in inequitable sharing of the gains from technological progress. The existence of monopolistic power in industry and labor unions in industrial societies results in distribution of gains to these units through higher profits and wages. Numerous specific actions such as the call for special and differential treatment in all phases of international market organization plus various forms of direct income and technological transfer from industrial to developing countries have been proposed to correct the situation. Some of the proposals for correction are similar in concept to those used to generate income transfers to farmers in industrial countries and would have similar kinds of effects on the operation of markets.

Economic Security. The extent of dependency by the United States and most other industrial countries on foreign sources of raw materials, including food, has increased sharply and creates a vulnerability not heretofore experienced. For some LDCs, food security and the ability to pay for needed energy imports have become crucial. The creation of reserve stocks and expansion of production in food-deficit countries involve a very complex set of policies related to trade and foreign aid. Another significant issue concerns the extent to which privately directed marketing systems can be relied on to deal with food security problems as opposed to the need for direct government intervention. A third dimension of the security goal is the nature of changes in international power relationships and whether political objectives will complicate security issues. The search for food security often causes governments to institute excessive protection programs for agriculture and leads to attempts to develop assured sources of supply through programs to stimulate domestic production, bilateral supply, and purchase guarantee agreements or through investment in foreign production capacity. All these factors have a significant impact on international markets and the market position of individual trading nations.

Market Stability. Conditions of interdependence and uncertainty in world markets have led to a greatly expanded concern with market stability. Issues related to domestic food industries center around the form of price supports, establishing domestic reserves, providing outlook information, and in general creating a flexible and informed market system. Internationally, these same issues exist but are exacerbated by variations in exchange rates, vagaries of weather around the world, and in some cases political considerations. Establishing market mechanisms to deal with stabilization of prices and market flows is a complex process. In the case of international food reserves, conflict arises as to how large reserves should be, where they should be maintained, how wide price bands should be, who should cover the costs involved, and numerous other issues. International marketing agreements to deal with instability also have been very difficult to arrange for a number of technical, economic, and political reasons. These agreements, often seeking to deal with both stabilization of price and commodity flows, must include mechanisms to remove and add supplies to the market in the short run and must adjust to longer run market equilibrium conditions to avoid creation of production shortages or surpluses.

Despite these and other difficulties, much effort is being expended to devise methods of reducing instability in international markets. This is being sought through establishment of international institutions and through efforts to modify domestic policies followed by individual countries. The linkage between domestic markets and international markets is very direct and requires simultaneous action.

Besides modifying some of the traditional performance goals, it is also necessary to introduce some additional criteria when dealing with the international marketing system. Although it may be argued that these objectives are subsumed under other classifications, separate treatment seems appropriate in light of the attention given them by many governments. These goals include eradicating hunger and famine, promoting economic development, improving the quality of life, and seeking consistency in overall foreign economic and political objectives. In addition, adaptability and flexibility in the agricultural sector have become important performance objectives as economic interdependence has increased throughout the world. Interdependence subjects agricultural industries to adjustment pressures since the effects of inflation or changes in relative prices are not confined to individual nations. Dependence on energy-intensive technologies and the increased size and specialization of farms retard adaptation of U.S. agriculture to such international developments as the rise in petroleum prices. Since adjustment is difficult under these conditions, increased pressure is brought to bear on governments to raise protection levels. These plus other performance criteria merit explicit consideration so that efforts can be directed to their precise specification.

DIRECTIONS FOR FUTURE RESEARCH

In turning to the research needs in international agricultural marketing, eight broad areas are identified; they are interrelated and reflect the linkages between domestic and international markets. The perspective used in presenting them is that of the U.S. agricultural sector, and no attempt is made to define the issues and problems facing other nations such as the LDCs. It should be emphasized that much descriptive research is needed in each of these areas to lay the foundation for more in-depth analysis and policy recommendations.

Interdependence and Instability. Two primary concerns are related to the links between international and domestic markets and price stability. The first involves our understanding of the mechanisms through which price impacts operate, and the second is related to the distributional and adjustment implications of price patterns.

Further understanding of the way in which international prices are determined is needed to differentiate between the impacts from stochastic supply and demand changes and those from policy decisions. Johnson (1977) has noted that price increases in the early 1970s may provide the basic data needed to resolve differing interpretations of the importance of domestic policy as it relates to international prices. While some writers believe that the rapid price increases were due to increased demand and production shortfalls, Johnson feels that they were caused by the inability of producers and consumers to adjust as a result of government policies to stabilize domestic prices. Greater understanding of the mechanisms through which international prices are established is essential if we are to develop alternative policies and institutions for international marketing.

We will have more to say about institutional research in a later section. However, it should be noted here that insight is needed on the potential effects of such institutional arrangements as international commodity agreements on the stability of both international and domestic prices. In the study of domestic markets, agricultural economists are frequently called upon to analyze the nature and implications of contractual arrangements for price discovery and market adjustment. The methods employed in these analyses possibly could be extended to the study of the impact of bilateral and multilateral commodity agreements in international markets.

There is also a need to know more about the rationale for considering price stability to be a desirable performance objective. This requires work on the distributional and adjustment effects of price instability. In particular, we need to know what the trade-offs are between domestic and international stability as they relate to different groups of producers and commodities. There are both short-term welfare considerations and long-term efficiency and resource allocation aspects to these questions. For example, the sacrifice of domestic stability in favor of international stability could impose fairly immediate welfare losses on particular groups of producers or marketing firms, while leading to more efficient long-run resource allocation. A fuller understanding of the distributional and adjustment implications of alternative policies relating to price stability would enhance our ability to develop recommendations for improved international market performance.

International Market Coordination and Efficiency. As noted earlier, efficient operation of international markets implies that functions such as processing, distribution, and financing should be performed on a least-cost basis and that the price mechanism should lead to optimum resource allocation. Using the free-trade framework, a common focus of past research has been on the efficiency and welfare losses caused by trade barriers. This emphasis has clouded the trade-offs between efficiency and other performance goals. High priority should be given to description of the structure of international markets and the transfer mechanisms through which they operate. In developing greater understanding of the behavioral and performance implications of existing and proposed international market structures, the industrial organization models of Sirhan and Johnson (1971) and Freebairn (1968) suggest promising approaches.

With well-defined structural parameters, a variety of issues could be explored. The interdependence of international and domestic markets raises issues in internal market coordination; e.g., movement of massive grain shipments to the Soviet Union caused serious disturbances in the domestic marketing and transportation systems. Avoiding disruptions of this nature provides incentives to organize international markets through such contractual arrangements as the U.S./U.S.S.R. grain agreement.

Another area of research concerns the potential for further adjustment according to international comparative advantage and the barriers to this adjustment. These questions are of practical as well as theoretical interest; e.g., cheaper air transportation has opened the possibility for some LDCs to export labor-intensive, perishable products such as strawberries to areas where labor costs are high. The potential for U.S. exporters to develop new export arrangements to further exploit their comparative advantage is also an important question. Determining whether gains could be achieved through multicommodity cooperative export efforts is an example of this type of research.

Additional issues in this area include the problems of vertical coordination in international markets, the potential for vertical integration across national boundaries by multinational firms, the implications for market coordination and efficiency stemming from the fact that large amounts of agricultural trade are handled through monolithic state trading organizations, and many others.

Implications of Present and Alternative Trade Policies. Agricultural trade policies are directly linked to domestic farm policy. Traditionally, economists have focused on tariffs as the central element of trade policy. While research on the degree of protection occasioned by tariffs is still necessary, the interrelation of farm and trade policies has raised the need for expanded efforts to measure levels of effective protection. Since nontariff barriers have become increasingly prominent for agricultural commodities, there is a need to determine the level and degree of protection provided by the combination of domestic farm policies and trade policies. This basic empirical research is a prerequisite to any assessment of the impact of present policies on production, consumption, trade, and resource allocation.

In addition to improved measures of protection levels, there is increased need for estimates of supply and demand functions for the principal temperate zone commodities in the major trading areas of the world. With more accurate assessment of the supply and demand conditions in other countries, it would be possible to begin an evaluation of the trade restricting effects of alternative policy mixes. If trade barriers are successfully reduced, it will have important consequences for the U.S. farm sector. Research related to the adjustment that would result from freer trade is important as a means to reduce resistance to trade liberalization from the domestic agricultural sector and as a foundation for intelligent trade negotiating positions.

Several issues need to be examined in studying the adjustments that would result from alternative policies. Policy changes will not affect all producers and commodities in the same way. Some industries such as tomatoes may be highly vulnerable to foreign competition, while others such as the feed grain sector would benefit from a general lowering of protection levels. The distribution of adjustment burdens among various U.S. producers is likely to be very uneven. A prerequisite for determining the distributional effects of various policies is a clear understanding of the impact of these policies on resource use, particularly employment, in agriculture and food industries. In addition, since capital assets are often industry-specific and relatively immobile, research on the time path of adjustment and the hardships that could be imposed on particular groups of farmers with high levels of investment in fixed assets is essential. Research on the distributional effects of policy among U.S. farmers should be complemented with research on alternative policies for adjustment assistance. It is important to be able to identify the groups that will be adversely affected by policy changes as well as to contribute to the design of programs that will meet their needs.

Another set of issues is related to the distributional effects between consumers and producers of alternative policies. Here it is necessary to understand the macrolevel impact of lowered protection on prices. If more liberal trade is achieved, to what extent will consumers benefit from lowered prices? Would increased foreign demand raise the prices of commodities in which the United States has a competitive advantage? Answers to these and other distributional questions can only be obtained with better estimates of international supply and demand conditions.

Institutional Change and International Trade. Many of the institutions governing international commercial relations have changed or have been challenged by various groups of nations. This is particularly true of the international monetary system, which is evolving rapidly. In addition, LDCs have called for a "new international economic order," new institutional arrangements that would facilitate an international redistribution of income and wealth. Generally, international institutions are created as a response to a particular need or opportunity with little thought given to their overall impact on international commercial relations. Clearly, there is room for creative thinking about international institutions, with the goal of understanding present arrangements as well as contributing to the design of new systems.

The multinational corporation (MNC) is an important international institution that presents unique problems in understanding its role and methods of operation. These firms frequently are able to circumvent the policies of countries in which they operate. Faced with the economic power of these firms, many countries (particularly the LDCs) have called for regulation through an international code of conduct. In designing such a code, much information is needed on the expected behavior of MNCs and the impact of this behavior on performance of international markets.

Along with the growth of large private firms in international trade, there has been a proliferation of centralized state trading organizations. An interesting question is the extent to which MNCs provide a form of countervailing power to these state agencies. Conversely, small exporting firms may be at a disadvantage in dealing with the centralized decision making of large state trading organizations. Similar problems may exist for small importing firms in dealing with centralized exporters such as the marketing boards found in many countries. In the United States, agricultural cooperatives are becoming increasingly interested in expanding their activities in international trade. The mix of organizations operating in particular international markets (including MNCs, small firms, cooperatives, state trading organizations and marketing boards) constitutes the primary component of that market's structure. The changing nature of international market structure has implications for the behavior of participants and the resulting market performance. To what extent, for example, would U.S. producers benefit from the creation of marketing boards to deal with state trading companies on their behalf? Could a cooperative become the sole international bargaining and selling agent for producers of particular commodities?

Institutional arrangements are closely related to policy. It would appear that the ideal of free trade is being progressively eroded in favor of policies of organized trade. Noting the special nature of agriculture, the EC have called for organized markets. Developing countries do not feel that international institutions allow them to capture the gains from trade. One result of these attitudes is a renewed interest in international commodity agreements, both multilateral and bilateral. Given the nature of trade with centrally planned economies, the United States has little choice but to enter into bilateral arrangements with such countries as the Soviet Union. Multilateral agreements have been established for such commodities as wheat and sugar and may be implemented for other commodities as well.

A number of issues surround the proliferation of interest in international commodity agreements. First, there are differences in expectations concerning the performance objectives of these arrangements. Developing countries see them as a means to stabilize prices and assure markets. Price discovery in thin international markets may become difficult if all existing proposals are implemented. Finally, adoption of international commodity agreements will affect the direction and volume of trade flows and will have an impact on particular domestic sectors.

Another important area for institutional research concerns the impact of regional integration on trade. Studies have been done on the impact of creation of the EC (Sorenson and Hathaway 1968). In addition, there have been efforts to evaluate the implications for U.S. producers of enlargement of the EC (Ferris et al. 1971). However, there is room for more research on the effects of regional groupings on trade and domestic production, particularly as the number of regional blocks among developing countries grows and the EC is further enlarged by the entry of Spain, Greece, and Portugal.

International Monetary System and Agricultural Trade. It has already been noted that international monetary institutions are in a state of rapid evolution. Dobbins and Smeal (1977), Kost (1976), and other writers have begun to investigate the relationships between exchange rate changes and agricultural trade. Dobbins and Smeal argue that most existing studies lack the detailed information on policies and foreign competitiveness needed for a full analysis of the effects of changing exchange rates on agricultural trade. Thus one focus of research on the international monetary system and agricultural trade should concern the mechanisms through which the mix of macroeconomic, farm, and trade policies interact with supply and demand to determine international prices and exchange rates. Closely related to this kind of question is the study of implications for trade flows and international marketing of exchange rate fluctuations and the resulting increased uncertainty in world markets.

Another concern is the development of new monetary and financial phenomena. Growth in the amount of Euro-currencies outstanding has altered the nature of international liquidity. Historically, this has been assured by movements of gold or dominant national currencies such as the British pound or the U.S. dollar. Since the Euro-currencies are outside the control of central banks and national governments (unlike dollars, pounds, or gold), their increased importance on international financial markets represents a fundamental change in the way nations settle their accounts and international capital movements occur. In addition, special drawing rights may become more important as an international reserve. As the agreements adopted at the Bretton-Woods Conference have broken down, new financial arrangements have developed. The move from fixed to floating exchange rates was a major change. The uncertainty resulting from abandonment of fixed exchange rates, the problem of recycling OPEC petroleum revenues, and the increasingly complicated interrelationships among the world's economies have all contributed to experiments with new institutions. The new European Monetary Union and past efforts to manage European currencies within the "snake" are obvious examples of changing institutional arrangements. Much work remains to be done simply to understand the relationships between the evolving international monetary system and agricultural trade.

Studies of Foreign Economies. In previous sections we have discussed a variety of subject areas where research is needed; study of foreign economies cuts across all of them. There are two sides to this question. First, insight is needed concerning the way in which the policies and institutions of other countries affect the performance of international markets. An example of this, as already noted, is understanding the implications of state trading for international agricultural markets. A second area concerns the determinants of policy within other countries. Not only is it necessary to develop our understanding of the basic economic, political, and social forces that influence the level and direction of trade, it is also necessary to understand the goals of foreign policy.

For example, the Japanese desire for food security is at the root of many of their policies. Information on the ways in which other countries view international equity, the role of trade in development, feasible policies and institutional arrangements, and increased knowledge of their resource bases and economic structures would be useful in predicting the positions they are likely to take in international negotiations. Finally, research on foreign economies could provide insights into the functioning of alternative institutions and suggest new arrangements that might be adapted to U.S. needs.

Market Growth and the Dynamics of Comparative Advantage. Most analyses of international trade have been carried out within the static framework of comparative advantage. Little has been done on the performance implications of such dynamic phenomena as inflation and changing labor and energy costs. Rising energy costs, in particular, may have a major impact on the international competitiveness of highly energy-intensive U.S. agricultural industries. The relatively high levels of inflation point to the need for research on the effects of rising prices on the competitive position of U.S. agriculture as well as the potential for reducing inflation through improving efficiency of the international marketing chain.

Growth and development of international markets constitutes another dynamic area where research is needed. In addition to forecasts of world supply and demand for various commodities, basic research on the sources and causes of market growth would provide useful insights into the potential for expanded exports. Technological progress, population growth, and rising incomes will have important implications for the direction, volume, and commodity composition of international trade. In addition to the sources and causes of market growth, assessment of the potential development of supply by foreign competitors is also needed.

Theory and Conceptual Framework. A final area requiring further research concerns the intellectual framework within which agricultural economists analyze international trade. In today's circumstances the free-trade framework is not an adequate approach to international market analysis. There would appear to be potential for additional theoretical work aimed at development of improved institutions and policies in a world where some of the marginal conditions assumed in the competitive model cannot be met and where adjustment may be slow and painful. Of fundamental importance to this approach is the definition and elaboration of a set of performance criteria for international agricultural markets. Just as analysis of domestic markets requires insights beyond those of neoclassical economics and perfect competition, study of international markets also requires the construction of an improved conceptual framework. Specification of international market performance criteria would be a good starting point.

ORGANIZATION OF RESEARCH

A final question is whether land-grant colleges and agricultural economists can gear up to undertake marketing research in a broadened domestic-international framework. This will depend partly on attitude and interest, partly on whether we are willing to make the investment in intellectual capital required to deal with problems of this kind, and partly on whether we can establish the appropriate operational linkages needed to implement this kind of research.

Clearly, more can be done, particularly if we define research and its objectives to include all activities involved in acquiring the knowledge needed for classroom and extension purposes and in the applied research designed to provide information needed by public or private decision makers. As a starter, increased library research to generate familiarization with conditions, problems, and issues is an activity all agricultural economists can undertake. Literature abounds from such sources as the UNCTAD, Organization for Economic Cooperation and Development, EC, USDA, and other sources.

Data are increasingly available from international organizations and the USDA to do quantitative analyses within individual countries of the effects of economic and policy factors that influence production, consumption, prices, and trade flows. In some cases, analysis of this kind can be extended to a multicountry or even worldwide framework. We should keep in mind that not all good analysis requires sophisticated model building. Good descriptive analysis with numbers that have not been processed by a computer probably will have to suffice more often than we would like.

Greater difficulty in implementing research arises where we try to deal with the intricacies of how international markets operate, why individual countries act as they do, or what kinds of policy changes are needed to achieve a given result. Much more insight is needed than can be obtained from available theory and secondary data. There is no way, for example, of assessing the policies Brazil should follow to optimize its export earnings from soybeans or of estimating what that total might be without knowing a lot more about Brazil than is available from secondary sources. This means that a basis must be provided for on-site study and research. This can be done only through increased cooperation among land-grant colleges and government and private foundations. Long-term commitments will be required, and research will have to be planned jointly. A starting point would be to seek increased cooperation between the USDA and the states. Little of this kind of work has been built into Agency for International Development programs, and it represents another avenue that should be explored. Other institutions such as the World Bank and private foundations might be interested if appropriate initiatives were forthcoming.

REFERENCES

Abel, Martin E., and Frederick V. Waugh. 1966. Measuring changes in international trade. *Am. J. Agric. Econ.* 48:847–61.
Bawden, D. Lee. 1966. A spatial price equilibrium model of international trade. *Am. J. Agric. Econ.* 48:862–74.
Bergsten, C. Fred, Robert O. Keohane, and Joseph Nye, Jr. 1975. International economics and international politics: A framework for analysis. In C. F. Bergsten and L. B. Krause, eds.,

World Politics and International Economics, p. 12. Washington, D.C.: Brookings Institution.

Dean, Gerald W., and N. R. Collins. 1966. Trade and welfare effects of EEC tariff policy: A case study of oranges. *Am. J. Agric. Econ.* 48:826–46.

Dobbins, Paul, and Gary L. Smeal. 1977. Exchange Rates and U.S. Agricultural Exports. OS/Policy Research, U.S. Treasury Department.

Epp, Donald J. 1968. Changes in Regional Grain and Livestock Prices under EEC Policies. Institute of International Agriculture Res. Rep. 4, Michigan State University, East Lansing.

Ferris, John, Timothy Josling, Brian Davey, Paul Weightman, Denis Lucey, Liam O'Callaghan, and Vernon Sorenson. 1971. The Impact on U.S. Agricultural Trade of the Accession of the U.K., Ireland, Denmark and Norway to the EEC. Institute of International Agriculture Res. Rep. 11, Michigan State University, East Lansing.

Freebairn, J. W. 1968. Competitive behavior in the international wheat market. *Rev. Mark. Agric. Econ.* 36:111–24.

Houthakker, H. S. 1971. Domestic farm policy and international trade. *Am. J. Agric. Econ.* 53: 762–66.

Johnson, D. Gale. 1977. Post-war policies relating to trade in agricultural products. In L. P. Martin, ed., *A Survey of Agricultural Economics Literature,* vol. 1. Minneapolis: University of Minnesota Press.

Josling, Timothy. 1976. Agricultural Protection and Stabilization Policies: An Analysis of Current Neo-Mercantilist Practices. Mimeo.

Kost, William E. 1976. Effects of an exchange rate change on agricultural trade. *Agric. Econ. Res.* 28:99–106.

McCalla, A. F. 1977. Strategies in International Agricultural Marketing: Public vs. Private Sector. Paper presented at Symposium on International Trade and Agriculture, Tucson, Ariz., April.

Rojko, Anthony S., and Martin W. Schwartz. 1976. Modeling the world grain-oilseeds-livestock economy to assess world food prospects. *Agric. Econ. Res.* 28:89–98.

Schuh, Edward G. 1977. Problems Involved in Doing Research on International Trade in the Agricultural Sector. Paper presented at the Symposium on International Trade and Agriculture, Tucson, Ariz., April.

Sirhan, Ghazi, and Paul R. Johnson. 1971. A market-share approach to the foreign demand for U.S. cotton. *Am. J. Agric. Econ.* 53:593–99.

Sorenson, Vernon L., and Dale E. Hathaway. 1968. The Grain-Livestock Economy and Trade Patterns of the EEC. Institute of International Agriculture Res. Rep. 5, Michigan State University, East Lansing.

Tontz, Robert L. 1966. Foreign market prospects and potentials. *Am. J. Agric. Econ.* 48:1348–78.

U.S. Department of Agriculture. 1970. World Demand Prospects for Agricultural Exports of LDCs in 1980. Foreign Agric. Econ. Rep. 60.

———. 1972. Food Marketing in West Germany. Foreign Agric. Econ. Rep. 76.

———. 1978. Alternative Futures for World Food. Foreign Agric. Econ. Rep. 152.

Valdés, Alberto. 1973. Trade policy and its effect on the external agricultural trade of Chile 1945–1965. *Am. J. Agric. Econ.* 55:154–64.

Vellianitis-Fidas, Amalia. 1976. The Impact of Devaluation on U.S. Agricultural Exports. USDA, AER.

Vermeer, James, D. W. Culver, and J. B. Penn. 1975. Effects of Trade Liberalization on U.S. Agriculture. USDA, AER.

Warley, T. K. 1967. *Agriculture: The Cost of Joining the Common Market.* European Ser. 3. London: Chatham House.

———. 1973. Trade liberalization in agricultural products. *Am. J. Agric. Econ.* 55:280–303.

Wipf, Larry J. 1971. Tariffs, nontariff distortions, and effective protection in U.S. agriculture. *Am. J. Agric. Econ.* 53:423–30.

Young, S. 1975. Trade creation and trade diversion in the EEC. *J. Agric. Econ.* 26:197–208.

Zarembka, Paul. 1966. Manufacturing and agricultural production functions and international trade. *Am. J. Agric. Econ.* 48:952–66.

HAROLD M. RILEY
MICHAEL T. WEBER

16

MARKETING IN DEVELOPING COUNTRIES

SINCE THE 1960s there has been a significant and growing involvement of U.S. agricultural economists in foreign assistance programs directed toward the less developed countries (LDCs). Through programs financed by the U.S. Agency for International Development (USAID), foundations, foreign governments, and multilateral international agencies, agricultural marketing economists have had opportunities to conduct research and provide advisory inputs into the development of programs to improve marketing systems. Concurrently, there has been an expanding flow of students from the LDCs through the graduate programs of U.S. universities (Stevenson 1979). To be effective in their roles as teachers, researchers, and advisors, economists have found it necessary to adapt their conceptual and analytical tools for use in different institutional, political, and social environments. In some instances marketing problems confronted in the LDCs appear to be similar to those experienced in the United States some 50–75 years ago when agricultural and industrial changes began to transform the economy in a relatively rapid and irreversible manner. However, the economic, political, and social conditions in the LDCs pose problems of market organization that require solutions carefully tailored to the needs of particular country situations. The paucity of basic information and data about existing LDC market systems and the lack of trained professionals in those countries have been major constraints to research and development activities.

This chapter is directed toward a primary audience composed of U.S. university faculty and students concerned with agricultural marketing problems in the LDCs. Marketing specialists with international agencies and LDC institutions are another audience group. We begin with a brief conceptual point of

HAROLD M. RILEY is Professor and **MICHAEL T. WEBER** is Assistant Professor, Department of Agricultural Economics, Michigan State University, East Lansing.

view regarding the role of marketing in the development process. This is followed by an assessment of past research activities. A major part of the chapter is devoted to the organization and conduct of future research.

CONCEPTUALIZING AGRICULTURAL MARKETING IN A DEVELOPMENT CONTEXT

Economic development should be viewed as a long-term process that occurs over decades and generations. Through technological innovation and economic organization, output per person increases and the material well-being of the population is raised to higher levels. Increased specialization of productive effort, industrialization, and urbanization are important elements in the growth and development process. These forces contribute to a growing demand for marketing services. In agriculture there is a transformation process, as relatively small-scale, predominantly self-sufficient family farming units become larger, more specialized, and increasingly dependent on marketing arrangements for the sale of their agricultural products and for purchased inputs. Rural markets emerge as local trading centers hierarchically interconnected within a larger regional and national market network.

In most developing countries there is a steady and sometimes relatively rapid migration of people from the rural areas to urban centers, with many of the capital cities growing at rates in excess of 5 percent per year. The buildup of urban population and rising levels of consumer income place great pressures on the marketing system to expand and undertake an increasingly complex set of activities that link the rural and urban sectors of the economy (Mittendorf 1978). Marketing services become a larger portion of the consumer food bill and the composition of the market basket shifts from low-cost, starchy foods toward higher cost livestock products, fruits, and vegetables. Major investments are required for transportation equipment, highways, and other physical facilities. Government agencies usually assume leadership in planning and financing much of the market infrastructure and frequently perform major roles in facilitating and regulating development of marketing institutions and in some instances actually organizing and managing marketing enterprises.

There is a wide range of viewpoints on the role of agricultural marketing institutions in economic development and the appropriate function of the public sector in bringing about desired changes. There are those who hold the view that marketing is an adaptive set of activities to be given secondary consideration in development planning strategies, with primary consideration being directed toward expansion of agricultural and industrial production. This view has been challenged by marketing economists who argue that marketing is a critical and dynamic component of development. Abbott (1967) and others in the Food and Agriculture Organization (FAO) marketing group (1970) have stressed the incentive role of effective product marketing systems that can reduce risks and lower costs for farmers and other market participants. The local availability of reasonably priced agricultural inputs and consumer goods is also seen as having a stimulating impact on economic activity in both rural and urban areas.

Collins and Holton (1964) have also challenged the view that marketing firms and institutions will automatically spring up in response to price incentives to provide the services most appropriate for new production situations. They argue that effective planning for economic development should give a great deal of attention to facilitating development of marketing institutions to complement programs for expanding physical production.

There seems to be a growing consensus among agricultural economists that aligns with the broader, more dynamic view of marketing as a major element in development of the agricultural sector and in coordinating agriculture with growth and development in other sectors. Hence food production, processing, and distribution activities are seen as a closely interrelated set of activities that operate in a "systems" context. The system includes the familiar components of farm production, rural assembly, processing, distribution (both rural and urban), and flow of industrially produced agricultural inputs and consumer goods to rural markets. In the more rurally based economies these activities take place within rural towns and their hinterlands; but as development progresses, the influence of larger urban centers becomes more important.

A simplified conceptual model was developed by a Michigan State University research team to illustrate a particular application of a "food system" approach to a marketing development program in Northeast Brazil (Table 16.1). The left-hand column lists five system components that are potential points of public sector intervention into a regional or national food system where the program objective is to stimulate economic growth and development. The vertical ordering of the system components gives emphasis to a demand-driven system, but there are important supply and demand interactions that link the various components into a semiclosed system. It is semiclosed because it does not include an export market component or an explicit linkage to the other sectors of the domestic economy. These components could be added to the model, but for the purposes of this chapter they have been set aside. The middle column lists a series of actions that might be taken by the public sector to bring about desired changes in the food system, giving emphasis to actions that will affect market organization and performance. An interrelated set of impacts on costs, demand, and output are summarized in the right-hand column. The model illustrates a particular sequence of actions that work back from the urban food market toward farm producers. However, there are many alternative sequences that could begin with any of the system components as long as there is adequate consideration of the pattern of repercussions that will be likely to occur. For example, demand from the rural purchased food and other basic consumer goods component has a direct pull effect on rural production and assembly components. And to the extent that there is regional specialization of agricultural production in a country, there is a direct linkage between rural demand and urban food distribution components that serve as concentration and redistribution mechanisms for the more specialized rural regions.

In the past there has been a strong tendency for agricultural planners to emphasize farm production expansion without sufficient consideration of market incentives and constraints, whether these come from rural or urban de-

TABLE 16.1. A conceptual model showing a series of interrelated food system reforms and expected effects on economic growth and development

Potential points of public sector intervention in regional or national food system processes	Possible actions to bring about desired changes	Postulated effects on economic growth and development
Urban food distribution components	Capital and technical assistance to stimulate improvements in efficiency of traditional urban marketers	Reduced marketing costs in urban areas for locally produced food products
	Timely introduction of infrastructures as a tool to stimulate improvement in market channel performance	Lowered food prices, increased effective income
	More effective public facilitative and regulatory programs	Increased effective urban demand for food and consumer goods and related marketing services
Rural food production components	Appropriate agricultural production extension efforts	Increased food production and agricultural production specialization
	Development of appropriate packages of inputs	Increased rural incomes and market participation on the supply and demand sides
	Effective market information and price stabilization programs	
	Supervised credit programs	
Rural assembly market components	Promotion of backward vertical coordination of food marketing	Increased rural and urban demand for organization and coordination services of commodity subsystems
	Capital and technical assistance to rural assemblers and transporters	Increased rural demand for improved physical distribution services, i.e., assembly activities
	Improvement of public storage, roads, exchange rules, grades	
Rural distribution components Purchased food Farm inputs Consumer goods	Improvement of rural distribution services and lower costs for farm inputs, purchased food, consumer goods	Increased rural demand for farm inputs, purchased food, rural- and urban-produced consumer goods, marketing services related to the above three
Rural and urban industrial and services components	Use of appropriate technologies in production processes	Increased demand and employment in industry and related services sectors
	Development of more appropriate products for local market demand characteristics	Increased income leading to increased demand for food and consumer goods
	Lower costs of mass distribution to rural and urban areas	

Source: Adapted from Figure 1.1 in Slater et al. (1969).

mand sources. Thus Table 16.1 illustrates a more comprehensive, market-oriented approach to agricultural development, emphasizing the dynamic interactions between agriculture and industry and between rural- and urban-based activities.

PAST RESEARCH

Research by U.S. scholars on agricultural marketing problems in developing countries can be categorized into three broad groupings: descriptive studies, feasibility studies, and broader diagnostic assessments.

Descriptive Studies. Descriptive studies have been conducted by individuals from various social science disciplines on existing arrangements for marketing specific commodities or carrying out selected marketing functions. Most have been carried out by professionals in academic institutions and their students. The studies have provided useful factual information about existing marketing arrangements, but limited accessibility has been a major factor restricting their use by government agencies and the private sector. In addition, many of the studies done by economists and agricultural economists are based upon conceptual perspectives of market organization dominated by the perfectly competitive theoretical model of economics. Much of the research has been concerned with issues involving the testing for conditions of structure, conduct, and performance predicted by the perfectly competitive model. A major problem with this relatively static framework is that it underplays the potential dynamic impacts of marketing institutions in achieving development goals regarding efficiency, equity, growth, and employment.

There have also been useful and insightful descriptive studies carried out by researchers who represent other social science perspectives. Geographers, with their interest in the location of economic activities, have undertaken a large number of descriptive studies of marketplaces, periodic markets, and itinerant traders in rural areas of developing countries. This research is important for the development process because it provides knowledge of how these traditional trading institutions function. Unfortunately, by the geographers' own assessments, much of this research suffers from the inability to offer normative solutions to questions concerning policy and planning of marketing systems (Ghosh and McNulty 1978; Smith 1978).

Anthropologists and sociologists have observed and described rural household behavior relative to combinations of production, consumption, storage, and sales decisions. Anthropologists also have a tradition of conducting individual village studies. Although these provide valuable descriptive information about rural populations and economic processes, they rarely contain analyses that lead directly to policy recommendations. A group of economic anthropologists are seeking to use concepts from regional science and geographical models to put their village studies into a more useful framework for understanding and promoting development. In a review article, C. Smith (1976) concludes that without the regional system context that geographical models can provide, anthropological marketing studies will not tell us a great deal more than we already know about the economic determinants of peasant behavior.

Feasibility Studies. Feasibility studies have been done to provide information needed by government agencies, international financial institutions, and private sector investors regarding capital investments in marketing infrastructure (e.g., processing plants, wholesale markets, grain storages, transportation facilities). These studies have varied widely in scope and quality of analysis. Most have been carried out by private consulting firms or professionals associated with university-based research institutes. The analyses are typically focused on the economic feasibility of a proposed project involving a large capital investment. Because of severe time constraints, heavy reliance is usually placed on the use of available secondary data, engineering estimation procedures, and qualitative information that can be obtained through interviews with informed local business leaders and professionals. Looking back at some of these studies and the recommendations that were subsequently carried out, several concerns can be identified. First, there has been a tendency toward unrealistic optimism regarding the transferability of technologies from the more highly developed countries to the LDCs. The analyses have tended to endorse capital-intensive technologies in situations where labor-capital costs are such that more labor-intensive technologies would be more appropriate (Timmer 1972). Second, the feasibility studies have sometimes misjudged the compatibility of new capital-intensive infrastructure with existing patterns of production, distribution, and consumption. As a result, there have been examples of serious underutilization of the new facilities (e.g., grain storage) (Lele 1975). Third, the lack of a trained labor force and a local capacity for the continued development of both skilled labor and management ability is either underplayed in the reports or not taken seriously by those responsible for local project implementation. Because of the problems mentioned above, there is a growing demand for better preparation of professionals who conduct feasibility studies. This is reflected in the increased interest in short-term training programs on project development and evaluation such as those offered by the World Bank and similar offerings by USAID, universities, and private consulting firms.

Broader Diagnostic Assessments. Broader diagnostic assessments of food system organization in developing countries have provided inputs to policy and program development and to an evolving conceptual and analytical framework for future research and development efforts. Several groups of U.S. university researchers have carried out these broader based studies of agricultural marketing processes in LDCs (Mellor 1966; Mellor et al. 1968; Jones 1970; King 1973; Lele 1976; CRED 1977). Mellor et al. (1968) have studied marketing in India. On the basis of extensive field surveys they have challenged the validity of several widely held views regarding the exploitive and unproductive activities of rural traders (Mellor 1966). Lele (1976) could find little evidence in her field studies to support the view that the monopolistic nature of private trade leads to excessively high marketing margins or that wide seasonal price variations were caused by speculative hoarding and profiteering practices of traders (Mellor et al. 1968). Price differentials among major wholesale centers were found to be closely related to expected price patterns based upon transportation cost differences. Indications are that entry into traditional trade is generally open and that there is overcrowding and significant competition at each level of marketing. Even in instances where a few traders are handling a large share of the market volume it was observed that they are unable to influence prices appreciably through collusive action as long as there is effective market intelligence and transportation among markets. Lele (1976) observes that public sector efforts to facilitate efficiency in traditional trade is necessary, since rural traders perform a number of important functions that cannot be replaced by government or cooperative agencies without incurring substantially greater costs in administrative manpower and finances than is implicit in allowing the private sector to operate. A broad-based and positive role for public sector involvement in marketing has been outlined by Lele.

Jones (1970) and his colleagues at Stanford University have conducted extensive studies of agricultural marketing in several African countries. The characteristics of existing marketing systems were compared with the requirements of a purely competitive model, and actual pricing relationships were checked against what would be expected in a perfect market. Their conclusions were "that average seasonal price movements correspond rather well with the cost of storage; that intermarket price correlations were somewhat less than might be hoped for; that year-to-year price movements were generally in accord with supply and marketing conditions; but that week-to-week price changes showed signs of serious random disturbances consistent with the hypothesis that traders were poorly informed about episodic changes in the conditions of supply and transport. . . . In terms of the tasks that marketing systems are asked to perform, the African ones that we studied are not performing badly." Despite this assessment of the existing marketing system, Jones points out the critical need for attention to marketing in economic development planning where major technological and institutional changes are being contemplated. Jones closes his article with the observation that "the invisible hand cannot be trusted completely to guide economies in socially acceptable directions, nor can the state rely on the marketing system to perform the tasks assigned to it without appropriate facilitating services best provided by government."

A major problem with the research framework developed in most of these diagnostic assessments is the lack of concern for the dynamic impacts that marketing services can have on production and consumption. The static focus of the research has been on whether prices and cost relationships over space and time behave as predicted by the perfectly competitive model. Relatively little effort has been made to better understand how the effectiveness of marketing services influences supply and demand functions, especially for small-scale farmers and low-income consumers.

Harriss (1979) has reviewed the methodology used in the Stanford-Cornell type approach to the analysis of market performance and makes a similar observation:

In dealing with easily available, even if qualitatively poor, data on agricultural commodity prices the analysis of market performance has been diverted away from the consideration of interrelationships—between the control of commodities and money; between exchange and production essential for the identification of the role of the marketing system in economic development. In this sense not only is the methodology itself usually statistically and interpretatively spurious but also the fetishism of competition in agricultural commodity markets (as revealed by price and commodity analysis here) has led agricultural marketing economists to overnarrow at least a decade of a substantial part of our research.

Another problem with much of the broader diagnostic research is the tendency to use secondary and usually macrolevel data in testing for conditions of structure, conduct, and performance predicted by the perfectly competitive model. Commodity studies of market flows, margins, elasticities, concentration, competition, and policies are generally based on industry or regional data that do not permit focusing on the microlevel behavior of marketing agents, including farmers' marketing decisions in the rural areas. These studies frequently include assumptions of homogeneous behavior on the part of farmers and marketing agents and use data that are averages of many observations (e.g., monthly price data) and thus obscure important variations in market behavior. Results are often inadequate for making specific recommendations for improvements in rural and/or urban markets, especially if the objective is to extend improved services to specific target groups such as small-scale farmers, other low-income rural residents or low-income urban consumers.

Still another problem of much of the broader diagnostic research for guiding overall marketing policy stems from its focus on few if any of the levels of interaction in the vertical marketing channels between farmers and ultimate consumers, whether they are located within rural areas themselves or in large urban areas. Even in semisubsistence economies there are interdependencies in the various stages of the farm production-assembly-processing, distribution, and consumption process. And even the "equity with growth" type of rural development being advocated in much of the literature involves constant structural transformation of the rural and urban economy, which leads to greater interdependencies among agricultural production, distribution, and consumption processes. The most important marketing problems related to achieving the desired structural transformation are in the design and promotion of new technologies and new institutional arrangements, which may be unprofitable or unavailable to individual market participants but if adopted by all participants could yield substantial system improvements.

Pritchard (1969) has stressed the importance of developing a broad analytical framework for studying and solving agricultural marketing problems in developing countries. He has cautioned against using a narrowly defined market structure framework that limits analysis to characteristics of the organization of a market that seem to influence strategically the nature of competition and pricing within the market. Pritchard has outlined an eclectic set of analytical procedures, bound together into a useful framework by the concept of agricultural marketing as an organized, operating behavioral system within the national economy. He has emphasized the need to use the framework to search for basic economic, technological, and social constraints in the environment in which marketing systems function and change.

A number of U.S. university researchers have approached broader diagnostic assessments from such a perspective. Researchers at Harvard University have extended their "agribusiness commodity systems" approach to problems of export market development in Central America and other areas (Goldberg et al. 1974). Phillips (1973) and his colleagues in the Food and Feed Grain Institute at Kansas State University have conducted a number of diagnostic assessments of grain marketing systems in LDCs, using a broad food chain conceptual approach. Physical handling at all stages in the farm-to-consumer chain has been examined and recommendations for improvement programs have been presented to government agencies. Pricing, storage, and regulatory policies have also been an important part of the country studies.

A group of Michigan State University researchers have developed a food system approach to conducting diagnostic studies of agriculture and food marketing systems linking large urban centers in selected Latin American countries with their rural supply areas. Field studies in northeast Brazil, Bolivia, Colombia, and Costa Rica were carried out collaboratively with local professionals representing universities and government agencies (Harrison et al. 1974). The diagnostic studies were the basis for the development of broad-based market improvement programs with specific project recommendations. A modified market structure-conduct-performance framework of analysis generally guided the organization of these diagnostic investigations. Such a framework is oriented toward the evaluation of system performance when judged against broad economic and social goals.

The basic thrust of the Michigan research was toward use of a descriptive-diagnostic procedure for identifying constraints and unexploited opportunities as perceived by marketing system participants and local political leaders and as identified through use of a wide array of standard economic analysis tools. The approach is pragmatic and eclectic and emphasizes the need to identify managerial, technological, and institutional innovations, which are unprofitable or unavailable to individuals within existing marketing channels but, if adopted across all stages of these interrelated production-marketing processes, could lead to substantial, channelwide improvements.

ORGANIZING AND CONDUCTING FUTURE RESEARCH

General Considerations. The nature of marketing problems varies widely with the degree to which a particular economy has been transformed from an agriculture-based rural economy toward a more urban-based, market-oriented economy. In countries that are still predominantly rural, marketing problems center around improvements in the functioning of local markets as providers of simple farm inputs and household necessities and as trading centers for basic food commodities produced and consumed within the local area or region. As an economy becomes more urbanized, food production and distribution takes on a higher priority in development plans with greater attention to improving physical infrastructure (transportation, processing, storage) and to policies and programs designed to stimulate production and facilitate system coordination. As the industrialization process continues, new technologies for processing and distributing food, more complex logistic and institutional arrangements, and increased participation of government agencies in planning and carrying out marketing programs usually occur.

In many developing countries there are dual agricultural production-marketing subsystems, one oriented toward export markets and the other toward domestic food needs. The export-oriented subsystem is typically better organized in terms of pricing and handling procedures and often involves large-scale parastatal agencies or multinational corporations with vertically coordinated production-marketing programs. The export subsystems play an important role in bringing new technologies and management innovations into LDC agricultural sectors.

Whatever the level of industrialization and urbanization, there is a need to approach marketing research in LDCs from a food systems perspective where the interdependencies of the various stages in the farm production-assembly-processing and distribution process can be taken into consideration. The food systems perspective that has evolved in U.S. agricultural marketing research provides a useful background, especially when viewed in a long-term (50–75 years) context.

National Goals and Development Planning. Nearly all the LDCs are continually preparing general development planning statements and project documentation for consideration by external funding agencies and their own domestic government agencies. The planning documents usually reflect basic underlying goals of increasing gross national product per capita, maintaining full employment of the labor force, and achieving an acceptable degree of equality in distribution of income and economic opportunities. National planning goals have given increased emphasis to improving the relative well-being of rural people and to measures that will slow the migration from over-populated rural areas to the cities. This shift toward greater concern for the rural poor has been reinforced by the policies and programs of international development agencies (McNamara 1977). As indicated later in this chapter, there is serious need for research to give direction to marketing programs that will benefit small farmers and rural communities. But in a broader context, agricultural marketing research should support the design and promotion of new technologies and new institutional arrangements that will contribute to achievement of a broad set of economic development goals.

Planning and Conducting Research. Lack of basic information about organization and functioning of the food system and a general distrust of "middlemen" are common characteristics among the LDCs. However, social and political pressures dictate that development programs and public policies be made on the basis of available but usually very inadequate information and analyses of alternative courses of action. Policymakers and the small contingent of professionals who staff the planning units and ministries of agriculture desperately need applied research to identify the most urgent marketing problems and the actions that might be taken to improve existing conditions. But there is also need for a more comprehensive understanding of marketing processes and a long-term view of the desired role of market organization and institutions in national economic development.

Probably the most fundamental issue to capture the attention of the international development community is one that centers around the observed ineffectiveness of past development programs to improve the relative or, in some cases, actual well-being of the poor majority in the LDCs (McNamara 1977; Seers 1977; Thiesenhusen 1977). In the poorest countries there is high concentration of the poor in rural areas, and rapid migration of the rural poor toward urban centers creates serious employment and related social and political difficulties. As a result, there has been a significant policy reorientation in international development assistance agencies and in many of the LDCs. While there is continuing debate over the appropriate strategy for promoting the desired rural development, there is general agreement that the fundamental issue is how to promote both growth and equity.

In this context the question arises as to how changes in marketing institutions might make greater contributions toward improvements in economic and social conditions in rural areas, while also contributing to broader goals of holding down food prices for families in rural and urban areas. To support such an objective, additional field research is needed to identify alternative opportunities for improving effectiveness of rural marketing systems within more comprehensive rural development programs designed for particular country situations.

The emergence and development of an applied marketing research program will follow different patterns, depending upon the circumstances within individual countries. The experience of the Michigan State University group in Latin America suggests that a small task force unit created to carry out applied research and assist in the formulation of programs and policies can contribute substantially to development of a progressive and efficient agricultural marketing system (Harrison et al. 1974). Such a task-oriented group can develop a data base on food marketing and an approach to market system analysis that not only will help to identify opportunities for marketing improvements but will also examine alternatives and make recommendations to appropriate action agencies.

To develop the broad analytical framework needed in a task force approach to understanding equity and growth concerns of development, it is necessary to focus on operation of the marketing system in terms of distribution of wealth and income, access to government services and political power, social status and organization, geographic considerations, and technical performance.

There is growing interest among other social science researchers (geography, anthropology, sociology, political science) in exploring various aspects of the rural community, which might be relevant to more realistic and effective development efforts. This suggests an opportunity for increased collaboration or at least a greater degree of communication and coordination between marketing economists and other social science researchers as they attempt to deal with very complex rural development issues, including marketing institutions.

Whether or not a task force unit is created and institutionalized on a more permanent basis, there is usually a need for broad descriptive-diagnostic research. Depending upon the size of the country and the available resources, these studies can be organized on a regional or national basis. It is important that the geographic region to be studied includes both urban and rural areas so that the rural-urban marketing linkages can be considered in a longer term development context.

During the actual conduct of the studies, preliminary reports and selected pieces of information should be transmitted to key individuals in government and the private sector. When sufficient research output is available to support major recommendations for an interrelated set of marketing improvement programs, a high-level seminar can contribute to further development and eventual adoption of programs and policies consistent with long-term national goals.

If resources make it impractical to carry out such a comprehensive study as a concentrated effort, an alternative is to establish a research agenda and arrange for contributing studies that might be carried out by local university students under faculty supervision, graduate students preparing theses for foreign universities, or private consulting firms.

TOWARD A RESEARCH AGENDA

A research agenda can be developed around a set of interrelated questions that address the basic information needs for diagnosing marketing problems and assessing program needs to achieve desired development goals. Suggested questions and interspersed comments follow.

What is the organizational structure at the farmer, assembly, and processor levels? What services are provided at these different stages, and what are the prevailing price spreads, costs, and investments at each stage? What are the procedures for arranging transactions and coordinating product flow in these stages of the marketing channels for the major food products? Is there evidence that market instability and poor market coordination have resulted in high and costly levels of risk and uncertainty for farmers as well as other market participants? What are the major causes of market-related risks and uncertainties?

Primary emphasis in the above question areas is needed on understanding the microlevel relations of agricultural production and marketing in rural areas and exploring the equity and efficiency implications of alternative marketing arrangements. It is essential to understand how the effectiveness of marketing services influences supply response of different types of farmers. For example, what effect does market information have on market risk and uncertainty associated with both small and large farmers' adoption of new agricultural production technology and farm-level enterprise selection? There is the critical question of understanding the role of local input marketing services in reducing costs, risk, and uncertainty associated with obtaining and correctly using new agricultural production inputs. There is also the question of learning how to coordinate planned output expansion with market demand potential, even though there is generally very inadequate information regarding characteristics of demand. This is particularly important for many of the rural development schemes that seek to raise small farmer income via high-value crop production such as fruits, vegetables, and meats. High marketing costs, risky market transaction channels, and underspecified quality characteristics for these products can quickly dampen price and other demand incentives for farmers, especially for smaller ones who tend to have less individual control over marketing methods and higher marketing costs as a result of smaller unit sales.

What will be the trends in population growth, level and distribution of incomes, and urbanization patterns in rural towns and large cities over the next 10–20 years? What effects will these changes have on demand for food products and food marketing services? What are the existing characteristics and problems of the rural as well as urban consumer market for food with respect to quantities purchased by different income groups, shopping habits, and attitudes toward existing retailing services?

An area of relative neglect in marketing is the backflow (or within rural area flow) of food, other consumer goods, and agricultural inputs through a hierarchical set of trading arrangements that link individual farmers and small villages to larger villages and ultimately to major urban centers. The effectiveness of this portion of the agricultural marketing system can have a major impact on the well-being of rural people and on the growth and development of economic activity in rural areas. For example, there is need to understand how the type and effectiveness of marketing-distribution services influence the mix and quality of foods marketed in rural areas and how these services influence farmers' ability to specialize in fewer crop and livestock enterprises. Researchers have begun to focus on understanding rural demand and consumption linkages for nonfood inputs and consumer goods (King and Byerlee 1977). These are an important source of demand for industrialized products that are well matched to local, effective demand characteristics. However, more research is needed to understand how to promote the organization of lower cost mass distribution of these products to rural consumers.

Another related area for future research is exploration of the relationship between nutrition and marketing services and their combined effect on the welfare of different consumer and producer groups. Nutrition studies are increasingly being redefined to include a broader set of variables instead of the isolated factors of health or total food supply. When subsistence rural households are encouraged to increase their cash incomes by producing food and/or cash crops for sale, improved nutrition may be achieved only if necessary foods are available at reasonable cost and in nutritious and consumable forms for purchase by these households. If rural food distribution services are ineffective and of high cost, this will reduce the quantity and quality of products available locally as well as the real purchasing power of rural consumers.

What is the organizational structure at the retailing and wholesaling levels? What services are provided at these stages and what are the prevailing price spreads, costs, and investments at each stage? What are the procedures for arranging transactions and coordinating product flow among the wholesale and retail stages of the marketing channels for the major foods consumed? What are the major problems confronting the more progressive food marketing entrepreneurs in finance, government regulation, competition from other entrepreneurs (public or private), and market infrastructure?

Past research has shown that the large and rapidly growing group of low-income consumers in major urban centers of LDCs allocate high portions of their cash income to purchased food, which they tend to procure from small-scale urban retailers located in marketplace stalls and neighborhood shops (Harrison et al. 1974; Mellor 1978; Mittendorf 1978). Among all urban food retailers, however, these smaller scale merchants tend to have higher costs of operation and more difficulty coordinating with urban and rural wholesale suppliers for the provision of the mix of foods demanded by their customers. Unfortunately, this group of small urban retailers has not received adequate research and development program assistance, while the urban marketing reforms undertaken have tended to benefit larger scale retailers and wholesalers who serve middle- and high-income consumers (Silva 1976; Bucklin 1977; Weber 1977; Stevenson 1979). Therefore, there is special need for research to determine how small-scale food retailing in both rural and urban areas can be improved through managerial and technological innovations that reflect each country's labor and capital endowments and contribute to more effective vertical market channel coordination linking small-scale merchants and farmers.

In terms of overall evaluation of marketing channels: What evidence can be cited to indicate poor market performance with respect to costs of providing existing services; effectiveness of vertical coordination mechanisms in communicating consumer demands to marketing firms and ultimately to farmers; adequacy of the variety, quality, and condition of products reaching consumers; effectiveness of product distribution over space and time; progressiveness of public and private enterprises in adopting new marketing practices; and equitability of the system in distributing benefits of marketing improvements? Is it possible to identify potential innovators (i.e., individuals who have adopted improved management practices that could be transferred to others)? What are the problems and opportunities for encouraging improved distribution channel coordination through sequential introduction of new management practices and coordination arrangements? What are the alternative roles the public sector can play in taking leadership to promote improved performance of marketing practices? What are the costs and benefits of pursuing these alternative roles?

Public and semipublic enterprises play an important role in the agricultural marketing systems of many LDCs. Their functions often involve purchase, storage, and distribution of large volumes of domestically produced staple food commodities and importation of additional food supplies. In several instances, parastatal enterprises have a central role in development of export-oriented agricultural commodity systems. Because of the size and nature of these public enterprises, it is difficult to achieve and maintain high levels of economic performance. There are the usual internal organization and management problems that should be attended with a continuing applied research and educational program. But there are larger, more troublesome problems regarding the appropriate role of the public agency in relation to the private sector, the distortion of real price-cost relationships through the use of taxing and subsidy powers, and the ever present possibility of political manipulation and corruption where large amounts of money and commodities are being handled. This area needs careful evaluation through a series of case studies from which a set of guidelines might be derived for organization and operation of public sector agencies and enterprises within a more global food system development strategy.

SUMMARY

This chapter has outlined a conceptual view of the development process within which the economic organization of market relationships in the food system can play a dynamic and critical role in achieving national development goals. A general approach to the development of a marketing research program within LDCs has been proposed. Priority has been directed toward applied research carried out within a descriptive-diagnostic-prescriptive framework that is relevant to policy and program development needs in the developing countries. These procedures seem appropriate for use in a wide range of political economic systems, although the specific forms of public sector participation in the food system will vary among countries.

U.S. agricultural economists and those from other developed countries will have continuing opportunities to assist in development of agricultural marketing research programs in the LDCs. Past experience indicates that their most important contributions will be in development of young professionals who become the indigenous professional cadre that actually carries out research, teaching, administrative, and entrepreneural roles within their own countries. But effective professional development requires a combination of formal training and long-term involvement in applied problem solving research and related activities. This pattern of professional development can be facilitated by developed country Ph.D. level training programs with supervised thesis research back in the student's own country or region and by assistance in development of M.S. level training progams in LDC institutions, with emphasis on relevant field research experience dealing with marketing problems. Collaborative task force marketing research projects involving professionals from local institutions in LDCs and the more developed countries can contribute to professional development goals while making timely and important inputs to current planning and program needs. Over the longer run, gradual evolvement of professional networks reinforced by linkages among and between LDC institutions and similar institutions in more developed coun-

tries can greatly strengthen the overall effectiveness of marketing research programs in the LDCs. These kinds of professional development–applied research
efforts are being reinforced and promoted by international agencies such as
the FAO and the Inter-American Institute of Agricultural Sciences and by
bilateral foreign assistance agencies and private foundations.

REFERENCES

Abbott, J. C. 1967. The development of marketing institutions. In H. M. Southworth and B. F.
    Johnston, eds., *Agricultural Development and Economic Growth,* Chap. 10. Ithaca: Cornell
    University Press.
Bucklin, Louis P. 1977. Improving food marketing in developing Asian countries. *Food Policy* 2:
    114–22.
Center for Research on Economic Development (CRED). 1977. *Marketing, Price Policy and Stor-
    age of Food Grains in the Sahel,* vols. 1 and 2. University of Michigan, Ann Arbor.
Collins, N. R., and R. H. Holton. 1964. Programming changes in marketing in planned economic
    development. In C. Eicher and L. W. Witt, eds., *Agriculture in Economic Development.* New
    York: McGraw-Hill.
Food and Agriculture Organization. Marketing—A Dynamic Force in Agricultural Development,
    p. 5. World Food Problems, Publ. 10, Rome.
Ghosh, Auijit, and Michael L. McNulty. 1978. Locational Analysis and Spatial Planning of Mar-
    keting Systems in Developing Countries. Paper presented at International Geographical
    Union Working Group on Marketplace Exchange Systems Symposium, Zaria, Nigeria, July
    23–30.
Goldberg, R. A., et al. 1974. *Agribusiness Management for Developing Countries: Latin America.*
    Cambridge, Mass.: Ballinger.
Harrison, K., D. Henley, H. Riley, and J. Shaffer. 1974. Improving Food Systems in Developing
    Countries: Experiences from Latin America. Latin American Studies Center Res. Rep. 6,
    Michigan State University, East Lansing.
Harriss, Barbara. 1979. Relevant and Feasible Research—Annex 3—Methodology and Data in the
    Measurement of Market Performance. Paper presented at Workshop on Socioeconomic Con-
    straints to Development of Semi-Arid Tropical Agriculture, ICRISAT, Hyderabad, India.
    Feb. 19–22.
Jones, W. O. 1970. Measuring the effectiveness of agricultural marketing in contributing to eco-
    nomic development: Some African examples. *Food Res. Inst. Stud. Agric. Econ., Trade,
    Dev.* 9:175–96.
King, R. A. 1973. Mobilizing the concept of marketing orientation in space, time and form. In
    Building Viable Food Chains in the Developing Countries, pp. 55–73. Food and Feed Grain
    Institute, Spec. Rep. 1, Kansas State University, Manhattan.
King, Robert P., and Derek Byerlee. 1977. Income Distribution, Consumption Patterns and Con-
    sumption Linkages in Rural Sierra Leone. African Rural Economy Paper 16, Michigan State
    University, East Lansing.
Kriesberg, M., and H. Steele. 1972. Improving Marketing Systems in Developing Countries.
    USDA, FEDS Staff Pap. 7.
Lele, Uma J. 1975. Marketing of agricultural output. In *The Design of Rural Development—Les-
    sons from Africa,* Chap. 6. Baltimore: Johns Hopkins University Press.
_____. 1976. Considerations Related to Optimum Pricing and Marketing Strategies in Rural De-
    velopment. Decision-making and Agriculture. Papers and reports from the 16th International
    Conference of Agricultural Economists, pp. 488–514, Nairobi, Kenya.
McNamara, R. 1977. Address to the Board of Governors, World Bank, Washington, D.C.
Mellor, J. W. 1966. *The Economics of Agricultural Development,* Chap. 18. Ithaca: Cornell Uni-
    versity Press.
_____. 1978. Food price policy and income distribution in low-income countries. *Econ. Dev.
    Cult. Change* 27:1–26.

Mellor, J. W., T. F. Weaver, Uma J. Lele, and S. R. Simon. 1968. *Developing Rural India,* Chaps. 12–14. Ithaca: Cornell University Press.

Mittendorf, H. J. 1978. The challenge of organizing city food marketing systems in developing countries. *Z. aus. Landwirtsch.* 17:323–41.

Phillips, Richard, ed. 1973. Building Viable Food Chains in the Developing Countries. Food and Feed Grain Institute, Kansas State University, Manhattan.

Pritchard, Norris T. 1969. A framework for analysis of agricultural marketing systems in developing countries. *Agric. Econ. Res.* 21:78–85.

Seers, D. 1977. The new meaning of development. *Int. Dev. Rev.* 19:2–7.

Silva, A. 1976. Evaluation of Food Market Reform: Corabastos-Bogota. Ph.D. diss., Michigan State University, East Lansing.

Slater, C., H. Riley, V. Farace, K. Harrison, F. Neves, A Bogatay, M. Doctoroff, D. Larson, R. Nason, and T. Webb. 1969. Market Processes in the Recife Area of Northeast Brazil. Latin American Studies Center Res. Rep. 2, Michigan State University, East Lansing.

Smith, Carol A. 1976. Regional Economic Systems, Linking Geographical Models and Socioeconomic Problems. In C. A. Smith, ed., *Regional Analysis,* vol. 1. *Economic Systems,* Chap. 1. New York: Academic Press.

Smith, Robert H. T. 1978. Marketplace Trade—Periodic Markets, Hawkers and Traders in Africa, Asia, and Latin America. Center for Transportation Studies, University of British Columbia, Vancouver.

Stevenson, Russell. 1979. Graduate students from less developed countries: The continuing demand for U.S. training. *Am. J. Agric. Econ.* 61:104–6.

Thiesenhusen, W. 1977. Reaching the rural poor and the poorest: A goal unmet. In H. Newby, ed., *International Research in Rural Studies: Progress and Prospects.* New York: Wiley.

Timmer, C. Peter. 1972. Employment aspects of investments in rice marketing in Indonesia. *Food Res. Inst. Stud.* 11:59–88.

Weber, Michael T. 1977. Towards Improvement of Rural Food Distribution. Paper presented at Inter-American Institute of Agricultural Sciences (IICA) Seminar on Marketing Strategies for Small Farmers in Latin America, San Jose, Costa Rica, April 25–28.

INDEX

Research and development (*cont.*)
  private sector, 124–26, 132
  public, 126–28, 132
Research and Marketing Act of 1946, 4, 6, 8
Resource allocation by the firm, 38–39
Resources, exhaustion of low costs, 80–85
Risk, reduction in vertical coordination,
  200–201

School Breakfast Program, 280
Securities and Exchange Commission, 125
Senate Committee on Agriculture, Nutrition,
  and Forestry, 271
Senate Select Committee on Nutrition and
  Human Needs, 29, 264
Sherley Amendments, 269
Sherman Antitrust Act of 1890, 4
Special Milk Program, 280, 286
Special Supplemental Food Program for
  Women, Infants, and Children (WIC),
  279, 282, 285, 289, 291
Structural change
  capital and entry requirements, 57
  effects on food industry, 54
  factor markets in the food system, 90
  farm input markets, 57–58
  national consumer panel, 75
  number and size distribution of farms and
    farm incomes, 55–57
  ownership and control of agricultural pro-
    duction and sales, 58–61
  socioeconomic characteristics of farm
    operations, 55
  specialization, productivity, and organiza-
    tion of production activities, 61–62
Structure-conduct-performance (S-C-P),
  31–32, 228–29, 323. *See also* Industrial
  organization, paradigm
Sugar beets, 58, 59, 61
Sugarcane, 58, 59
Surveys, preference, 242–43

Talmadge, Herman, 271
Tariffs, 300, 302, 308
Tax
  implications of ownership forms, 203–4
  policy and capital flow in food system, 87
  provisions and farm ownership, 58
Tax Reform Act, 57
Technology
  adoption of new, 120–24
  assessment, 128–31, 132–33
  change associated with production func-
    tion, 61
  computer, 10
  defined, 117
  developing countries, 320
  development, 122
  development in a maturing economy,
    134–35
  diffusion studies, 21
  economic organization of farm production,
    185
  firm size and market power, 21

food manufacturing industries, 21
future research, 74
government policies, 127, 133
incentive for vertical coordination, 198–99
organization of industry, 79
post-harvest (PHT), 127, 128, 130
preference articulation, 241
production agriculture and marketing,
  64–68
public vs. private role in research, 128
reducing food losses, 134
research and development, 126–28, 131–35
technological imperative, 123–24, 190
Temporary National Economic Committee
  (TNEC), 6, 18
10-State Survey, 283
Theory of Monopolistic Competition, The, 6
Transactions, market, lowering cost, 201
Transportation
  agricultural demand, 103, 105–6
  agricultural dependence, 99–100
  defined, 98
  delivery of services, 107–8
  efficiency, 107
  energy issues, 109
  facilitator of trade, 100
  firm studies, 103–4, 107–8
  future research needs, 107–13
  international trade, 109–10
  ownership and control, 112–13
  past research categories, 103–7
  railroad, 108, 112
  regulation, 103, 104–5, 107, 110–11
  research, 101–3, 106–7
  road and highway, 111–12
  service pricing, 105
  spatial equilibrium studies, 103, 104–5
  subsidies, 111
  uncertainty, 108–9

Unionization, 32, 57, 69
United Nations Committee on Trade and
  Development (UNCTAD), 297, 301, 313
U.S. Agency for International Development
  (USAID), 315, 320
U.S. Department of Agriculture (USDA)
  administration of marketing research, 11
  Agricultural Marketing Service, 155, 165
  Bureau of Chemistry, 269
  Cooperative State Research Service
    (CSRS), 11
  data series, 165–66
  Economic Research Service, 64
  Economics and Statistics Service (ESS),
    165
  food inspection, 268, 275
  Food and Nutrition Service, 278, 283
U.S. Department of Commerce, 133, 166
  food inspection, 268
U.S. Department of Health and Human
  Services, 242, 269
  Food and Drug Administration (FDA),
    252, 264, 268, 269